# Studies in Computational Intelligence

Volume 738

**Series editor**

Janusz Kacprzyk, Polish Academy of Sciences, Warsaw, Poland
e-mail: kacprzyk@ibspan.waw.pl

*About this Series*

The series "Studies in Computational Intelligence" (SCI) publishes new developments and advances in the various areas of computational intelligence—quickly and with a high quality. The intent is to cover the theory, applications, and design methods of computational intelligence, as embedded in the fields of engineering, computer science, physics and life sciences, as well as the methodologies behind them. The series contains monographs, lecture notes and edited volumes in computational intelligence spanning the areas of neural networks, connectionist systems, genetic algorithms, evolutionary computation, artificial intelligence, cellular automata, self-organizing systems, soft computing, fuzzy systems, and hybrid intelligent systems. Of particular value to both the contributors and the readership are the short publication timeframe and the world-wide distribution, which enable both wide and rapid dissemination of research output.

More information about this series at http://www.springer.com/series/7092

Adam E. Gawęda · Janusz Kacprzyk
Leszek Rutkowski · Gary G. Yen
Editors

# Advances in Data Analysis with Computational Intelligence Methods

Dedicated to Professor Jacek Żurada

 Springer

*Editors*
Adam E. Gawęda
Department of Medicine, Division
  of Nephrology and Hypertension
University of Louisville
Louisville, KY
USA

Janusz Kacprzyk
Systems Research Institute
Polish Academy of Sciences
Warsaw
Poland

Leszek Rutkowski
Institute of Computational Intelligence,
  Department of Mechanical Engineering
  and Computer Science
Częstochowa University of Technology
Częstochowa
Poland

Gary G. Yen
School of Electrical and Computer
  Engineering
Oklahoma State University
Stillwater, OK
USA

ISSN 1860-949X             ISSN 1860-9503   (electronic)
Studies in Computational Intelligence
ISBN 978-3-319-88516-2       ISBN 978-3-319-67946-4   (eBook)
https://doi.org/10.1007/978-3-319-67946-4

Printed on acid-free paper

This Springer imprint is published by Springer Nature
The registered company is Springer International Publishing AG
The registered company address is: Gewerbestrasse 11, 6330 Cham, Switzerland

*This volume is an expression of gratitude to one of our professional colleagues and friends with whom we have had the pleasure to meet and work. It is dedicated to Dr. Jacek M. Zurada, one of the most prominent scientists and technical leaders in the field of computational intelligence, who has made many pioneering contributions to the field, notably to the theory and applications of neural networks. But, in addition to his widely recognized and cited works, he has equally distinguished himself through his career by his exceptional leadership and service to the profession and community. His illustrious academic and*

*professional career spans over three stages: the early doctoral years in Poland and postdoctoral training at ETH Zürich in Switzerland (1972–1980), his academic work in the USA (1980–present), and his intermittent visiting professorship positions, first during his sabbatical at Princeton University and then during his summer stays in Singapore and Japan.*

*It can be easily argued that the impact of Dr. Zurada's technical, professional, and educational accomplishments has been extraordinary in each of these aspects alone, but only their sum, broadened by his generous personality, has made a truly unique and influential career.*

*His work has heavily influenced the field, and it continues to inspire and benefit numerous researchers as it will for sure do for decades to come. For over 25 years, he has been one of the most recognizable scientists and personalities of the field of computational intelligence, notably in neural networks. This recognition is to be credited to his seminal works that continue to have lasting societal and technical impact. He has contributed to the fundamental understanding of the field through publishing many papers on significant theoretical advances and applications of tools and techniques developed in the field, to a large extent by himself and his collaborators, and through authoring a groundbreaking book that has been widely considered as a pioneering and standard reference of the field. His contributions have resulted in about 10500 citations.*

*One of his most widely recognized singular contributions is to the area of recurrent attractor networks that use complex-valued neurons (1996) which established a new and seminal paradigm of the Hopfield-type associative memories. Further, he has developed one of the most successful and widely applied methods to deal with the "black box nature" of neural networks through their sensitivity evaluations. This novel idea has allowed for an efficient and systematic reduction of oversized architectures, pruning of inputs and other simplifications. Based on this seminal concept of the perceptron networks sensitivity, other algorithms have been developed for network pruning, derivation of logic rules and explanation capabilities. Well over two dozen authors had continued extending the early concepts proposed by Dr. Zurada.*

*One of the most attractive applications of neural networks has been in the field of computer-assisted medicine. Here, Dr. Zurada's work in drug dosing with computational methods has opened new avenues and lines of inquiry for numerous researchers all over the world. Working with his colleagues in the University of Louisville's School of Medicine and his Ph.D. students, he has devised new pharmacokinetical models for renal clinic patients and for different drugs. These pioneering works have resulted in numerous articles in many highly respected journals and opened new vistas for both the theory and practice.*

*Last but not least, Dr. Zurada's signature contribution is his famous book "Introduction to Artificial Neural Systems" which is widely recognized as the first comprehensive and cohesive academic text of the field. It has ingeniously combined the scope and depth of coverage with the clarity of exposition. It has also been reprinted in Singapore, Poland, Egypt, and most recently in India, the latter of which underscores this pioneering work's outstanding longevity. This book has successfully bridged the gaps between the early multifaceted research by scientists from various fields including psychology, physics, information theory, computer science, electrical engineering, and others. In fact, this contribution of Dr. Zurada has laid the foundation for numerous neurocomputing courses in electrical/computer and other engineering and computer science departments throughout the world.*

*Dr. Zurada's academic teaching emphasizes in-depth, project-based learning that helps electrical/computer engineers to stay technically current as the technology evolves during their careers. He has also served the industry as a consultant and lecturer. He has advised 21 Ph.D. and many more M.Sc. students many of whom now hold leadership positions in industrial R&D centers, academia, and governmental agencies in the USA, Korea, and Poland. He has also delivered 170 invited, plenary, and keynote conference presentations and seminars throughout the world. He has served as IEEE Distinguished Speaker for the IEEE Systems, Man and Cybernetics Society, and served as a Distinguished Lecturer the IEEE Circuits and Systems and Computational Intelligence Societies.*

*It is also the editors' pleasure to cite
Dr. Zurada's distinguished career of service
to the profession, mostly to the IEEE all the
editors are strongly attached to. Since 1992,
he has served in many editorial roles. He was
an Associate Editor of the IEEE Transactions
on Circuits and Systems, Parts I and II, and a
Member of the Editorial Board of the
Proceedings of the IEEE and Senior Advisory
Editor of IEEE Computational Intelligence
Magazine. He also served as the
Editor-in-Chief of the IEEE Transactions on
Neural Networks (1998–2003). He has served
as a chair or member of about 140
conference committees.*

*He has made an extraordinary impact on the
IEEE Computational Intelligence Society
(formerly Neural Networks Society) and was
the Society President (2004–2005). More
recently, he has held several top IEEE
positions in the Publications, Products and
Services and Technical Activities Boards,
including Chair of IEEE TAB Periodicals
Committee (2010–2011) and of Periodical
Review and Advisory Committee
(2012–2013). He was also elected as the
2014 TAB Chair or Vice-President of IEEE,
Technical Activities (2014—VP Elect, and
2016—Past VP).*

*In recognition of his research
accomplishments and his unselfish service to
the profession, Dr. Zurada has received a
number of awards for distinction in research,
teaching, and service including the 1993
Presidential Award for Research,
Scholarship and Creative Activity, and the
2001 Presidential Distinguished Service
Award for Service to the Profession. He*

*received the Golden Jubilee IEEE Medal
from the Circuits and Systems Society in 2000
and the Meritorious Service Award from the
Computational Intelligence Society in 2008.
He is an IEEE Life Fellow. In 2003, he was
conferred the Title of Professor by the
President of Poland. In 2005, he was elected
a Foreign Member of the Polish Academy of
Sciences. He also holds four honorary
doctorates from European and Asian
universities.
We are the four coeditors who represent
hundreds of privileged individuals in our
community who have had the pleasure to
personally and professionally know
Dr. Zurada, work with him, or study under
his direction, and enjoy his friendship. We
have undertaken this editorial effort to honor
his dedication and impact he had on all of us.
Our intention is to present these select topical
papers in our research field as a token of
appreciation for his efforts that have
benefitted so many of us.*

Adam E. Gawęda
Janusz Kacprzyk
Leszek Rutkowski
Gary G. Yen
Spring 2017

# Preface

This volume is dedicated to Prof. Jacek Żurada, Full Professor at the Computational Intelligence Laboratory, Department of Electrical and Computer Engineering, J.B. Speed School of Engineering, University of Louisville, Kentucky, USA, as a token of appreciation for his scientific and scholarly achievements, and his longtime service to many communities, notably—from the point of view of research interests topics—those of computational intelligence, in particular neural networks, machine learning, data analyses, and data mining, but also fuzzy logic, evolutionary computation, to just mention a few. On the other hand, from an institutional and organizational point of view, this is also a small token of appreciation for his longtime dedication and service to so many scientific, scholarly, and professional communities and societies, notably those of IEEE (Institute of Electrical and Electronics Engineers), the world largest professional technical professional organization dedicated to advancing science and technology in a broad spectrum of areas and fields related to its scope of interest.

Dr. Żurada's illustrious scientific and scholarly career spans over so many fields and areas of science and technology exemplified primarily by neural networks, the area he has been for years an iconic personality, but also many other areas from the broadly perceived fields of data sciences, machine learning, knowledge engineering, and—to put it most generally, maybe by using too general a name—for all kinds of intelligent systems. In a more applied direction, his influential works in the field of computer-assisted medicine deserve much appreciation, both because of their scientific quality and—which is maybe even more important—for their crucial relevance and value to the society.

The volume is divided into five parts that cover main issues related to the topic of the volume. Part I deals with theoretic, algorithmic, and implementation problems related to an intelligent use of data in the sense of how to get from data information and knowledge which can be in general useful for solving some relevant tasks, such as, data mining, machine learning, and knowledge discovery.

In his paper on "Tensor Networks for Dimensionality Reduction, Big Data and Deep Learning," Andrzej Cichocki provides a comprehensive and critical state-of-the-art survey, complemented with a deep vision on some innovative links between low-rank tensor network decompositions and deep neural networks. This survey and analysis is motivated by the fact that large-scale multidimensional data are often provided as multiway arrays or higher-order tensors, and they can be approximately represented in distributed forms via low-rank tensor decompositions and tensor networks. Due to the underlying low-rank approximations, tensor networks may help reduce the dimensionality and alleviate the infamous curse of dimensionality in many real-life cases, exemplified buy large-scale optimization problems and deep learning. A novel view of links between the low-rank tensor network decompositions and the deep neural networks is provided and graphically illustrated. It is shown in an intuitively appealing way that due to low-rank tensor approximations and sophisticated contractions of core tensors, tensor networks attain a remarkable ability to perform distributed computations on otherwise prohibitively large volume of data/parameters. The approach is mainly related to the Hierarchical Tucker tensor train (TT) decompositions and the MERA tensor networks in some specific applications.

Jerzy Błaszczyński and Jerzy Stefanowski ("Local Data Characteristics in Learning Classifiers from Imbalanced Data") deal with a very important yet difficult and challenging problem of learning classifiers from imbalanced data. Standard classifiers do not usually show a good performance due to many factors, notably those related to data difficulty related to internal and local characteristics of class distributions. Many of these difficulties can be alleviated by some approximation through an analysis of some neighborhoods of learning examples and the identification of different types of examples from the minority class. The authors assume a recent research direction for the evaluation of the types of examples that are based on the use of either the $k$-nearest neighbor or kernel-based methods. Some approaches are shown for tuning the size of both kinds of neighborhoods depending on the data set characteristics as well as for the evaluation of their usefulness in a series of both benchmark type and real data. Then, a claim is considered and analyzed that a proper analysis of these neighborhoods could be a basis for the development of new specialized algorithms for dealing with imbalanced data. For illustration, some generalizations of oversampling in preprocessing methods and neighborhood-based ensembles are discussed.

Paweł Szmeja, Maria Ganzha, Marcin Paprzycki, and Wiesław Pawłowski ("Similarity dimensions of semantic ontologies") deal with a very important, yet difficult, problem of semantic similarity which is usually meant in the sense of tools, models, and methods applied in knowledge bases, semantic graphs, text disambiguation, and ontology matching, to just name a few more relevant problem classes. Many models and algorithms have been proposed for that purpose, and— though they are usually very different both with respect to the very idea, algorithm, and implementation—they are all meant to produce a single numerical score evaluation, termed a "semantic similarity" that is meant to capture all aspects of similarity. The authors claim that there are many ways in which semantic entities can be similar, and a single score may not be the best option. In their approach, a division of knowledge (and, consequently, the similarity) into categories (dimensions) of semantic relationships is performed, with each dimension representing a different "type" of similarity, with this process guided by an interpretation of the meaning (semantics) of a similarity score in a particular dimension. Therefore, an add extra information to a similarity score can be added to emphasize differences and similarities between results obtained by using different methods.

Ryszard Tadeusiewicz ("Some interesting phenomenon occurring during self-learning process with its psychological interpretation") discusses some interesting and general issues related to neural networks and artificial intelligence. The point of departure is that neural networks are very often useful for solving many practical problems but this usefulness can be viewed limited in the sense that it can be interesting and valid for a limited number of readers who are concerned with similar problems and applications. Therefore, a reasonable approach may be that some more interesting observations, which are related to phenomena observed, are selected during the neural network self-learning process. Since there is some intrinsic similarity to psychological processes that can be observed during a natural activity in the human mind, such phenomena are called "artificial dreams" meant here as spontaneous and unexpected processes emerging from natural self-learning procedures. These phenomena are very interesting and exciting, even mysterious, yet are rarely considered in a sufficient depth by the artificial intelligence or computational intelligence communities. The main reason may be viewed to be die to the fact that most contributions presenting methods and results of self-learning, even in neural networks which are main tool considered in this work, are mainly goal-oriented, and authors of almost all papers first try to obtain the best result in terms of solving a specified problem, for instance, by building a neural network based model of some process or finding a solution of a pattern recognition problem. Therefore, in the discussion of the self-learning results, the authors usually take into account only the final result exemplified by the value of a measure of the quality of the model or the correctness of classification. Issues discussed in this paper occur when the self-learning system has not been learned enough, and emphasis is on a rarely considered issue of a detailed analysis of behavior of a network, or other self-learning system, during the learning process, as well as some unexpected outcomes.

Part II is devoted to various aspects of neural networks and connectionist systems. Filippo Maria Bianchi, Lorenzo Livi, and Cesare Alippi, in their paper "On the interpretation and characterization of echo state networks dynamics: A complex systems perspective," discuss some relevant, recently developed methods for characterizing the dynamics of recurrent neural network using some concepts and tools and techniques of complex systems theory. They focus on the so-called echo state networks which are a class of recurrent networks. They show a method for the characterization and analysis of the evolution of internal states, which makes it possible to provide a qualitative interpretation of the network dynamics, as well as to assess the very important problem, for theoretical and practical points of view, of stability of the system. Then, the identification of the onset of criticality in such networks is dealt with. The authors discuss an unsupervised method based on Fisher information which can be used to tune the network hyperparameters. It is shown that as compared to standard supervised techniques, the proposed approach is effective and efficient for many problems, and shows better results.

Martha Pulido, Patricia Melin, and Olivia Mendoza ("Optimization of Ensemble Neural Networks with Type-1 and Interval Type-2 Fuzzy Integration for Forecasting the Taiwan Stock Exchange") describe an optimization method based on the PSO (particle swarm optimization) for ensemble neural networks with type-1 and type-2 fuzzy aggregation for the forecasting complex time series, notably related to financial data. Notably, the optimization of the structure of the ensemble neural network with type-1 and type-2 fuzzy integration is concerned. For the comparison of the new hybrid method proposed with traditional methods, the data from the Taiwan Stock Exchange (TAIEX) are used, and the simulation results show that the ensemble approach produces good prediction results.

In his paper "Deep Neural Networks—A Brief History," Krzysztof J. Cios provides a description on and insight into Deep Neural Networks (DNN), their history, and some related concepts and works. Basically, the DNNs—which are one of the most efficient tools that belong to the so-called deep learning—process input information in a hierarchical way in that each subsequent level of processing extracts more abstract/global/invariant features so that the DNNs (semi) automatically learn key features from data and then aggregate them for some purpose, such as the recognition of objects in the images. To be more specific, the author illustrates how the DNNs using some example from face recognition where the inputs are images from which at the first level (the first hidden layer) of processing simple image characteristics such as edges are extracted, then—at the second and subsequent levels—more complex parts of an image are formed, and—finally, at the output layer—human faces are recognized. Then, the author concentrates on the fully unsupervised DNNs, the field in which little progress has been reported so far. The focus here is on the DNNs, including those that use spiking neuron models and the corresponding learning rules.

Part III deals with broadly perceived tools and techniques for intelligent technologies in systems modeling. Grzegorz J. Nalepa ("Techniques for Construction and Integration of Rule Bases") discusses issues related to the use of rules for capturing and executing knowledge. He deals with rule-based shells, software frameworks that support knowledge engineers by providing a rule language for encoding the rule base and a generic inference engine. One of the best known shells, CLIPS (C Language Integrated Production System) is now a multiparadigm programming language that provides support for rule-based, object-oriented, and procedural programming. This wide acceptance of CLIPS has implied the development of Jess which, although being similar, has been entirely written in Java which improved its integration capabilities. The development of intelligent systems in last decades shows that the rule-based systems (RBS) are still a technology with a great potential and many applications. However, it is also clear that rules, while very useful, need to be integrated with other paradigms, including those related to data and knowledge processing, software development, implementation, etc. In this paper, the author presents an identification of some issues that are relevant for the integration of rule-based systems, notably: high-level modeling techniques for rule bases, integration architectures for rule-based systems, and rule interoperability. A human assisted and an automatic derivation of rules are discussed, and some challenging common problems, notably the handling of large rules sets through structuring, integration of rule-based components, as well as rule interoperability issues, are discussed.

Krystian Łapa, Krzysztof Cpałka, and Leszek Rutkowski ("New Aspects of Interpretability of Fuzzy Systems for Nonlinear Modeling") discuss fuzzy systems as a well suited tool for modeling nonlinear systems. The authors emphasize that the fuzzy systems can be effectively and efficiently used if their structure and structure parameters are properly chosen, and the rules are clear and interpretable. A new algorithm for the automatic learning of fuzzy systems and new interpretability criteria of fuzzy systems are proposed. The interpretability criteria are related to all aspects of those systems, not only their fuzzy sets and rules, and also concern the choice and analysis of parameterized triangular norms, discretization points and weights of importance from the rules. Such a comprehensive solution is novel. The proposed criteria are taken into account in the learning process which proceeds using a new learning algorithm that combines the genetic algorithm and the firework algorithms, which makes it possible to automatically choose not only the parameters but also the structure of the system. The new approach is tested on some relevant simulation problems of nonlinear modeling.

Krassimir T. Atanassov and Peter Vassilev discuss in their paper "On the Intuitionistic Fuzzy Sets of $n$-th Type" the use of various extensions of the concept of a fuzzy set introduced by Zadeh, notably some extensions along the line of Atanassov's intuitionistic fuzzy set that makes it possible not only to express imprecision of information but a very important problem related to the fact that the human beings tend to use in their everyday discourse, judgments, reasoning, etc., aspects for and against. The author clarifies some misconceptions and introduces a unified framework for such approaches.

In Part IV, "Intelligent Technologies in Decision Making, Optimization and Control," the first paper by Jacek Mańdziuk ("MCTS/UCT in solving real-life problems") deals with the Monte Carlo Tree Search (MCTS) supported by the Upper Confidence Bounds Applied to Trees (UCT) method, i.e., the so-called MCTS/UCT which is one of the state-of-the-art techniques in the game-playing domain. In particular, it is emphasized the spectacular success of this method (combined with the use of deep neural networks trained with the reinforcement learning algorithm) in the game of Go. The author summarizes his works and experience in the application of MCTS/UCT to domains other than games, with a particular emphasis on hard real-life problems with a large degree of uncertainty due to the existence of some stochastic factors in their definition, exemplified by the Capacitated Vehicle Routing Problem with Traffic Jams, and the Risk-Aware Project Scheduling Problem. It is shown how MCTS/UCT is a viable method in these two domains, notably due its ability to effectively and efficiently deal with uncertainty by online adaptation of the core MCTS simulations to the current situation.

Miłosz Kadziński, Michał K. Tomczyk, and Roman Słowiński ("Interactive Cone Contraction for Evolutionary Multiple Objective Optimization") present a new interactive evolutionary algorithm for Multiple Objective Optimization (MOO) which combines the NSGA-II method with a cone contraction method. The new approach requires the Decision Maker (DM) to provide the preference information as a reference point and pairwise comparisons of solutions from a current population. This information is represented using a compatible Achievement Scalarizing Function (ASF) which is used to guide the evolutionary search toward the most preferred region of the Pareto front. The proposed algorithm is tested on a set of benchmark problems, and the results show its quick convergence to the DM's most preferred region. Moreover, it also indicated the advantage of the new algorithm of the well-known NEMO-0, in particular when the DM provides a richer preference information composed of a greater number of pairwise comparisons of solutions.

Oscar Castillo, Carlos Soto, and Fevrier Valdez ("A Review of Fuzzy and Mathematic Methods for Dynamic Parameter Adaptation in the Firefly Algorithm") are concerned with some issues related to the design and use of the firefly algorithm, a well-known meta-heuristic. The authors concentrate on the choice of parameters of the firefly algorithm, its analysis, and dynamic adjustments. Some relevant traditional and fuzzy logic-based approaches are analyzed and numerically compared.

In Part V, "Applications of Intelligent Technologies," in the first paper by Adam E. Gawęda and Michael E. Brier ("Computational Intelligence Methods in Personalized Pharmacotherapy"), the authors are concerned with a pharmacologic therapy of chronic diseases that remains a big challenge to physicians, notably because individual dose-response characteristics of patients may vary significantly across patient populations, and—due to a chronic nature of the process—they may change over time within individual patients as well. Current state-of-the-art protocols for dose adjustment of pharmacologic agents rely heavily on data from the drug approval process and a physician's expertise but they do not fuzzy utilize the wealth of knowledge hidden in patient data collected during his or her treatment. The authors review the application of two computational intelligence methods: the artificial neural networks and fuzzy sets theory, to personalized pharmacologic treatment of a chronic condition using patient data. As an example, the authors use data on patients with anemia and renal failure.

Zdzisław Kowalczuk and Michał Czubenko ("Embodying Intelligence in Autonomous and Robotic Systems with the Use of Cognitive Psychology and Motivation Theories") discuss a coherent anthropological approach for the control of autonomous robots or agents. This modern approach is based on an appropriate modeling of the human mind using the available psychological knowledge. One of the main reasons that have inspired the authors is the lack of available and effective top-down approaches resulting from the some known results from the area of autonomous robotics. On the other hand, a system for a comprehensive and effective and efficient modeling of human psychology for the purpose of constructing autonomous systems is lacking. The authors review the recent progress in the understanding of the mechanisms of cognitive computations underlying decision-making and existing challenges, notably those founded on cognitive ideas such as LIDA, CLARION, SOAR, MANIC, DUAL, and OpenCog. In particular, the idea of an Intelligent System of Decision-making (ISD) is emphasized that is based on the results of cognitive psychology (using the aspect of "information path"), motivation theory (where the needs and emotions serve as the main drives, or motivations, in the mechanism of governing autonomous systems), and several other detailed theories, which concern memory, categorization, perception, and decision-making. In the ISD system, in particular, an xEmotion subsystem is focused on that covers the psychological theories on emotions, including the appraisal, evolutionary, and somatic theories.

Krystian Łapa and Krzysztof Cpałka ("Evolutionary Approach for Automatic Design of PID Controllers") present a new approach to an automatic design of the well-known and widely used PID controllers. It is based on a meta-heuristic hybrid algorithm which combines the genetic algorithm and the imperialist one. The main characteristic of the proposed approach is its capability to design the structure of the controller and the structure of its parameters. This eliminates the need for a trial-and-error process during the design of the controller structure. Moreover, in the proposed approach, various control criteria can be reflected.

Marcin Zalasiński, Krzysztof Cpałka, and Leszek Rutkowski ("Fuzzy-genetic Approach to Identity Verification Using a Handwritten Signature") discuss a relevant biometric problem of the verification of the dynamic signature. There are many methods for the signature verification using dynamics of the signing process often based on the so-called global features. In this paper, a new approach to the signature verification using global features is proposed. Basically, it involves the classification of the signature which is performed using a fuzzy-genetic system; the selection of an individual set of features for each signer which uses a genetic algorithm with an appropriately designed evaluation function and works without access to the signatures called skilled forgeries; and the determination of weights of importance for evolutionarily selected features which are taken into account in the classification process. The main advantages of this new approach is that the feature selection via a fuzzy-genetic systems works with access to the signatures called skilled forgeries, and also that the proposed classifier can do without machine learning with respect to its work interpretation and possibility of an analytical determination of its parameters. Simulation results for the BioSecure signature database, distributed by the BioSecure Association, are performed and confirm the above mentioned good results.

S. Piasecki, R. Szmurlo, J. Rabkowski, and M.P. Kaźmierkowski ("A Method of Design and Optimization for SiC-based Grid-connected AC-DC Converters") present a method of design and optimization for three-phase AC-DC converters. The main idea of presented work is to provide a tool which supports the design process and helps to achieve the main desired properties: efficiency, volume, weight, and cost. The proposed design method is described with a special attention paid to calculations regarding the power section of the converter. The authors concentrate on the new technology of SiC power devices. The method is illustrated on three SiC-based laboratory models rated at 10 and 20 kVA, respectively.

Each model is a result of an optimization process performed for different input requirements related to the volume and efficiency. Finally, the performance of all models is verified during the operation with a 3x400V AC grid.

We would like to express our gratitude to all the authors for their interesting, novel, and inspiring contributions. Peer-reviewers also deserve a deep appreciation, because their insightful and constructive remarks and suggestions have considerably improved many contributions.

And last but not least, we wish to thank Dr. Tom Ditzinger, Dr. Leontina di Cecco, and Mr. Holger Schaepe for their dedication and help to implement and finish this large publication project on time maintaining the highest publication standards.

Louisville, USA
Warsaw, Poland
Częstochowa, Poland
Stillwater, USA
Spring 2017

Adam E. Gawęda
Janusz Kacprzyk
Leszek Rutkowski
Gary G. Yen

# Contents

Part III   Intelligent Technologies in Systems Modeling

Part IV   Intelligent Technologies in Decision Making, Optimization
            and Control

Part V   Applications of Intelligent Technologies

# Part I
# Data Mining, Machine Learning, Knowledge Discovery

# Tensor Networks for Dimensionality Reduction, Big Data and Deep Learning

Andrzej Cichocki

**Abstract** Large scale multidimensional data are often available as multiway arrays or higher-order tensors which can be approximately represented in distributed forms via low-rank tensor decompositions and tensor networks. Our particular emphasis is on elucidating that, by virtue of the underlying low-rank approximations, tensor networks have the ability to reduce the dimensionality and alleviate the curse of dimensionality in a number of applied areas, especially in large scale optimization problems and deep learning. We briefly review and provide tensor links between low-rank tensor network decompositions and deep neural networks. We elucidating, through graphical illustrations, that low-rank tensor approximations and sophisticated contractions of core tensors, tensor networks have the ability to perform distributed computations on otherwise prohibitively large volume of data/parameters. Our focus is on the Hierarchical Tucker, tensor train (TT) decompositions and MERA tensor networks in specific applications.

## 1 Introduction and Objectives

This paper aims to present some new ideas and methodologies related to tensor decompositions (TDs) and tensor networks models (TNs), especially in applications to deep neural networks (DNNs) and dimensionality reduction. The resurgence of artificial neural systems, especially deep learning neural networks has formed an active frontier of machine learning, signal processing and data mining [1–6, 13, 14]. Tensor decompositions (TDs) decompose complex data tensors of exceedingly high volume into their factor (component) matrices, while tensor networks (TNs)

A. Cichocki (✉)
Systems Research Institute, Polish Academy of Science, Warsaw, Poland
e-mail: a.cichocki@riken.jp

A. Cichocki
RIKEN Brain Science Institute, Tokyo, Japan

A. Cichocki
SKOLTECH, Moscow, Russia

© Springer International Publishing AG 2018
A.E. Gawęda et al. (eds.), *Advances in Data Analysis with Computational Intelligence Methods*, Studies in Computational Intelligence 738,
https://doi.org/10.1007/978-3-319-67946-4_1

decompose higher-order tensors into sparsely interconnected small-scale factor matrices and/or low-order core tensors [7–14]. These low-order core tensors are called "components", "blocks", "factors" or simply "cores". In this way, large-scale data can be approximately represented in highly compressed and distributed formats.

In this paper, the TDs and TNs are treated in a unified way, by considering TDs as simple tensor networks or sub-networks; the terms "tensor decompositions" and "tensor networks" will therefore be used interchangeably. Tensor networks can be thought of as special graph structures which break down high-order tensors into a set of sparsely interconnected low-order core tensors, thus allowing for both enhanced interpretation and computational advantages [12–14].

Tensor networks offer a theoretical and computational framework for the analysis of computationally prohibitive large volumes of data, by "dissecting" such data into the "relevant" and "irrelevant" information. In this way, tensor network representations often allow for super-compression of data sets as large as $10^8$ entries, down to the affordable levels of $10^5$ or even less entries [15–25].

**Challenges in Big Data Processing**. Extreme-scale volumes and variety of modern data are becoming ubiquitous across the science and engineering disciplines. In the case of multimedia (speech, video), remote sensing and medical/biological data, the analysis also requires a paradigm shift in order to efficiently process massive data sets within tolerable time (velocity). Such massive data sets may have billions of entries and are typically represented in the form of huge block matrices and/or tensors. This has spurred a renewed interest in the development of tensor algorithms that are suitable for extremely large-scale data sets.

Apart from the huge Volume, the other features which characterize big data include Veracity, Variety and Velocity (see Fig. 1a and b). Each of the "V features" represents a research challenge in its own right. For example, high volume implies the need for algorithms that are scalable; high Velocity requires the processing of big data streams in near real-time; high Veracity calls for robust and predictive algorithms for noisy, incomplete and/or inconsistent data; high Variety demands the fusion of different data types, e.g., continuous, discrete, binary, time series, images, video, text, probabilistic or multi-view. Some applications give rise to additional "V challenges", such as Visualization, Variability and Value. The Value feature is particularly interesting and refers to the extraction of high quality and consistent information, from which meaningful and interpretable results can be obtained.

Our objective is to show that tensor networks provide a natural sparse and distributed representation for big data, and address both established and emerging methodologies for tensor-based representations and optimization. Our particular focus is on low-rank tensor network representations, which allow for huge data tensors to be approximated (compressed) by interconnected low-order core tensors [10, 11, 14].

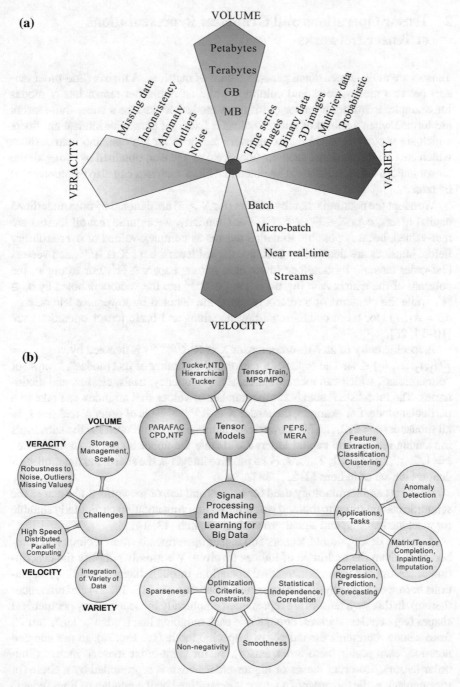

**Fig. 1** **a** The 4 V challenges for big data. **b** A framework for extremely large-scale data analysis and the potential applications based on tensor decomposition approaches

## 2 Tensor Operations and Graphical Representations of Tensor Networks

Tensors are multi-dimensional generalizations of matrices. A matrix (2nd-order tensor) has two modes, rows and columns, while an $N$th-order tensor has $N$ modes for example, a 3rd-order tensor (with three-modes) looks like a cube. Sub-tensors are formed when a subset of tensor indices is fixed. Of particular interest are fibers which are vectors obtained by fixing every tensor index but one, and matrix slices which are two-dimensional sections (matrices) of a tensor, obtained by fixing all the tensor indices but two. It should be noted that block matrices can also be represented by tensors.

We adopt the notation whereby tensors (for $N \geq 3$) are denoted by bold underlined capital letters, e.g., $\underline{\mathbf{X}} \in \mathbb{R}^{I_1 \times I_2 \times \cdots \times I_N}$. For simplicity, we assume that all tensors are real-valued, but it is possible to define tensors as complex-valued or over arbitrary fields. Matrices are denoted by boldface capital letters, e.g., $\mathbf{X} \in \mathbb{R}^{I \times J}$, and vectors (1st-order tensors) by boldface lower case letters, e.g., $\mathbf{x} \in \mathbb{R}^J$. For example, the columns of the matrix $\mathbf{A} = [\mathbf{a}_1, \mathbf{a}_2, \dots, \mathbf{a}_R] \in \mathbb{R}^{I \times R}$ are the vectors denoted by $\mathbf{a}_r \in \mathbb{R}^I$, while the elements of a matrix (scalars) are denoted by lowercase letters, e.g., $a_{ir} = \mathbf{A}(i, r)$ (for more details regarding notations and basic tensor operations see [10–14, 26].

A specific entry of an $N$th-order tensor $\underline{\mathbf{X}} \in \mathbb{R}^{I_1 \times I_2 \times \cdots \times I_N}$ is denoted by $x_{i_1, i_2, \dots, i_N} = \underline{\mathbf{X}}(i_1, i_2, \dots, i_N) \in \mathbb{R}$. The order of a tensor is the number of its "modes", "ways" or "dimensions", which can include space, time, frequency, trials, classes, and dictionaries. The term "size" stands for the number of values that an index can take in a particular mode. For example, the tensor $\underline{\mathbf{X}} \in \mathbb{R}^{I_1 \times I_2 \times \cdots \times I_N}$ is of order $N$ and size $I_n$ in all modes-$n$ ($n = 1, 2, \dots, N$). Lower-case letters e.g., $i, j$ are used for the subscripts in running indices and capital letters $I, J$ denote the upper bound of an index, i.e., $i = 1, 2, \dots, I$ and $j = 1, 2, \dots, J$. For a positive integer $n$, the shorthand notation $<n>$ denotes the set of indices $\{1, 2, \dots, n\}$.

Notations and terminology used for tensors and tensor networks differ across the scientific communities to this end we employ a unifying notation particularly suitable for machine learning and signal processing research [13, 14].

A precise description of tensors and tensor operations is often tedious and cumbersome, given the multitude of indices involved. We grossly simplify the description of tensors and their mathematical operations through diagrammatic representations borrowed from physics and quantum chemistry (see [13, 14, 27] and references therein). In this way, tensors are represented graphically by nodes of any geometrical shapes (e.g., circles, squares, dots), while each outgoing line ("edge", "leg", "arm") from a node represents the indices of a specific mode (see Fig. 2a). In our adopted notation, each scalar (zero-order tensor), vector (first-order tensor), matrix (2nd-order tensor), 3rd-order tensor or higher-order tensor is represented by a circle (or rectangular), while the order of a tensor is determined by the number of lines (edges) connected to it. According to this notation, an $N$th-order tensor $\underline{\mathbf{X}} \in \mathbb{R}^{I_1 \times \cdots \times I_N}$ is represented by a circle (or any shape) with $N$ branches each of size $I_n$, $n = 1, 2, \dots, N$

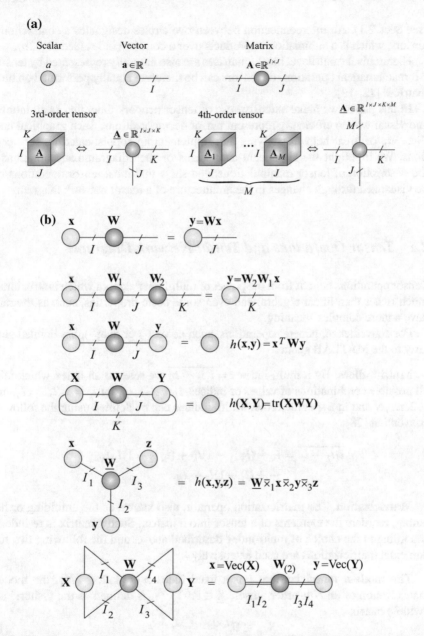

**Fig. 2** Graphical representation of tensor operations. **a** Basic building blocks for tensor network diagrams. **b** Tensor network diagrams for matrix-vector and tensor-vectors multiplications

(see Sect. 2.1). An interconnection between two circles designates a contraction of tensors, which is a summation of products over a common index (see Fig. 2b).

Hierarchical (multilevel block) matrices are also naturally represented by tensors. All mathematical operations on tensors can be therefore equally performed on block matrices [12, 13].

In this paper, we make extensive use of tensor network diagrams as an intuitive and visual way to efficiently represent tensor decompositions. Such graphical notations are of great help in studying and implementing sophisticated tensor operations. We highlight the significant advantages of such diagrammatic notations in the description of tensor manipulations, and show that most tensor operations can be visualized through changes in the architecture of a tensor network diagram.

## 2.1 Tensor Operations and Tensor Network Diagrams

Tensor operations benefit from the power of multilinear algebra which is structurally much richer than linear algebra, and even some basic properties, such as the rank, have a more complex meaning.

For convenience, general operations, such as vec($\cdot$) or diag($\cdot$), are defined similarly to the MATLAB syntax.

**Multi-indices**: By a multi-index $i = \overline{i_1 i_2 \cdots i_N}$ we refer to an index which takes all possible combinations of values of indices, $i_1, i_2, \ldots, i_N$, for $i_n = 1, 2, \ldots, I_n$, $n = 1, 2, \ldots, N$ and in a specific order. Multi–indices can be defined using the following convention [28]:

$$\overline{i_1 i_2 \cdots i_N} = i_N + (i_{N-1} - 1)I_N + (i_{N-2} - 1)I_N I_{N-1} + \cdots + (i_1 - 1)I_2 \cdots I_N.$$

**Matricization**. The matricization operator, also known as the unfolding or flattening, reorders the elements of a tensor into a matrix. Such a matrix is re-indexed according to the choice of multi-index described above, and the following two fundamental matricizations are used extensively.

**The mode-$n$ matricization**. For a fixed index $n \in \{1, 2, \ldots, N\}$, the mode-$n$ matricization of an $N$th-order tensor, $\underline{\mathbf{X}} \in \mathbb{R}^{I_1 \times \cdots \times I_N}$, is defined as the ("short" and "wide") matrix

$$\mathbf{X}_{(n)} \in \mathbb{R}^{I_n \times I_1 I_2 \cdots I_{n-1} I_{n+1} \cdots I_N}, \tag{1}$$

with $I_n$ rows and $I_1 I_2 \cdots I_{n-1} I_{n+1} \cdots I_N$ columns, the entries of which are

$$(\mathbf{X}_{(n)})_{i_n, \overline{i_1 \cdots i_{n-1} i_{n+1} \cdots i_N}} = x_{i_1, i_2, \ldots, i_N}.$$

Note that the columns of a mode-$n$ matricization, $\mathbf{X}_{(n)}$, of a tensor $\underline{\mathbf{X}}$ are the mode-$n$ fibers of $\underline{\mathbf{X}}$.

**The mode-$\{n\}$ canonical matricization**. For a fixed index $n \in \{1, 2, \ldots, N\}$, the mode-$(1, 2, \ldots, n)$ matricization, or simply mode-$n$ canonical matricization, of a tensor $\underline{\mathbf{X}} \in \mathbb{R}^{I_1 \times \cdots \times I_N}$ is defined as the matrix

$$\mathbf{X}_{<n>} \in \mathbb{R}^{I_1 I_2 \cdots I_n \times I_{n+1} \cdots I_N}, \tag{2}$$

with $I_1 I_2 \cdots I_n$ rows and $I_{n+1} \cdots I_N$ columns, and the entries

$$(\mathbf{X}_{<n>})_{\overline{i_1 i_2 \ldots i_n}, \overline{i_{n+1} \ldots i_N}} = x_{i_1, i_2, \ldots, i_N}.$$

The matricization operator in the MATLAB notation (reverse lexicographic) is given by

$$\mathbf{X}_{<n>} = \text{reshape} \left( \underline{\mathbf{X}}, I_1 I_2 \cdots I_n, I_{n+1} \cdots I_N \right). \tag{3}$$

As special cases we immediately have

$$\mathbf{X}_{<1>} = \mathbf{X}_{(1)}, \quad \mathbf{X}_{<N-1>} = \mathbf{X}_{(N)}^{\text{T}}, \quad \mathbf{X}_{<N>} = \text{vec}(\mathbf{X}). \tag{4}$$

The tensorization of a vector or a matrix can be considered as a reverse process to the vectorization or matricization (see Fig. 3) [14].

The following symbols are used for most common tensor multiplications: ∘ for the outer product $\otimes$ for the Kronecker product, $\odot$ for the Khatri–Rao product, ⊛ for the Hadamard (componentwise) product, and $\times_n$ for the mode-$n$ product. We refer to [13, 14, 26, 29] for more detail regarding the basic notations and tensor operations (Figs. 4 and 5).

**Outer product**. The central operator in tensor analysis is the outer or tensor product, which for the tensors $\underline{\mathbf{A}} \in \mathbb{R}^{I_1 \times \cdots \times I_N}$ and $\underline{\mathbf{B}} \in \mathbb{R}^{J_1 \times \cdots \times J_M}$ gives the tensor $\underline{\mathbf{C}} = \underline{\mathbf{A}} \circ \underline{\mathbf{B}} \in \mathbb{R}^{I_1 \times \cdots \times I_N \times J_1 \times \cdots \times J_M}$ with entries $c_{i_1, \ldots, i_N, j_1, \ldots, j_M} = a_{i_1, \ldots, i_N} \, b_{j_1, \ldots, j_M}$.

Note that for 1st-order tensors (vectors), the tensor product reduces to the standard outer product of two nonzero vectors, $\mathbf{a} \in \mathbb{R}^I$ and $\mathbf{b} \in \mathbb{R}^J$, which yields a rank-1 matrix, $\mathbf{X} = \mathbf{a} \circ \mathbf{b} = \mathbf{a}\mathbf{b}^{\text{T}} \in \mathbb{R}^{I \times J}$. The outer product of three nonzero vectors, $\mathbf{a} \in \mathbb{R}^I$, $\mathbf{b} \in \mathbb{R}^J$ and $\mathbf{c} \in \mathbb{R}^K$, gives a 3rd-order rank-1 tensor (called pure or elementary tensor), $\underline{\mathbf{X}} = \mathbf{a} \circ \mathbf{b} \circ \mathbf{c} \in \mathbb{R}^{I \times J \times K}$, with entries $x_{ijk} = a_i \, b_j \, c_k$.

The outer (tensor) product has been generalized to the nonlinear outer (tensor) products, as follows

$$\left( \underline{\mathbf{A}} \circ_\rho \underline{\mathbf{B}} \right)_{i_1, \ldots, i_N j_1, \ldots, J_M} = \rho \left( a_{i_1, \ldots, i_N}, b_{j_1, \ldots, j_M} \right), \tag{5}$$

where $\rho$ is, in general, nonlinear suitably chosen function (see [30] and Sect. 7 for more detail).

In a similar way, we can define the generalized Kronecker and the Khatri-Rao products. Generalized Kronecker product of two tensors $\underline{\mathbf{A}} \in \mathbb{R}^{I_1 \times I_2 \times \cdots \times I_N}$ and

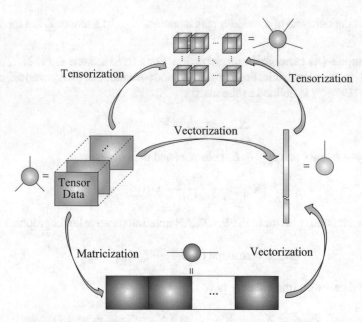

**Fig. 3** Tensor reshaping operations: matricization, vectorization and tensorization. Matricization refers to converting a tensor into a matrix, vectorization to converting a tensor or a matrix into a vector, while tensorization refers to converting a vector, a matrix or a low-order tensor into a higher-order tensor

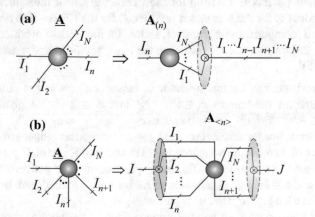

**Fig. 4** Matricization (flattening, unfolding) used in tensor reshaping. **a** Tensor network diagram for the mode-$n$ matricization of an $N$th-order tensor, $\underline{\mathbf{A}} \in \mathbb{R}^{I_1 \times I_2 \times \cdots \times I_N}$, into a short and wide matrix, $\mathbf{A}_{(n)} \in \mathbb{R}^{I_n \times I_1 \cdots I_{n-1} I_{n+1} \cdots I_N}$. **b** Mode-$\{1, 2, \ldots, n\}$th (canonical) matricization of an $N$th-order tensor, $\underline{\mathbf{A}}$, into a matrix $\mathbf{A}_{<n>} = \mathbf{A}_{\overline{(i_1 \ldots i_n} \; ; \; \overline{i_{n+1} \ldots i_N)}} \in \mathbb{R}^{I_1 I_2 \cdots I_n \times I_{n+1} \cdots I_N}$

$\underline{\mathbf{B}} \in \mathbb{R}^{J_1 \times J_2 \times \cdots \times J_N}$ yields a tensor $\underline{\mathbf{C}} = \underline{\mathbf{A}} \otimes_\rho \underline{\mathbf{B}} \in \mathbb{R}^{I_1 J_1 \times \cdots \times I_N J_N}$, with entries $c_{\overline{i_1 j_1}, \dots, \overline{i_N j_N}} = \rho(a_{i_1, \dots, i_N}, b_{j_1, \dots, j_N})$.

Analogously, we can define a generalized Khatri–Rao product of two matrices $\mathbf{A} = [\mathbf{a}_1, \dots, \mathbf{a}_J] \in \mathbb{R}^{I \times J}$ and $\mathbf{B} = [\mathbf{b}_1, \dots, \mathbf{b}_J] \in \mathbb{R}^{K \times J}$ is a matrix $\mathbf{C} = \mathbf{A} \odot_\rho \mathbf{B} \in \mathbb{R}^{IK \times J}$, with columns $\mathbf{c}_j = \mathbf{a}_j \otimes_\rho \mathbf{b}_j \in \mathbb{R}^{IK}$.

**CP decomposition, Kruskal tensor**. Any tensor can be expressed as a finite sum of rank-1 tensors, in the form

$$\underline{\mathbf{X}} = \sum_{r=1}^{R} \mathbf{b}_r^{(1)} \circ \mathbf{b}_r^{(2)} \circ \cdots \circ \mathbf{b}_r^{(N)} = \sum_{r=1}^{R} \left( \overset{N}{\underset{n=1}{\circ}} \mathbf{b}_r^{(n)} \right), \quad \mathbf{b}_r^{(n)} \in \mathbb{R}^{I_n}, \tag{6}$$

which is exactly the form of the Kruskal tensor, also known under the names of CANDECOMP/PARAFAC, Canonical Polyadic Decomposition (CPD), or simply the CP decomposition in (23). We will use the acronyms CP and CPD.

**Tensor rank**. The tensor rank, also called the CP rank, is a natural extension of the matrix rank and is defined as a minimum number, $R$, of rank-1 terms in an exact CP decomposition of the form in (6).

**Multilinear products**. The mode-$n$ (multilinear) product, also called the tensor-times-matrix product (TTM), of a tensor, $\underline{\mathbf{A}} \in \mathbb{R}^{I_1 \times \cdots \times I_N}$, and a matrix, $\mathbf{B} \in \mathbb{R}^{J \times I_n}$, gives the tensor

$$\underline{\mathbf{C}} = \underline{\mathbf{A}} \times_n \mathbf{B} \in \mathbb{R}^{I_1 \times \cdots \times I_{n-1} \times J \times I_{n+1} \times \cdots \times I_N}, \tag{7}$$

with entries $c_{i_1, i_2, \dots, i_{n-1}, j, i_{n+1}, \dots, i_N} = \sum_{i_n=1}^{I_n} a_{i_1, i_2, \dots, i_N} b_{j, i_n}$. An equivalent matrix representation is $\mathbf{C}_{(n)} = \mathbf{B} \mathbf{A}_{(n)}$, which allows us to employ established fast matrix-by-vector and matrix-by-matrix multiplications when dealing with very large-scale tensors. Efficient and optimized algorithms for TTM are, however, still emerging [31–33].

**Full Multilinear Product**. A full multilinear product, also called the Tucker product,[1] of an $N$th-order tensor, $\underline{\mathbf{G}} \in \mathbb{R}^{R_1 \times R_2 \times \cdots \times R_N}$, and a set of $N$ factor matrices, $\underline{\mathbf{B}}^{(n)} \in \mathbb{R}^{I_n \times R_n}$ for $n = 1, 2, \dots, N$, performs the multiplications in all the modes and can be compactly written as

$$\underline{\mathbf{C}} = \underline{\mathbf{G}} \times_1 \mathbf{B}^{(1)} \times_2 \mathbf{B}^{(2)} \cdots \times_N \mathbf{B}^{(N)} \in \mathbb{R}^{I_1 \times I_2 \times \cdots \times I_N}.$$

Observe that this format corresponds to the Tucker decomposition [26, 34, 35] (see also Sect. 3.1).

**Multilinear product of a tensor and a vector (TTV)**. In a similar way, the mode-$n$ multiplication of a tensor, $\underline{\mathbf{G}} \in \mathbb{R}^{R_1 \times \cdots \times R_N}$, and a vector, $\mathbf{b} \in \mathbb{R}^{R_n}$ (tensor-times-vector, TTV) yields a tensor

---

[1] The standard multilinear product can be generalized to nonlinear multilinear product as $\underline{\mathbf{C}} = \underline{\mathbf{G}} \times_1^\sigma \mathbf{B}^{(1)} \times_2^\sigma \mathbf{B}^{(2)} \cdots \times_N^\sigma \mathbf{B}^{(N)}$, where $\underline{\mathbf{G}} \times_n^\sigma \mathbf{B} = \sigma(\underline{\mathbf{G}} \times_n \mathbf{B})$, and $\sigma$ is a suitably chosen nonlinear activation function.

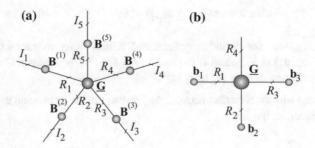

**Fig. 5** Generalized (nonlinear) multilinear tensor products used in deep learning in a compact tensor network notation. **a** Generalized multilinear product of a tensor, $\underline{\mathbf{G}} \in \mathbb{R}^{R_1 \times R_2 \times \cdots \times R_5}$, and five factor (component) matrices, $\mathbf{B}^{(n)} \in \mathbb{R}^{I_n \times R_n}$ ($n = 1, 2, \ldots, 5$), yields the tensor $\underline{\mathbf{C}} = (((((\underline{\mathbf{G}} \times_1^\sigma \mathbf{B}^{(1)}) \times_2^\sigma \mathbf{B}^{(2)}) \times_3^\sigma \mathbf{B}^{(3)}) \times_4^\sigma \mathbf{B}^{(4)}) \times_5^\sigma \mathbf{B}^{(5)}) \in \mathbb{R}^{I_1 \times I_2 \times \cdots \times I_5}$. This corresponds to the generalized Tucker format. **c** Generalized multi-linear product of a 4th-order tensor, $\underline{\mathbf{G}} \in \mathbb{R}^{R_1 \times R_2 \times R_3 \times R_4}$, and three vectors, $\mathbf{b}_n \in \mathbb{R}^{R_n}$ ($n = 1, 2, 3$), yields the vector $\mathbf{c} = (((\underline{\mathbf{G}} \bar{\times}_1^\sigma \mathbf{b}_1) \bar{\times}_2^\sigma \mathbf{b}_2) \bar{\times}_3^\sigma \mathbf{b}_3) \in \mathbb{R}^{R_4}$, where, in general, $\sigma$ is a nonlinear activation function

$$\underline{\mathbf{C}} = \underline{\mathbf{G}} \bar{\times}_n \mathbf{b} \in \mathbb{R}^{R_1 \times \cdots \times R_{n-1} \times R_{n+1} \times \cdots \times R_N}, \tag{8}$$

with entries $c_{r_1, \ldots, r_{n-1}, r_{n+1}, \ldots, r_N} = \sum_{r_n=1}^{R_n} g_{r_1, \ldots, r_{n-1}, r_n, r_{n+1}, \ldots, r_N} \, b_{r_n}$.

Note that the mode-$n$ multiplication of a tensor by a matrix does not change the tensor order, while the multiplication of a tensor by vectors reduces its order, with the mode $n$ removed.

Multilinear products of tensors by matrices or vectors play a key role in deterministic methods for the reshaping of tensors and dimensionality reduction, as well as in probabilistic methods for randomization/sketching procedures and in random projections of tensors into matrices or vectors. In other words, we can also perform reshaping of a tensor through random projections that change its entries, dimensionality or size of modes, and/or the tensor order. This is achieved by multiplying a tensor by random matrices or vectors, transformations which preserve its basic properties [36–43].

**Tensor contractions**. Tensor contraction is a fundamental and the most important operation in tensor networks, and can be considered a higher-dimensional analogue of matrix multiplication, inner product, and outer product.

In a way similar to the mode-$n$ multilinear product,[2] the mode-$\binom{m}{n}$ product (tensor contraction) of two tensors, $\underline{\mathbf{A}} \in \mathbb{R}^{I_1 \times I_2 \times \cdots \times I_N}$ and $\underline{\mathbf{B}} \in \mathbb{R}^{J_1 \times J_2 \times \cdots \times J_M}$, with common modes, $I_n = J_m$, yields an $(N + M - 2)$-order tensor, $\underline{\mathbf{C}} \in \mathbb{R}^{I_1 \times \cdots \times I_{n-1} \times I_{n+1} \times \cdots \times I_N \times J_1 \times \cdots \times J_{m-1} \times J_{m+1} \times \cdots \times J_M}$, in the form (see Fig. 6a)

$$\underline{\mathbf{C}} = \underline{\mathbf{A}} \times_n^m \underline{\mathbf{B}}, \tag{9}$$

---

[2] In the literature, sometimes the symbol $\times_n$ is replaced by $\bullet_n$.

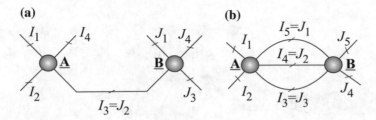

**Fig. 6** Examples of contractions of two tensors. **a** Tensor contraction of two 4th-order tensors, along mode-3 in $\underline{\mathbf{A}}$ and mode-2 in $\underline{\mathbf{B}}$, yields a 6th-order tensor, $\underline{\mathbf{C}} = \underline{\mathbf{A}} \times_3^2 \underline{\mathbf{B}} \in \mathbb{R}^{I_1 \times I_2 \times I_4 \times J_1 \times J_3 \times J_4}$, with entries $c_{i_1,i_2,i_4,j_1,j_3,j_4} = \sum_{i_3} a_{i_1,i_2,i_3,i_4} b_{j_1,i_3,j_3,j_4}$. **b** Tensor contraction of two 5th-order tensors along the modes 3, 4, 5 in $\underline{\mathbf{A}}$ and 1, 2, 3 in $\underline{\mathbf{B}}$ yields a 4th-order tensor, $\underline{\mathbf{C}} = \underline{\mathbf{A}} \times_{5,4,3}^{1,2,3} \underline{\mathbf{B}} \in \mathbb{R}^{I_1 \times I_2 \times J_4 \times J_5}$. Nonlinear contraction can be also performed similar to formula (5)

for which the entries are computed as $c_{i_1,\ldots,i_{n-1},i_{n+1},\ldots,i_N,j_1,\ldots,j_{m-1},j_{m+1},\ldots,j_M} = \sum_{i_n=1}^{I_n} a_{i_1,\ldots,i_{n-1},i_n,i_{n+1},\ldots,i_N} b_{j_1,\ldots,j_{m-1},i_n,j_{m+1},\ldots,j_M}$. This operation is referred to as a contraction of two tensors in single common mode.

Tensors can be contracted in several modes (or even in all modes), as illustrated in Fig. 6. Often, the super- or sub-index, e.g., $m, n$, will be omitted in a few special cases. For example, the multilinear product of the tensors, $\underline{\mathbf{A}} \in \mathbb{R}^{I_1 \times I_2 \times \cdots \times I_N}$ and $\underline{\mathbf{B}} \in \mathbb{R}^{J_1 \times J_2 \times \cdots \times J_M}$, with common modes, $I_N = J_1$, can be written as

$$\underline{\mathbf{C}} = \underline{\mathbf{A}} \times_N^1 \underline{\mathbf{B}} = \underline{\mathbf{A}} \times^1 \underline{\mathbf{B}} = \underline{\mathbf{A}} \bullet \underline{\mathbf{B}} \in \mathbb{R}^{I_1 \times I_2 \times \cdots \times I_{N-1} \times J_2 \times \cdots \times J_M}, \qquad (10)$$

for which the entries $c_{\mathbf{i}_{2:N},\mathbf{j}_{2:M}} = \sum_{i=1}^{I_1} a_{i,\mathbf{i}_{2:N}} b_{i,\mathbf{j}_{2:M}}$ by using the MATLAB notation $\mathbf{i}_{p:q} = \{i_p, i_{p+1}, \ldots, i_{q-1}, i_q\}$.

In this notation, the multiplications of matrices and vectors can be written as, $\mathbf{A} \times_2^1 \mathbf{B} = \mathbf{A} \times^1 \mathbf{B} = \mathbf{AB}$, $\mathbf{A} \times_2^2 \mathbf{B} = \mathbf{AB}^T$, $\mathbf{A} \times_{1,2}^{1,2} \mathbf{B} = \mathbf{A} \bar{\times} \mathbf{B} = \langle \mathbf{A}, \mathbf{B} \rangle$, and $\mathbf{A} \times_2^1 \mathbf{x} = \mathbf{A} \times^1 \mathbf{x} = \mathbf{Ax}$.

In practice, due to the high computational complexity of tensor contractions, especially for tensor networks with loops, this operation is often performed approximately [44–47].

## 3 Mathematical and Graphical Representation of Basic Tensor Networks

Tensor networks (TNs) represent a higher-order tensor as a set of sparsely interconnected lower-order tensors (see Fig. 7), and in this way provide computational and storage benefits. The lines (branches, edges) connecting core tensors correspond to the contracted modes while their weights (or numbers of branches) represent the

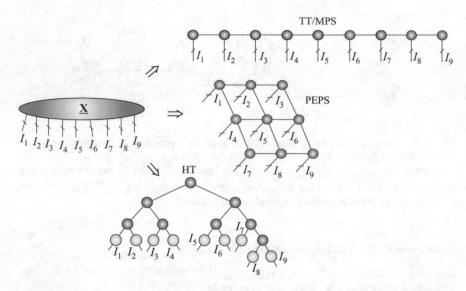

**Fig. 7** Illustration of the decomposition of a 9th-order tensor, $\underline{\mathbf{X}} \in \mathbb{R}^{I_1 \times I_2 \times \cdots \times I_9}$, into different forms of tensor networks (TNs). In general, the objective is to decompose a very high-order tensor into sparsely (weakly) connected low-order and small size core tensors, typically 3rd-order and 4th-order cores. Top: The Tensor Train (TT) model, which is equivalent to the Matrix Product State (MPS) with closed boundary conditions (CBC). Middle: The Projected Entangled-Pair States (PEPS). Bottom: The Hierarchical Tucker (HT)

rank of a tensor network,[3] whereas the lines which do not connect core tensors correspond to the "external" physical variables (modes, indices) within the data tensor. In other words, the number of free (dangling) edges (with weights larger than one) determines the order of a data tensor under consideration, while set of weights of internal branches represents the TN rank.

## 3.1 The CP and Tucker Tensor Formats

The CP and Tucker decompositions have long history. For recent surveys and more detailed information we refer to [12–14, 26, 48–50]. Compared to the CP decomposition, the Tucker decomposition provides a more general factorization of an $N$th-order tensor into a relatively small size core tensor and factor matrices, and can be expressed as follows:

---

[3]Strictly speaking, the minimum set of internal indices $\{R_1, R_2, R_3, \dots\}$ is called the rank (bond dimensions) of a specific tensor network.

$$\underline{\mathbf{X}} \cong \sum_{r_1=1}^{R_1} \cdots \sum_{r_N=1}^{R_N} g_{r_1 r_2 \dots r_N} \left( \mathbf{b}_{r_1}^{(1)} \circ \mathbf{b}_{r_2}^{(2)} \circ \cdots \circ \mathbf{b}_{r_N}^{(N)} \right)$$

$$= \underline{\mathbf{G}} \times_1 \mathbf{B}^{(1)} \times_2 \mathbf{B}^{(2)} \cdots \times_N \mathbf{B}^{(N)}$$

$$= [\![ \underline{\mathbf{G}}; \mathbf{B}^{(1)}, \mathbf{B}^{(2)}, \dots, \mathbf{B}^{(N)} ]\!], \tag{11}$$

where $\underline{\mathbf{X}} \in \mathbb{R}^{I_1 \times I_2 \times \cdots \times I_N}$ is the given data tensor, $\underline{\mathbf{G}} \in \mathbb{R}^{R_1 \times R_2 \times \cdots \times R_N}$ is the core tensor, and $\mathbf{B}^{(n)} = [\mathbf{b}_1^{(n)}, \mathbf{b}_2^{(n)}, \dots, \mathbf{b}_{R_n}^{(n)}] \in \mathbb{R}^{I_n \times R_n}$ are the mode-$n$ factor (component) matrices, $n = 1, 2, \dots, N$ (see Fig. 8). The core tensor (typically, $R_n \ll I_n$) models a potentially complex pattern of mutual interaction between the vectors in different modes. The model in (11) is often referred to as the Tucker-$N$ model.

Using the properties of the Kronecker tensor product, the Tucker-$N$ decomposition in (11) can be expressed in an equivalent vector form as

$$\text{vec}(\underline{\mathbf{X}}) \cong [\mathbf{B}^{(N)} \otimes \mathbf{B}^{(N-1)} \otimes \cdots \otimes \mathbf{B}^{(1)}] \, \text{vec}(\underline{\mathbf{G}}), \tag{12}$$

where the multi-indices are ordered in a reverse lexicographic order (little-endian).

Note that the CP decomposition can be considered as a special case of the Tucker decomposition, whereby the cube core tensor has nonzero elements only on the main diagonal. In contrast to the CP decomposition, the unconstrained Tucker decomposition is not unique. However, constraints imposed on all factor matrices and/or core tensor can reduce the indeterminacies to only column-wise permutation and scaling, thus yielding a unique core tensor and factor matrices [51].

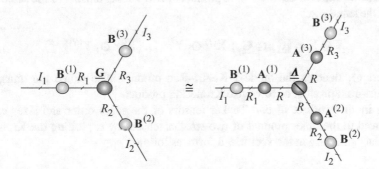

**Fig. 8** Illustration of the standard Tucker and Tucker-CP decompositions, where the objective is to compute the factor matrices, $\mathbf{B}^{(n)}$, and the core tensor, $\underline{\mathbf{G}}$. Tucker decomposition of a 3rd-order tensor, $\underline{\mathbf{X}} \cong \underline{\mathbf{G}} \times_1 \mathbf{B}^{(1)} \times_2 \mathbf{B}^{(2)} \times_3 \mathbf{B}^{(3)}$. In some applications, the core tensor can be further approximately factorized using the CP decomposition as $\underline{\mathbf{G}} \cong \sum_{r=1}^{R} \mathbf{a}_r \circ \mathbf{b}_r \circ \mathbf{c}_r$, or alternatively using TT/HT decompositions. Graphical representation of the Tucker-CP decomposition for a 3rd-order tensor, $\underline{\mathbf{X}} \cong \underline{\mathbf{G}} \times_1 \mathbf{B}^{(1)} \times_2 \mathbf{B}^{(2)} \times_3 \mathbf{B}^{(3)} = [\![ \underline{\mathbf{G}}; \mathbf{B}^{(1)}, \mathbf{B}^{(2)}, \mathbf{B}^{(3)} ]\!] \cong (\underline{\mathbf{\Lambda}} \times_1 \mathbf{A}^{(1)} \times_2 \mathbf{A}^{(2)} \times_3 \mathbf{A}^{(3)}) \times_1 \mathbf{B}^{(1)} \times_2 \mathbf{B}^{(2)} \times_3 \mathbf{B}^{(3)} = [\![ \underline{\mathbf{\Lambda}}; \mathbf{B}^{(1)} \mathbf{A}^{(1)}, \mathbf{B}^{(2)} \mathbf{A}^{(2)}, \mathbf{B}^{(3)} \mathbf{A}^{(3)} ]\!]$

## 3.2 Operations in the Tucker Format

If very large-scale data tensors admit an exact or approximate representation in their TN formats, then most mathematical operations can be performed more efficiently using the so obtained much smaller core tensors and factor matrices.

As illustrative example, consider the $N$th-order tensors $\underline{\mathbf{X}}$ and $\underline{\mathbf{Y}}$ in the Tucker format, given by

$$\underline{\mathbf{X}} = [\![\underline{\mathbf{G}}_X; \mathbf{X}^{(1)}, \dots, \mathbf{X}^{(N)}]\!] \quad \text{and} \quad \underline{\mathbf{Y}} = [\![\underline{\mathbf{G}}_Y; \mathbf{Y}^{(1)}, \dots, \mathbf{Y}^{(N)}]\!], \tag{13}$$

for which the respective multilinear ranks are $\{R_1, R_2, \dots, R_N\}$ and $\{Q_1, Q_2, \dots, Q_N\}$, then the following mathematical operations can be performed directly in the Tucker format, which admits a significant reduction in computational costs [13, 52–54]:

- **The addition** of two Tucker tensors of the same order and sizes

$$\underline{\mathbf{X}} + \underline{\mathbf{Y}} = [\![\underline{\mathbf{G}}_X \oplus \underline{\mathbf{G}}_Y; [\mathbf{X}^{(1)}, \mathbf{Y}^{(1)}], \dots, [\mathbf{X}^{(N)}, \mathbf{Y}^{(N)}]]\!], \tag{14}$$

  where $\oplus$ denotes a direct sum of two tensors, and $[\mathbf{X}^{(n)}, \mathbf{Y}^{(n)}] \in \mathbb{R}^{I_n \times (R_n + Q_n)}$, $\mathbf{X}^{(n)} \in \mathbb{R}^{I_n \times R_n}$ and $\mathbf{Y}^{(n)} \in \mathbb{R}^{I_n \times Q_n}$, $\forall n$.

- **The Kronecker product** of two Tucker tensors of arbitrary orders and sizes

$$\underline{\mathbf{X}} \otimes \underline{\mathbf{Y}} = [\![\underline{\mathbf{G}}_X \otimes \underline{\mathbf{G}}_Y; \mathbf{X}^{(1)} \otimes \mathbf{Y}^{(1)}, \dots, \mathbf{X}^{(N)} \otimes \mathbf{Y}^{(N)}]\!]. \tag{15}$$

- **The Hadamard** or element-wise product of two Tucker tensors of the same order and the same sizes

$$\underline{\mathbf{X}} \circledast \underline{\mathbf{Y}} = [\![\underline{\mathbf{G}}_X \otimes \underline{\mathbf{G}}_Y; \mathbf{X}^{(1)} \odot_1 \mathbf{Y}^{(1)}, \dots, \mathbf{X}^{(N)} \odot_1 \mathbf{Y}^{(N)}]\!], \tag{16}$$

  where $\odot_1$ denotes the mode-1 Khatri–Rao product, also called the transposed Khatri–Rao product or row-wise Kronecker product.

- **The inner product** of two Tucker tensors of the same order and sizes can be reduced to the inner product of two smaller tensors by exploiting the Kronecker product structure in the vectorized form, as follows

$$\langle \underline{\mathbf{X}}, \underline{\mathbf{Y}} \rangle = \text{vec}(\underline{\mathbf{X}})^{\mathrm{T}} \, \text{vec}(\underline{\mathbf{Y}}) \tag{17}$$

$$= \text{vec}(\underline{\mathbf{G}}_X)^{\mathrm{T}} \left( \bigotimes_{n=1}^{N} \mathbf{X}^{(n)\,\mathrm{T}} \right) \left( \bigotimes_{n=1}^{N} \mathbf{Y}^{(n)} \right) \text{vec}(\underline{\mathbf{G}}_Y)$$

$$= \text{vec}(\underline{\mathbf{G}}_X)^{\mathrm{T}} \left( \bigotimes_{n=1}^{N} \mathbf{X}^{(n)\,\mathrm{T}} \mathbf{Y}^{(n)} \right) \text{vec}(\underline{\mathbf{G}}_Y)$$

$$= \langle [\![\underline{\mathbf{G}}_X; (\mathbf{X}^{(1)\,\mathrm{T}} \mathbf{Y}^{(1)}), \dots, (\mathbf{X}^{(N)\,\mathrm{T}} \mathbf{Y}^{(N)})]\!], \underline{\mathbf{G}}_Y \rangle.$$

- **The Frobenius norm** can be computed in a particularly simple way if the factor matrices are orthogonal, since then all products $\mathbf{X}^{(n)\,\mathrm{T}}\mathbf{X}^{(n)}$, $\forall n$, become the identity matrices, so that

$$
\begin{aligned}
\|\mathbf{X}\|_F &= \langle \underline{\mathbf{X}}, \underline{\mathbf{X}} \rangle \\
&= \mathrm{vec}\left( [\![ \underline{\mathbf{G}}_X; (\mathbf{X}^{(1)\,\mathrm{T}}\mathbf{X}^{(1)}), \dots, (\mathbf{X}^{(N)\,\mathrm{T}}\mathbf{X}^{(N)}) ]\!] \right)^{\mathrm{T}} \mathrm{vec}(\underline{\mathbf{G}}_X) \\
&= \mathrm{vec}(\underline{\mathbf{G}}_X)^{\mathrm{T}}\,\mathrm{vec}(\underline{\mathbf{G}}_X) = \|\underline{\mathbf{G}}_X\|_F.
\end{aligned}
\tag{18}
$$

- **The $N$-D discrete convolution** of tensors $\underline{\mathbf{X}} \in \mathbb{R}^{I_1 \times \cdots \times I_N}$ and $\underline{\mathbf{Y}} \in \mathbb{R}^{J_1 \times \cdots \times J_N}$ in their Tucker formats can be expressed as

$$
\underline{\mathbf{Z}} = \underline{\mathbf{X}} * \underline{\mathbf{Y}} = [\![ \underline{\mathbf{G}}_Z; \mathbf{Z}^{(1)}, \dots, \mathbf{Z}^{(N)} ]\!]
\tag{19}
$$
$$
\in \mathbb{R}^{(I_1+J_1-1) \times \cdots \times (I_N+J_N-1)}.
$$

If $\{R_1, R_2, \dots, R_N\}$ is the multilinear rank of $\underline{\mathbf{X}}$ and $\{Q_1, Q_2, \dots, Q_N\}$ the multilinear rank $\underline{\mathbf{Y}}$, then the core tensor $\underline{\mathbf{G}}_Z = \underline{\mathbf{G}}_X \otimes \underline{\mathbf{G}}_Y \in \mathbb{R}^{R_1 Q_1 \times \cdots \times R_N Q_N}$ and the factor matrices

$$
\mathbf{Z}^{(n)} = \mathbf{X}^{(n)} \boxdot_1 \mathbf{Y}^{(n)} \in \mathbb{R}^{(I_n+J_n-1) \times R_n Q_n},
\tag{20}
$$

where $\mathbf{Z}^{(n)}(:,s_n) = \mathbf{X}^{(n)}(:,r_n) * \mathbf{Y}^{(n)}(:,q_n) \in \mathbb{R}^{(I_n+J_n-1)}$ for $s_n = \overline{r_n q_n} = 1, 2, \dots, R_n Q_n$.

- **Super Fast discrete Fourier transform** (MATLAB functions fftn($\underline{\mathbf{X}}$) and fft($\mathbf{X}^{(n)}$, [], 1)) of a tensor in the Tucker format

$$
\mathcal{F}(\underline{\mathbf{X}}) = [\![ \underline{\mathbf{G}}_X; \mathcal{F}(\mathbf{X}^{(1)}), \dots, \mathcal{F}(\mathbf{X}^{(N)}) ]\!].
\tag{21}
$$

Note that if the data tensor admits low multilinear rank approximation, then performing the FFT on factor matrices of relatively small size $\mathbf{X}^{(n)} \in \mathbb{R}^{I_n \times R_n}$, instead of a large-scale data tensor, decreases considerably computational complexity. This approach is referred to as the super fast Fourier transform in Tucker format.

Similar operations can be performed in other TN formats [13].

## 4 Curse of Dimensionality and Separation of Variables for Multivariate Functions

The term curse of dimensionality was coined by Bellman [55] to indicate that the number of samples needed to estimate an arbitrary function with a given level of accuracy grows exponentially with the number of variables, that is, with the dimensionality of the function. In a general context of machine learning and the underlying optimization problems, the "curse of dimensionality" may also refer to an exponentially increasing number of parameters required to describe the data/system or an extremely large number of degrees of freedom. The term "curse of dimensionality", in the context of tensors, refers to the phenomenon whereby the number of elements, $I^N$, of an $N$th-order tensor of size ($I \times I \times \cdots \times I$) grows exponentially with the tensor order, $N$. Tensor volume can therefore easily become prohibitively big for multiway arrays for which the number of dimensions ("ways" or "modes") is very high, thus requiring huge computational and memory resources to process such data. The understanding and handling of the inherent dependencies among the excessive degrees of freedom create both difficult to solve problems and fascinating new opportunities, but comes at a price of reduced accuracy, owing to the necessity to involve various approximations.

The curse of dimensionality can be alleviated or even fully dealt with through tensor network representations; these naturally cater for the excessive volume, veracity and variety of data (see Fig. 1) and are supported by efficient tensor decomposition algorithms which involve relatively simple mathematical operations. Another desirable aspect of tensor networks is their relatively small-scale and low-order core tensors, which act as "building blocks" of tensor networks. These core tensors are relatively easy to handle and visualize, and enable super-compression of the raw, incomplete, and noisy huge-scale data sets. This suggests a solution to a more general quest for new technologies for processing of exceedingly large data sets within affordable computation times [13, 18, 56–58].

To address the curse of dimensionality, this work mostly focuses on approximative low-rank representations of tensors, the so-called low-rank tensor approximations (LRTA) or low-rank tensor network decompositions.

A tensor is said to be in a full or raw format when it is represented as an original (raw) multidimensional array [59], however, distributed storage and processing of high-order tensors in their full format is infeasible due to the curse of dimensionality. The sparse format is a variant of the full tensor format which stores only the nonzero entries of a tensor, and is used extensively in software tools such as the Tensor Toolbox [60] and in the sparse grid approach [61–63].

As already mentioned, the problem of huge dimensionality can be alleviated through various distributed and compressed tensor network formats, achieved by low-rank tensor network approximations.

The underpinning idea is that by employing tensor networks formats, both computational costs and storage requirements may be considerably reduced through distributed storage and computing resources. It is important to note that, except for very special data structures, a tensor cannot be compressed without incurring some compression error, since a low-rank tensor representation is only an approximation of the original tensor.

The concept of compression of multidimensional large-scale data by tensor network decompositions can be intuitively explained as follows [13]. Consider the approximation of an $N$-variate function $h(\mathbf{x}) = h(x_1, x_2, \ldots, x_N)$ by a finite sum of products of individual functions, each depending on only one or a very few variables [64–67]. In the simplest scenario, the function $h(\mathbf{x})$ can be (approximately) represented in the following separable form

$$h(x_1, x_2, \ldots, x_N) \cong h^{(1)}(x_1)\, h^{(2)}(x_2) \cdots h^{(N)}(x_N). \tag{22}$$

In practice, when an $N$-variate function $h(\mathbf{x})$ is discretized into an $N$th-order array, or a tensor, the approximation in (22) then corresponds to the representation by rank-1 tensors, also called elementary tensors (see Sect. 2.1). Observe that with $I_n$, $n = 1, 2, \ldots, N$ denoting the size of each mode and $I = \max_n\{I_n\}$, the memory requirement to store such a full tensor is $\prod_{n=1}^{N} I_n \leq I^N$, which grows exponentially with $N$. On the other hand, the separable representation in (22) is completely defined by its factors, $h^{(n)}(x_n)$, $(n = 1, 2, \ldots, N)$, and requires only $\sum_{n=1}^{N} I_n \ll I^N$ storage units.

If $x_1, x_2, \ldots, x_N$ are statistically independent random variables, their joint probability density function is equal to the product of marginal probabilities, $p(\mathbf{x}) = p^{(1)}(x_1)p^{(2)}(x_2)\ldots p^{(N)}(x_N)$, in an exact analogy to outer products of elementary tensors. Unfortunately, the form of separability in (22) is rather rare in practice.

It should be noted that a function $h(x_1, x_2)$ is a continuous analogue of a matrix, say $\mathbf{H} \in \mathbb{R}^{I_1 \times I_2}$, while a function $h(x_1, \ldots, x_N)$ in $N$ dimensions is a continuous analogue of an $N$-order grid tensor $\underline{\mathbf{H}} \in \mathbb{R}^{I_1 \times \cdots \times I_N}$. In other words, the discretization of a continuous score function $h(x_1, x_2, \ldots, x_N)$ on a hyper-cube leads to a grid tensor of order $N$. Specifically, we make use of a grid tensor that approximates and/or interpolates $h(x_1, \ldots, x_N)$ on a grid of points.

The concept of tensor networks rests upon generalized (full or partial) separability of the variables of a high dimensional function. This can be achieved in different tensor formats, including:

1. The Canonical Polyadic (CP) format, where

$$h(x_1, x_2, \ldots, x_N) \cong \sum_{r=1}^{R} h_r^{(1)}(x_1)\, h_r^{(2)}(x_2) \cdots h_r^{(N)}(x_N), \tag{23}$$

in an exact analogy to (22). In a discretized form, the above CP format can be written as an $N$th-order tensor

$$\underline{\mathbf{H}} \cong \sum_{r=1}^{R} \mathbf{h}_r^{(1)} \circ \mathbf{h}_r^{(2)} \circ \cdots \circ \mathbf{h}_r^{(N)} \in \mathbb{R}^{I_1 \times I_2 \times \cdots \times I_N}, \tag{24}$$

where $\mathbf{h}_r^{(n)} \in \mathbb{R}^{I_n}$ denotes a discretized version of the univariate function $h_r^{(n)}(x_n)$, symbol $\circ$ denotes the outer product, and $R$ is the tensor rank.

2. The Tucker format, given by (see Sect. 3.1)

$$h(x_1, \ldots, x_N) \cong \sum_{r_1=1}^{R_1} \cdots \sum_{r_N=1}^{R_N} g_{r_1, \ldots, r_N} \, h_{r_1}^{(1)}(x_1) \cdots h_{r_N}^{(N)}(x_N), \tag{25}$$

and its distributed tensor network variants,
3. The Tensor Train (TT) format (see Sect. 6.2), in the form

$$h(x_1, x_2, \ldots, x_N) \cong \sum_{r_1=1}^{R_1} \sum_{r_2=1}^{R_2} \cdots \sum_{r_{N-1}=1}^{R_{N-1}} h_{r_1}^{(1)}(x_1) h_{r_1 r_2}^{(2)}(x_2) \cdots$$
$$\cdots h_{r_{N-2} r_{N-1}}^{(N-2)}(x_{N-1}) \, h_{r_{N-1}}^{(N)}(x_N), \tag{26}$$

with the equivalent compact matrix representation

$$h(x_1, x_2, \ldots, x_N) \cong \mathbf{H}^{(1)}(x_1) \, \mathbf{H}^{(2)}(x_2) \cdots \mathbf{H}^{(N)}(x_N), \tag{27}$$

where $\mathbf{H}^{(n)}(x_n) \in \mathbb{R}^{R_{n-1} \times R_n}$, with $R_0 = R_N = 1$.

All the above approximations adopt the form of "sum-of-products" of single-dimensional functions, a procedure which plays a key role in all tensor factorizations and decompositions.

Indeed, in many applications based on multivariate functions, a relatively good approximations are obtained with a surprisingly small number of factors; this number corresponds to the tensor rank, $R$, or tensor network ranks, $\{R_1, R_2, \ldots, R_N\}$ (if the representations are exact and minimal). However, for some specific cases this approach may fail to obtain sufficiently good low-rank TN approximations [67]. The concept of generalized separability has already been explored in numerical methods for high-dimensional density function equations [22, 66, 67] and within a variety of huge-scale optimization problems [13, 14].

To illustrate how tensor decompositions address excessive volumes of data, if all computations are performed on a CP tensor format in (24) and not on the raw $N$th-order data tensor itself, then instead of the original, exponentially growing, data dimensionality of $I^N$, the number of parameters in a CP representation reduces to $NIR$, which scales linearly in the tensor order $N$ and size $I$. For example, the discretization of a 5-variate function over 100 sample points on each axis would yield the difficulty to manage $100^5 = 10,000,000,000$ sample points, while a rank-2 CP representation would require only $5 \times 2 \times 100 = 1000$ sample points.

In contrast to CP decomposition algorithms, TT tensor network formats in (26) exhibit both very good numerical properties and the ability to control the error of approximation, so that a desired accuracy of approximation is obtained relatively easily [13, 68–70]. The main advantage of the TT format over the CP decomposition is the ability to provide stable quasi-optimal rank reduction, achieved through, for example, truncated singular value decompositions (tSVD) or adaptive cross-approximation [64, 71, 72]. This makes the TT format one of the most stable and simple approaches to separate latent variables in a sophisticated way, while the associated TT decomposition algorithms provide full control over low-rank TN approximations.[4] We therefore, make extensive use of the TT format for low-rank TN approximations and employ the TT toolbox software for efficient implementations [68]. The TT format will also serve as a basic prototype for high-order tensor representations, while we also consider the Hierarchical Tucker (HT) and the Tree Tensor Network States (TTNS) formats (having more general tree-like structures) whenever advantageous in applications [13].

Furthermore, the concept of generalized separability of variables and the tensorization of structured vectors and matrices allows us to to convert a wide class of huge-scale optimization problems into much smaller-scale interconnected optimization sub-problems which can be solved by existing optimization methods [11, 14].

The tensor network optimization framework is therefore performed through the two main steps:

- Tensorization of data vectors and matrices into a high-order tensor, followed by a distributed approximate representation of a cost function in a specific low-rank tensor network format.
- Execution of all computations and analysis in tensor network formats (i.e., using only core tensors) that scale linearly, or even sub-linearly (quantized tensor networks), in the tensor order $N$. This yields both the reduced computational complexity and distributed memory requirements.

The challenge is to extend beyond the standard Tucker and CP tensor decompositions, and to demonstrate the perspective of TNs in extremely large-scale data analytic, together with their role as a mathematical backbone in the discovery of hidden structures in prohibitively large-scale data. Indeed, TN models provide a framework for the analysis of linked (coupled) blocks of tensors with millions and even billions of non-zero entries [13, 14].

## 5   Tensor Networks Approaches for Deep Learning

Revolution (breakthroughs) in the fields of Artificial Intelligence (AI) and Machine Learning triggered by class of deep convolutional neural networks (DCNNs), often simply called CNNs, has been a vehicle for a large number of practical applications and commercial ventures in computer vision, speech recognition, language processing, drug discovery, biomedical informatics, recommender systems, robotics, games, and artificial creativity, to mention just a few.

---

[4]Although similar approaches have been known in quantum physics for a long time, their rigorous mathematical analysis is still a work in progress (see [27, 69] and references therein).

The renaissance of deep learning neural networks [5, 6, 73, 74] has both created an active frontier of machine learning and has provided many advantages in applications, to the extent that the performance of DNNs in multi-class classification problems can be similar or even better than what is achievable by humans.

Deep learning is highly interesting in very large-scale data analysis for many reasons, including the following [14]:

1. High-level representations learnt by deep NN models, that are easily interpretable by humans, can also help us to understand the complex information processing mechanisms and multi-dimensional interaction of neuronal populations in the human brain;
2. Regarding the degree of nonlinearity and multi-level representation of features, deep neural networks often significantly outperform their shallow counterparts;
3. In big data analytic, deep learning is very promising for mining structured data, e.g., for hierarchical multi-class classification of a huge number of images.

It is well known that both shallow and deep NNs are universal function approximators in the sense that they are able to approximate arbitrarily well any continuous function of $N$ variables on a compact domain, under the condition that a shallow network has an unbounded width (i.e., the size of a hidden layer), that is, an unlimited number of parameters. In other words, a shallow NN may require a huge (intractable) number of parameters (curse of dimensionality), while DNNs can perform such approximations using a much smaller number of parameters.

Universality refers to the ability of a deep learning network to approximate any function when no restrictions are imposed on its size. On the other hand, depth efficiency refers to the case when a function realized by polynomially-sized deep neural network requires shallow networks to have super-polynomial (exponential) size for the same accuracy of approximation (course of dimensionality). This is often referred to as the expressive power of depth.

Despite recent advances in the theory of DNNs, there are several open fundamental challenges (or open problems) related to understanding high performance DNNs, especially the most successful and perspective DCNNs [13, 14]:

- Theoretical and practical bounds on the expressive power of a specific architecture, i.e., quantification of the ability to approximate or learn wide classes of unknown nonlinear functions;
- Ways to reduce the number of parameters without a dramatic reduction in performance;
- Ability to generalize while avoiding overfitting in the learning process;
- Fast learning and the avoidance "bad" local and spurious minima, especially for highly nonlinear score (objective) functions;
- Rigorous investigation of the conditions under which deep neural networks are "much better" the shallow networks (i.e., NNs with one hidden layer).

The aim of this section is to discuss the many advantages of tensor networks in addressing the first two of the above challenges and to build up both intuitive and mathematical links between DNNs and TNs. Revealing such links and inherent connections will both cross-fertilize deep learning and tensor networks and provide new insights.

In addition to establishing the existing and developing new links, this will also help to optimize existing DNNs and/or generate new architectures with improved performances.

We shall first present an intuitive approach using a simplified hierarchical Tucker (HT) model, followed by alternative simple but efficient, tensor train/tensor chain (TT/TC) architectures. We also propose to use more sophisticated TNs, such as MERA tensor network models in order to enable more flexibility, improved performance, and/or higher expressive power of the next generation of DCCNs.

## 5.1 Why Tensor Networks Are Important in Deep Learning?

Several research groups have recently investigated the application of tensor decompositions to simplify DNNs and to establish links between the deep learning and low-rank tensor networks [14, 75–80]. For example, [80] presented a general and constructive connection between Restricted Boltzmann Machines (RBM), which is a fundamental basic building block in class of Deep Boltzmann Machines, and (TNs) together with the correspondence between general Boltzmann machines and TT/MPS. In a series of research papers [30, 79, 81, 82] the authors analyze the expressive power of a class of DCNNs using simplified Hierarchal Tucker (HT) models (see the next sections). Particularly, Convolutional Arithmetic Circuits (ConvAC), also known as Sum-Product Networks, and Convolutional Rectifier Networks (CRN) have been considered as HT model. They claim that a shallow (single hidden layer) network realizes the classic CP decomposition, whereas a deep network with $\log_2 N$ hidden layers realizes Hierarchical Tucker (HT) decomposition (see the next section). Some researchers also argued that the "unreasonable success" of deep learning can be explained by inherent law of physics, such as the theory of TNs that often employ physical constraints locality, symmetry, compositional hierarchical functions, entropy, and polynomial log-probability, imposed on measurements or input training data [77, 80, 83]. In fact, a very wide spectrum of tensor networks can be potentially used to model and analyze some specific classes of DNNs, in order to obtain simpler and/or more efficient neural networks in the sense that they could provide more expressive power or reduced complexity. Such an approach not only promises to open the door to various mathematical and algorithmic tools for enhanced analysis of DNNs, but also allows us to design novel multi-layer architectures and practical implementations of various deep learning systems. In other words, the consideration of tensor networks in this context may give rise to new NN architectures which could be even potentially superior to the existing ones, but have so far been overlooked by practitioners. Furthermore, methods used for reducing or approximating TNs could be a vehicle to achieve more efficient DNNs, with a reduced number of parameters. This follows from the facts that redundancy

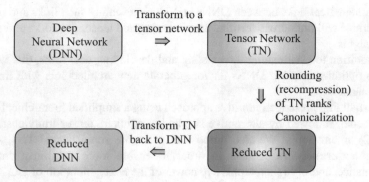

**Fig. 9** Optimization of Deep Neural Networks (DNNs) using The Tensor Networks (TNs) approach. In order to simplify a specific DNN and reduce the number of its parameters, we first transform the DNN into a specific TN, e.g., TT/MPS, then transform the approximated (with reduced rank) TN back to a new optimized DNN. Such transformation can performed in a layer by layer fashion, or globally for the whole DNN. For detailed discussions of such mappings for the Restricted Boltzmann Machine (RBM) see [80]. Optionally, we may choose to first construct and learn (e.g., via tensor completion) a tensor network and then transform it to an equivalent DNN

is inherent both in TNs and DNNs. Moreover, both TNs and DNNs are usually not unique. For example, two NNs with different connection weights and biases may result into the modeling the same nonlinear function. Therefore, the knowledge about redundancy in TNs can help simplify DNNs [14].

The general concept of optimization of DNNs via TNs is illustrated in Fig. 9. Given a specific DNN, we first construct an equivalent TN representation of the given DNN, then the TN is transformed into its reduced or canonical form by performing, e.g., the truncated SVD at each rank (bond). This will reduce the rank dimensions to the minimal requirement determined by a desired accuracy of approximation. Finally, we map back the reduced and optimized TN to another DNN. Since a rounded (approximated) TN has smaller ranks dimensions, a final DNN can be simpler than the original one,and with the same or slightly reduced performance.

It should be noted that, in practice, low-rank TN approximations have many potential advantages over a direct reduction of redundant DNNs, due to availability of many efficient optimization methods to reduce the number of parameters and achieve a pre-specified approximation error. Moreover, low-rank tensor networks are capable of avoiding the curse of dimensionality through low-order sparsely interconnected core tensors.

In the past two decades, quantum physicists and computer scientists have developed solid theoretical understanding and efficient numerical techniques for low-rank TN decompositions.

The entanglement entropy, Renyi's entropy, entanglement spectrum and long range correlations are four of the most widely used quantities (calculated from a spatial reduced density matrix) investigated in the theory of tensor networks. The spatial reduced density matrix is determined by splitting a TN into two parts, say, regions A and B, where a density matrix in region A is given by integrating out all the degrees of freedom in region B. The entanglement spectra are determined by the eigenvalues of the reduced density matrix [84, 85].

Entanglement is a physical phenomenon that occurs when pairs or groups of particles, such as photons, electrons, or qubits, are generated or interact in such way that the quantum state of each particle cannot be described independently of the others, so that a quantum state must be described for the system as a whole. Entanglement entropy is therefore a measure for the amount of entanglement. Strictly speaking, entanglement entropy is a measure of how quantum information is stored in a quantum state and it is mathematically expressed as the von Neumann entropy of the reduced density matrix. Entanglement entropy characterizes the information content of a bipartition of a specific TN. Furthermore, the entanglement area law explains that the entanglement entropy increases only proportionally to the boundary between the two tensor sub-networks. Also entanglement entropy characterizes the information content of the distribution of singular values of a matricized tensor, and can be viewed as a proxy for the correlations between the two partitions; uncorrelated data has zero entanglement entropy at any bipartition.

Note that TNs are usually designed to efficiently represent large systems which exhibit a relatively low entanglement entropy. In practice, we often need to only care about a small fraction of the input training data among a huge number of possible inputs similar to deep neural networks. This all suggest that certain guiding principles in DNNs correspond to the entanglement area law used in the theory of tensor networks. These may then used to quantify the expressive power of a wide class of DCNNs. Note that long range correlations also typically increase with the entanglement. We therefore conjecture that realistic data sets in most successful machine learning applications have relatively low entanglement entropies [86]. On the other hand, by exploiting the entanglement entropy bound of TNs, we can rigorously quantify the expressive power of a wide class of DNNs applied to complex and highly correlated data sets.

## 5.2 Basic Features of Deep Convolutional Neural Networks

Basic DCNNs are usually characterized by at least three features: locality, weight sharing (optional) and pooling explained below [14, 79]

- Locality refers to the connection of a (artificial) neuron only to neighboring neurons in the preceding layer, as opposed to being fed by the entire layer (this is consistent with biological NNs).
- Weight sharing reflects the property that different neurons in the same layer, connected to different neighborhoods in the preceding layer, often share the same weights. Note that weight sharing, when combined with locality, gives rise to standard convolution. However, it should noted that although weight sharing may reduce the complexity of a deep neural network, it is optional. However, the locality

at each layer is a key factor which gives DCNNs an exponential advantage over shallow NNs [77, 87, 88].

- Pooling, is essentially an operator that gradually decimates (reduces) layer sizes by replacing the local population of neural activations in a spatial window by a single value (e.g., by taking their maxima, average values or their scaled products). In the context of images, pooling induces invariance to translation, which often does not affect semantic content, and is interpreted as a way to create a hierarchy of abstractions in the patterns that neurons respond to [14, 79, 87].

Usually, DCNNs perform much better when dealing with compositional function approximations[5] and multi-class classification problems than shallow network architectures with one hidden layer. In fact, DCNNs can efficiently and conveniently select a subset of features for multiple classes, while for efficient learning a DCNN model can be pre-trained by first learning each DCNN layer, followed by fine tuning of the parameter of the entire model e.g., stochastic gradient descent. To summarize, the deep learning neural networks have the ability to exploit and approximate the complexity of compositional hierarchical functions arbitrarily well, whereas shallow networks are blind to them.

## 5.3 Score Functions for Deep Convolutional Neural Networks

Consider a multi-class classification task where the input training data, also called local structures or instances (e.g., input patches in images), are denoted by $X = (\mathbf{x}_1, \dots, \mathbf{x}_N)$, where $\mathbf{x}_n \in \mathbb{R}^S$ $n = 1, \dots, N$) belonging to one of $C$ categories (classes) denoted by $y_c \in \{y_1, y_2, \dots, y_C\}$. Such a representation is quite natural for many high-dimensional data—in images, the local structures represent patches consisting of $S$ pixels, while in audio data voice can be represented through spectrograms.

For this kind of problems, DCNNs aim is to model the following set of multivariate score functions:

$$h_{y_c}(\mathbf{x}_1, \dots, \mathbf{x}_N) = \sum_{i_1=1}^{I_1} \cdots \sum_{i_N=1}^{I_N} \underline{\mathbf{W}}_{y_c}(i_1, \dots, i_N) \, \Phi_{i_1, \dots, i_N}(\mathbf{x}_1, \dots, \mathbf{x}_N),$$

$$\Phi_{i_1, \dots, i_N}(\mathbf{x}_1, \dots, \mathbf{x}_N) = \prod_{n=1}^{N} f_{\theta_{i_n}}(\mathbf{x}_n), \tag{28}$$

$$\text{for} \quad y_c = y_1, y_2, \dots, y_C,$$

where $\underline{\mathbf{W}}_{y_c} \in \mathbb{R}^{I_1 \times \cdots \times I_N}$ is an $N$th-order coefficient tensor (typically, with all dimensions $I_n = I$, $\forall n$), $N$ is the number of (typically overlapped) input patches $\mathbf{x}_n$, $I_n$ is

---

[5]A compositional function can take, for example, the following form $h_1(\dots h_3(h_{21}(h_{11}(x_1, x_2)h_{12}(x_3, x_4)), h_{22}(h_{13}(x_5, x_6)h_{14}(x_7, x_8)) \dots)))$.

the size (dimension) of each mode $\underline{\mathbf{W}}_{y_c}$, and $f_{\theta_1}, \dots, f_{\theta_{I_n}}$ are referred to as the representation functions (in the representation layer) selected from a parametric family of nonlinear functions.[6]

In general, the one-dimensional basis functions could be polynomials, splines or other sets of basis functions. Natural choices for this family of nonlinear functions are also radial basis functions (Gaussian RBFs), wavelets, and affine functions followed by point-wise activations. Particularly interesting are Gabor wavelets, owing to their ability to induce features that resemble representations in the visual cortex of human brain.

Note that the representation functions in standard (artificial) neurons have the form

$$f_{\theta_i}(\mathbf{x}) = \sigma(\tilde{\mathbf{w}}_i^T \mathbf{x} + b_i), \tag{29}$$

for the set of parameters $\theta_i = \{\tilde{\mathbf{w}}_i, b_i\}$, where $\sigma(\cdot)$ is a suitably chosen activation function.

The representation layer play a key role to transform the inputs, by means of $I$ nonlinear functions, $f_{\theta_i}(\mathbf{x}_n)$ ($i = 1, 2, \dots, I$), to template input patches, thereby creating $I$ feature maps [81]. Note that the representation layer can be expressed by a feature vector defined as

$$\mathbf{f} = \mathbf{f}_\theta(\mathbf{x}_1) \otimes \mathbf{f}_\theta(\mathbf{x}_2) \otimes \cdots \otimes \mathbf{f}_\theta(\mathbf{x}_N) \in \mathbb{R}^{I_1 I_2 \cdots I_N}, \tag{30}$$

where $\mathbf{f}_\theta(\mathbf{x}_n) = [f_{\theta_1}(\mathbf{x}_n), f_{\theta_2}(\mathbf{x}_n), \dots, f_{\theta_{I_n}}(\mathbf{x}_n)]^T \in \mathbb{R}^{I_n}$ for $n = 1, 2, \dots, N$ and $i_n = 1, 2, \dots, I_n$.

Alternatively, the representation layer can be expressed as rank one tensor (see Fig. 10a)

$$\underline{\mathbf{F}} = \mathbf{f}_\theta(\mathbf{x}_1) \circ \mathbf{f}_\theta(\mathbf{x}_2) \circ \cdots \circ \mathbf{f}_\theta(\mathbf{x}_N) \in \mathbb{R}^{I_1 \times I_2 \times \cdots \times I_N}. \tag{31}$$

This allows us to represent the score function as an inner product of two tensors, as illustrated in Fig. 10a

$$h_{y_c}(\mathbf{x}_1, \dots, \mathbf{x}_N) = \langle \underline{\mathbf{W}}_{y_c}, \underline{\mathbf{F}} \rangle = \underline{\mathbf{W}}_{y_c} \bar{\times}_1 \mathbf{f}_\theta(\mathbf{x}_1) \bar{\times}_2 \mathbf{f}_\theta(\mathbf{x}_2) \dots \bar{\times}_N \mathbf{f}_\theta(\mathbf{x}_N). \tag{32}$$

To simplify the notations of a grid tensor, we can construct square matrices $\mathbf{F}_n$ ($n = 1, 2, \dots, N$), as follows

$$\mathbf{F}_n = \begin{bmatrix} f_{\theta_1}(\mathbf{x}_n^{(1)}) & f_{\theta_2}(\mathbf{x}_n^{(1)}) & \cdots & f_{\theta_{I_n}}(\mathbf{x}_n^{(1)}) \\ f_{\theta_1}(\mathbf{x}_n^{(2)}) & f_{\theta_2}(\mathbf{x}_n^{(2)}) & \cdots & f_{\theta_{I_n}}(\mathbf{x}_n^{(2)}) \\ \vdots & \vdots & \ddots & \vdots \\ f_{\theta_1}(\mathbf{x}_n^{(I_n)}) & f_{\theta_2}(\mathbf{x}_n^{(I_n)}) & \cdots & f_{\theta_{I_n}}(\mathbf{x}_n^{(I_n)}) \end{bmatrix} \in \mathbb{R}^{I_n \times I_n}, \tag{33}$$

---

[6]Note that the representation layer can be considered as a tensorization of input patches $\mathbf{x}_n$.

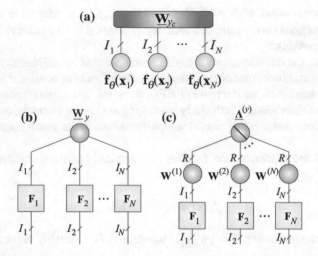

**Fig. 10** Various representations of the score function of a DCNN. **a** Direct Representation of the score function $h_{y_c}(\mathbf{x}_1, \mathbf{x}_2, \ldots, \mathbf{x}_N) = \underline{\mathbf{W}}_{y_c} \bar{\times}_1 \mathbf{f}_\theta(\mathbf{x}_1) \bar{\times}_2 \mathbf{f}_\theta(\mathbf{x}_2) \cdots \bar{\times}_N \mathbf{f}_\theta(\mathbf{x}_N)$. Note that the coefficient tensor $\underline{\mathbf{W}}_c$ can be represented in a distributed form by any suitable tensor network. **b** Graphical illustration of the $N$th-order grid tensor of the score function $h_c$. This model can be considered as a special case of Tucker-$N$ model where the representation matrix $\mathbf{F}_n \in \mathbb{R}^{I_n \times I_n}$ built up factor matrices; note that typically all the factor matrices are the same and $I_n = I$, $\forall n$. **c** CP decomposition of the coefficient tensor $\underline{\mathbf{W}}_{y_c} = \underline{\mathbf{\Lambda}}^{(y_c)} \times_1 \mathbf{W}^{(1)} \times_2 \mathbf{W}^{(2)} \ldots \times_N \mathbf{W}^{(N)} = \sum_{r=1}^{R} \lambda_r^{(y_c)} (\mathbf{w}_r^{(1)} \circ \mathbf{w}_r^{(2)} \circ \cdots \circ \mathbf{w}_r^{(N)})$, where $\mathbf{W}^{(n)} = [\mathbf{w}_1^{(n)}, \ldots, \mathbf{w}_R^{(n)}] \in \mathbb{R}^{I_n \times R}$. This CP model corresponds to a simple shallow neural network with one hidden layer, comprising weights $w_{ir}^{(n)}$, and the output layer comprising weights $\lambda_r^{(y_c)}$, $r = 1, \ldots, R$

which holding the values of taken by the nonlinear basis functions $\{f_{\theta_1}, \ldots, f_{\theta_{I_n}}\}$ on the selected fixed vectors, referred to as templates, $\{\mathbf{x}_n^{(1)}, \mathbf{x}_n^{(2)}, \ldots, \mathbf{x}_n^{(I_n)}\}$. Usually, we can assume that $I_n = I$, $\forall n$, and $\mathbf{x}_n^{(i_n)} = \mathbf{x}^{(i)}$ [30].

For discrete data values, the score function can be represented by a grid tensor, as graphically illustrated in Fig. 10b. The grid tensor of the nonlinear score function $h_{y_c}(\mathbf{x}_1, \ldots, \mathbf{x}_N)$ determined over all the templates $\mathbf{x}_n^{(1)}, \mathbf{x}_n^{(2)}, \ldots, \mathbf{x}_n^{(I)}$ can be expressed as a grid tensor

$$\underline{\mathbf{W}}(h_{y_c}) = \underline{\mathbf{W}}_{y_c} \times_1 \mathbf{F}_1 \times_2 \mathbf{F}_2 \cdots \times_N \mathbf{F}_N. \tag{34}$$

Of course, since the order $N$ of the coefficient (core) tensor is large, it cannot be implemented, or even saved on computer due to the curse of dimensionality.

The simplest model to represent coefficient tensors would be to apply the CP decomposition to reduce the number of parameters, as illustrated in Fig. 10c. This leads to a simple shallow network, however, this approach is associated with two problems: (i) the rank $R$ of the coefficient tensor $\underline{\mathbf{W}}_{y_c}$ can be very large (so compression of parameters cannot be very high), (ii) the existing CP decomposition algorithms are not very stable for very high-order tensors, and so an alternative promising approach would be to apply tensor networks such as HT that enable us to avoid the curse of dimensionality.

Following the representation layer, a DCNN may consists of a cascade of $L$ convolutional hidden layers with pooling in-between, where the number of layers $L$ should be at least two. In other words, each hidden layer performs 3D or 4D convolution followed by spatial window pooling, in order to reduce (decimate) feature maps by e.g., taking a product of the entries in sub-windows. The output layer is a linear dense layer.

Classification can then be carried out in a standard way, through the maximization of a set of labeled score functions, $h_{y_c}$ for $C$ classes, that is, the predicted label for the input instants $X = (\mathbf{x}_1, \dots, \mathbf{x}_N)$ will be the index $\hat{y}_c$ for which the score value attains a maximum, that is

$$\hat{y}_c = \arg \max_{y_c \in \{y_1, \dots, y_C\}} h_{y_c}(\mathbf{x}_1, \dots, \mathbf{x}_N). \tag{35}$$

Such score functions can be represented through their coefficient tensors which, in turn, can be approximated by low-rank tensor network decompositions [13].

The one restriction of the so formulated score functions (29) is that they allow for straightforward implementation of only a particular class of DCNNs, called convolutional Arithmetic Circuit (ConvAC). However, the score functions can be approximated indirectly and almost equivalently using more popular CNNs (see the next section). For example, it was shown recently how NNs with a univariate ReLU nonlinearity may perform multivariate function approximation [77].

The main idea is to employ a low-rank tensor network representation to approximate and interpolate a multivariate function $h_{y_c}(\mathbf{x}_1, \dots, \mathbf{x}_N)$ of $N$ variables by a finite sum of separated products of simpler functions (i.e., via sparsely interconnected core tensors) [13, 14].

# 6 Convolutional Arithmetic Circuits (ConvAC) Using Tensor Networks

Once the set of score functions has been formulated (29), we need to construct (design) a suitable multilayered or distributed representation for DCNN implementation. The objective is to estimate the parameters $\theta_1, \dots, \theta_I$ and coefficient tensors[7] $\underline{\mathbf{W}}_{y_1}, \dots, \underline{\mathbf{W}}_{y_C}$. Since the tensors are of $N$th-order and each with $I^N$ entries, in order to avoid the curse of dimensionality, we need to perform dimensionality reduction

---

[7]It should be noted that these tensors share the same entries, except for the parameters in the output layer.

through low-rank tensor network decompositions. Note that a direct implementation of (29) is intractable due to a huge number of parameters.

Conceptually, the ConvAC can be divided into three parts: (i) the first (input) layer is the representation layer which transforms input vectors $(\mathbf{x}_1, \ldots, \mathbf{x}_N)$ into $N \cdot I$ real valued scalars $\{f_{\theta_i}(\mathbf{x}_n)\}$ for $n = 1, \ldots, N$ and $i = 1, \ldots, I$. In other words, the representation functions, $f_{\theta_i} : \mathbb{R}^S \to \mathbb{R}$, $i = 1, \ldots, I$, map each local patch $\mathbf{x}_n$ into a feature space of dimension $I$. We can denote the feature vector by $\mathbf{f}_n = [f_{\theta_1}(\mathbf{x}_n), \ldots, f_{\theta_I}(\mathbf{x}_n)]^T \in \mathbb{R}^I$, $n = 1, \ldots, N$; (ii) the second, a key or kernel part, is a convolutional arithmetic circuits with many hidden layers that takes the $N \cdot I$ measurements (training samples) generated by the representation layer; (iii) the output layer represented by a full matrix $\mathbf{W}^{(L)}$, which computes $C$ different score functions $h_{y_c}$ [79].

## 6.1   Hierarchical Tucker (HT) and Tree Tensor Network State (TTNS) Models

The simplified HT tensor network [79] shown in Fig. 11 contains sparse 3rd-order core tensors $\underline{\mathbf{W}}^{(l,j)} \in \mathbb{R}^{R^{(l-1,2j-1)} \times R^{(l-1,2j-1)} \times R^{(l,j)}}$ for $l = 1, \ldots, L-1$ and matrices $\mathbf{W}^{(0,j)} = [\mathbf{w}_1^{(0,j)}, \ldots, \mathbf{w}_{R^{(0,j)}}^{(0,j)}] \in \mathbb{R}^{I_j \times R^{(0,j)}}$ for $l = 0$ and $j = 1, \ldots, N/2^l$, and a full matrix $\mathbf{W}^{(L)} = [\mathbf{w}_1^{(L)}, \ldots, \mathbf{w}_{y_C}^{(L)}] \in \mathbb{R}^{R^{(L-1)} \times y_C}$ with column vectors $\mathbf{w}_{y_c}^{(L)} = \mathrm{diag}(\mathbf{W}_{y_c}^{(L)}) = [\lambda_1^{(y_c)}, \ldots, \lambda_{R^{(L-1)}}^{(y_c)}]^T$ in the output $L$-layer (or equivalently the output sparse tensor $\underline{\mathbf{W}}^{(L)} \in \mathbb{R}^{R^{(L-1)} \times R^{(L-1)} \times y_C}$. The number of channels in the input layer is denoted by $I$, while for the $j$th node in the $l$th layer (for $l = 0, 1, \ldots, L-1$) is denoted by $R^{(l,j)}$. The $y_C$ different values of score functions are calculated in the output layer.

For simplicity, and in order to mimic basic features of the standard ConvAC, we assume that $R^{(l,j)} = R^{(l)}$ for all $j$, and that frontal slices of the core tensors $\underline{\mathbf{W}}^{(l,j)}$ are diagonal matrices with entries $w_{r^{(l-1)}, r^{(l)}}^{l,j}$. Note that such sparse core tensors can be represented by non-zero matrices defined as $\mathbf{W}^{(l,j)} \in \mathbb{R}^{R^{(l-1)} \times R^{(l)}}$.

The simplified HT tensor network can be mathematically described in the following recursive form

$$\mathbf{W}^{(0,j)} = [\mathbf{w}_1^{(0,j)}, \ldots, \mathbf{w}_{R^{(0)}}^{(0,j)}] \in \mathbb{R}^{I_j \times R^{(0)}}$$

$$\mathbf{W}_{r^{(1)}}^{(\leq 1,j)} = \sum_{r^{(0)}=1}^{R^{(0)}} w_{r^{(0)}, r^{(1)}}^{(1,j)} \cdot \left( \mathbf{w}_{r^{(0)}}^{(0,2j-1)} \circ \mathbf{w}_{r^{(0)}}^{(0,2j)} \right) \in \mathbb{R}^{I_{2j-1} \times I_{2j}}$$

$$\cdots \tag{36}$$

$$\underline{\mathbf{W}}_{r^{(l)}}^{(\leq l,j)} = \sum_{r^{(l-1)}=1}^{R^{(l-1)}} w_{r^{(l-1)}, r^{(l)}}^{(l,j)} \cdot \left( \underline{\mathbf{W}}_{r^{(l-1)}}^{(\leq l-1,2j-1)} \circ \underline{\mathbf{W}}_{r^{(l-1)}}^{(\leq l-1,2j)} \right)$$

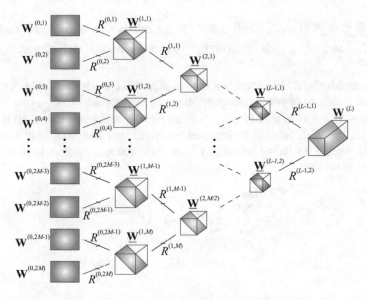

**Fig. 11** Architecture of a simplified Hierarchical Tucker (HT) network with sparse core tensors, which simulates the coefficient tensor for a ConvAC deep learning network with a pooling-2 window [79]. The HT tensor network consists of $L = \log_2(N)$ hidden layers and pooling-2 window. For simplicity, we assumed that we, $N = 2M = 2^L$ input patches, $R^{(l,j)} = R^{(l)}$, for $l = 0, 1, \ldots, L - 1$. The representation layer is not shown explicitly in this figure. Note that since all core tensors can be represented by matrices, we do not need to use tensors notation in this case

$$\underline{\mathbf{W}}^{(\leq L-1,j)}_{r^{(L-1)}} = \sum_{r^{(L-2)}=1}^{R^{(L-2)}} w^{(L-1,j)}_{r^{(L-2)},r^{(L-1)}} \cdot \left( \underline{\mathbf{W}}^{(\leq L-2,2j-1)}_{r^{(L-2)}} \circ \underline{\mathbf{W}}^{(\leq L-2,2j)}_{r^{(L-2)}} \right)$$

$$\underline{\mathbf{W}}_{y_c} \cong \sum_{r^{(L-1)}=1}^{R^{(L-1)}} \lambda^{(y_c)}_{r^{(L-1)}} \cdot \left( \underline{\mathbf{W}}^{(\leq L-1,1)}_{r^{(L-1)}} \circ \underline{\mathbf{W}}^{(\leq L-1,2)}_{r^{(l-1)}} \right) \in \mathbb{R}^{I_1 \times \cdots \times I_N},$$

where $\lambda^{(y_c)} = \operatorname{diag}(\lambda^{(y_c)}_1, \ldots, \lambda^{(y_c)}_{R^{(L-1)}}) = \underline{\mathbf{W}}^{(L)}(:, :, y_c)$.

In a special case when the weights in each layer are shared, i.e., $\mathbf{W}^{(l,1)} = \mathbf{W}^{(l,2)} = \cdots = \mathbf{W}^{(l)}$, the above equation can be considerably simplified to

$$\underline{\mathbf{W}}^{\leq l}_{r^{(l)}} = \sum_{r^{(l-1)}=1}^{R^{(l-1)}} w^{(l)}_{r^{(l-1)},r^{(l)}} \left( \underline{\mathbf{W}}^{\leq l-1}_{r^{(l-1)}} \circ \underline{\mathbf{W}}^{\leq l-1}_{r^{(l-1)}} \right) \tag{37}$$

for the layers $l = 1, \ldots, L - 1$, while for the output layer

$$\underline{\mathbf{W}}_{y_c} \cong \sum_{r^{(L-1)}=1}^{R^{(L-1)}} \lambda^{(y_c)}_{r^{(L-1)}} \left( \underline{\mathbf{W}}^{\leq L-1}_{r^{(L-1)}} \circ \underline{\mathbf{W}}^{\leq L-1}_{r^{(L-1)}} \right), \tag{38}$$

where $\underline{\mathbf{W}}^{\leq l}_{r^{(l)}} = \underline{\mathbf{W}}^{\leq l}(:,\ldots,:,r^{(l)}) \in \mathbb{R}^{I \times \cdots \times I}$ are sub-tensors of $\underline{\mathbf{W}}^{\leq l}$, for each $r^{(l)} = 1,\ldots,R^{(l)}$, and $w^{(l)}_{r^{(l-1)},r^{(l)}}$ is the $(r^{(l-1)},r^{(l)})$th entry of the weight matrix $\mathbf{W}^{(l)} \in \mathbb{R}^{R^{(l-1)} \times R^{(l)}}$.

However, it should be noted that the simplified HT model shown in Fig. 11 has a limited ability to approximate an arbitrary coefficient tensor, $\underline{\mathbf{W}}_{y_c}$, due to strong constraints imposed of core tensors. A more flexible and powerful model is shown in Fig. 12, in which constraints imposed on 3rd-order cores have been completely removed. Such a HT tensor network (with a slight abuse of notation) can be mathematically expressed as

$$\underline{\mathbf{W}}^{(\leq 1,j)}_{r^{(1)}} = \sum_{r_1=1}^{R^{(0,2j-1)}} \sum_{r_2=1}^{R^{(0,2j)}} w^{(1,j)}_{r_1,r_2,r^{(1)}} \cdot \left( \mathbf{w}^{(0,2j-1)}_{r_1} \circ \mathbf{w}^{(0,2j)}_{r_2} \right)$$

$$\cdots \tag{39}$$

$$\underline{\mathbf{W}}^{(\leq l,j)}_{r^{(l)}} = \sum_{r_1=1}^{R^{(l-1,2j-1)}} \sum_{r_2=1}^{R^{(l-1,2j)}} w^{(l,j)}_{r_1,r_2,r^{(l)}} \cdot \left( \underline{\mathbf{W}}^{(\leq l-1,2j-1)}_{r^{(l-1)}} \circ \underline{\mathbf{W}}^{(\leq l-1,2j)}_{r^{(l-1)}} \right)$$

$$\cdots$$

$$\underline{\mathbf{W}}^{(\leq L-1,j)}_{r^{(L-1)}} = \sum_{r_1=1}^{R^{(L-2,2j-1)}} \sum_{r_2=1}^{R^{(L-2,2j)}} w^{(L-1,j)}_{r_1,r_2,r^{(L-1)}} \cdot \left( \underline{\mathbf{W}}^{(\leq L-2,2j-1)}_{r^{(L-2)}} \circ \underline{\mathbf{W}}^{(\leq L-2,2j)}_{r^{(L-2)}} \right)$$

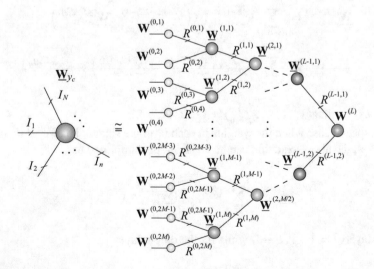

**Fig. 12** Hierarchical Tucker (HT) tensor network for the approximation of coefficient tensors, $\underline{\mathbf{W}}_{y_c}$, of the score functions $h_{y_c}(\mathbf{x}_1,\ldots,\mathbf{x}_N)$

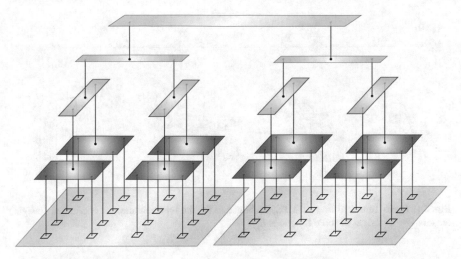

**Fig. 13** Tree Tensor Networks States (TTNS) with variable order of core tensors. The rectangles represent core tensors of orders 5 and 3 that allows pooling of window size 4 and 2, respectively

$$\underline{\mathbf{W}}^{(y_c)} \cong \sum_{r_1=1}^{R^{(L-1,2j-1)}} \sum_{r_2=1}^{R^{(L-1,2j)}} w_{r_1,r_2,y_c}^{(L)} \cdot \left( \underline{\mathbf{W}}_{r^{(L-1)}}^{(\leq L-1,1)} \circ \underline{\mathbf{W}}_{r^{(l-1)}}^{(\leq L-1,2)} \right).$$

The HT network can be further extended to the Tree Tensor Networks States (TTNS), as illustrated in Fig. 13. The use of TTNS, instead HT tensor networks, allows for more flexibility in the choice of size of pooling-window, the pooling size in each hidden layer cab be adjusted by applying core tensors with a suitable variable order in each layer. For example, if we use 5th-order (4rd-order) core tensors instead 3rd-order cores, then the pooling will employ a size-4 pooling window (size-3 pooling) instead of only size-2 pooling window when using 3rd-order core tensors in HT tensor networks. For more detail regrading HT networks and their generalizations to TTNS [13].

## 6.2 Alternative Tensor Network Model: Tensor Train (TT) Networks

We should emphasize that the HT/TTNS architectures are not the only one suitable TN decompositions which can be used to model DCNNs, and the whole family of powerful tensor networks can be employed to model individual hidden layers. In this section we discuss modified TT/MPS and TC models for this purpose for which efficient learning algorithms exist.

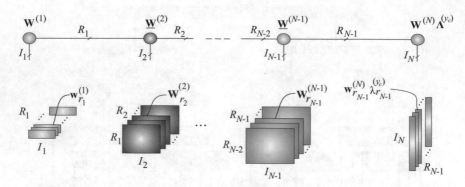

**Fig. 14** Basic Tensor Train (TT/MPS) architecture for the representation of coefficient (weight) tensors $\underline{\mathbf{W}}_{y_c}$ of the set of score functions $h_{y_c}$

The Tensor Train (TT) format can be interpreted as a special case of the HT format, where all nodes (TT-cores) of the underlying tensor network are connected in cascade (or train), i.e., they are aligned while factor matrices corresponding to the leaf modes are assumed to be identities and thus need not be stored. The TT format was first proposed in numerical analysis and scientific computing by Oseledets [15, 69].

Figure 14 presents the concept of TT decomposition for an $N$th-order tensor, the entries of which can be computed as a cascaded (multilayer) multiplication of appropriate matrices (slices of TT-cores). The weights of internal edges (denoted by $\{R_1, R_2, \ldots, R_{N-1}\}$) represent the TT-rank. In this way, the so aligned sequence of core tensors represents a "tensor train" where the role of "buffers" is played by TT-core connections. It is important to highlight that TT networks can be applied not only for the approximation of tensorized vectors but also for scalar multivariate functions, matrices, and even large-scale low-order tensors [13].

In the quantum physics community, the TT format is known as the Matrix Product State (MPS) representation with the Open Boundary Conditions (OBC). In fact, the TT/MPS was rediscovered several times under different names: MPS, valence bond states, and density matrix renormalization group (DMRG) (see [13, 27, 89–94] and references therein).

An important advantage of the TT/MPS format over the HT format is its simpler practical implementation, as no binary tree needs to be determined. Another attractive property of the TT-decomposition is its simplicity when performing basic mathematical operations on tensors directly in the TT-format (that is, employing only core tensors). These include matrix-by-matrix and matrix-by-vector multiplications, tensor addition, and the entry-wise (Hadamard) product of tensors. These operations produce tensors, also in the TT-format, which generally exhibit increased TT-ranks. A detailed description of basic operations supported by the TT format is given in [13]. Note that only TT-cores need to be stored and processed, which makes the number of TN parameters to scale linearly in the tensor order, $N$, of a data tensor and all mathematical operations are then performed only on the low-order and relatively small size core tensors.

The TT rank is defined as an $(N - 1)$-tuple of the form

$$\text{rank}_{TT}(\underline{\mathbf{X}}) = \mathbf{r}_{TT} = \{R_1, \dots, R_{N-1}\}, \quad R_n = \text{rank}(\mathbf{X}_{<n>}), \tag{40}$$

where $\mathbf{X}_{<n>} \in \mathbb{R}^{I_1 \cdots I_n \times I_{n-1} \cdots I_N}$ is an $n$th canonical matricization of the tensor $\underline{\mathbf{X}}$. Since the TT rank determines memory requirements of a tensor train, it has a strong impact on the complexity, i.e., the suitability of tensor train representation for a given raw data tensor.

The number of data samples to be stored scales linearly in the tensor order, $N$, and the size, $I$, and quadratically in the maximum TT rank bound, $R$, that is

$$\sum_{n=1}^{N} R_{n-1} R_n I_n \sim \mathcal{O}(N R^2 I), \quad R := \max_n \{R_n\}, \quad I := \max_n \{I_n\}. \tag{41}$$

This is why it is crucially important to have low-rank TT approximations.[8]

As illustrated in Fig. 14 the simplest possible implementation of the ConvAC network is via the standard tensor train (TT/MPS) (unbalanced binary tree), which can be represented by recursive formulas as

$$\underline{\mathbf{W}}^{\leq 1} = \mathbf{W}^{(1)}$$

$$\underline{\mathbf{W}}^{\leq 2} = \sum_{r_1=1}^{R_1} \underline{\mathbf{W}}^{(1)}_{r_1} \circ \mathbf{W}^{(2)}_{r_1} \in \mathbb{R}^{I_1 \times I_2 \times R_2}$$

$$\cdots$$

$$\underline{\mathbf{W}}^{\leq n} = \sum_{r_{n-1}=1}^{R_{n-1}} \underline{\mathbf{W}}^{\leq n-1}_{r_{n-1}} \circ \mathbf{W}^{(n)}_{r_{n-1}} \in \mathbb{R}^{I_1 \times \cdots \times I_n \times R_n}$$

$$\cdots \tag{42}$$

$$\underline{\mathbf{W}}^{\leq N-1} = \sum_{r_{N-2}=1}^{R_{N-2}} \underline{\mathbf{W}}^{\leq N-2}_{r_{N-2}} \circ \mathbf{W}^{(N-1)}_{r_{N-2}} \in \mathbb{R}^{I_1 \times \cdots \times I_{N-1} \times R_{N-1}}$$

$$\underline{\mathbf{W}}_{y_c} = \underline{\mathbf{W}}^{\leq N} = \sum_{r_{N-1}=1}^{R_{N-1}} \lambda^{(y_c)}_{r_{N-1}} (\underline{\mathbf{W}}^{\leq N-1}_{r_{N-1}} \circ \mathbf{w}^{(N)}_{r_{N-1},1}) \in \mathbb{R}^{I_1 \times \cdots \times I_N},$$

where $\mathbf{W}^{(n)}_{r_{n-1}} = \underline{\mathbf{W}}^{(n)}(r_{n-1}, :, :) \in \mathbb{R}^{I_n \times R_n}$ are lateral slices of the core tensor $\underline{\mathbf{W}}^{(n)} \in \mathbb{R}^{R_{n-1} \times I_n \times R_n}$ and $\underline{\mathbf{W}}^{\leq n}_{r_n} = \underline{\mathbf{W}}^{\leq n}(:, \dots, :, r_n) \in \mathbb{R}^{I_1 \times \cdots \times I_n}$ are sub-tensors of $\underline{\mathbf{W}}^{\leq n} \in \mathbb{R}^{I_1 \times \cdots \times I_n \times R_n}$ for $n = 1, \dots, N$ (Fig. 14).

The above recursive formulas for the TT network can be written in a compact form as

---

[8]In the worst case scenario the TT ranks can grow up to $I^{(N/2)}$ for an $N$th-order tensor.

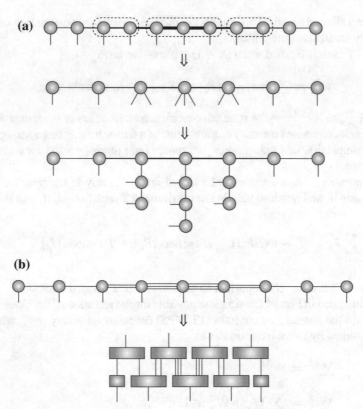

**Fig. 15** Extended (modified) Tensor Train (TT/MPS) architectures for the representation of coefficient (weight) tensors $\underline{\mathbf{W}}_{y_c}$ of the score function $h_{y_c}$. **a** TT-tucker network, also called fork tensor product states (FTPS) with reduced TT ranks. **b** Hierarchical TT network consisting of core tensors with different orders, where each high-order core tensor can be represented by a TT or HT sub-network. Rectangular boxes represent core tensors (sub-tensors) with variable orders. The ticker horizontal lines or double/tripple lines indicate relatively higher internal TT-ranks

$$\underline{\mathbf{W}}_{y_c} = \sum_{r_1=1}^{R_1} \cdots \sum_{r_{N-1}=1}^{R_{N-1}} \lambda_{r_{N-1}}^{(y_c)} (\mathbf{w}_{1,r_1}^{(1)} \circ \mathbf{w}_{r_1,r_2}^{(2)} \circ \cdots \circ \mathbf{w}_{r_{N-2},r_{N-1}}^{(N-1)} \circ \mathbf{w}_{r_{N-1},1}^{(N)}), \qquad (43)$$

where $\mathbf{w}_{r_{n-1},r_n}^{(n)} = \underline{\mathbf{W}}^{(n)}(:,i_n,:) \in \mathbb{R}^{I_n}$ are tubes of the core tensor $\underline{\mathbf{W}}^{(n)} \in \mathbb{R}^{R_{n-1} \times I_n \times R_n}$.

The TT-rank of the standard tensor train network with 3rd-order cores, shown in Fig. 15, can be very large for core tensors located in the middle of the chain. In the worst case scenario, the TT ranks can grow even, up to $I^{(N/2)}$ for an exact representation of $N$th-order tensor. In order to reduce the TT-rank and consequently reduce the number of parameters of a DCNN, we can apply two approaches, as explained in Fig. 14a and b. The main idea is to employ only a few core tensors with order larger than three, whereby each of these core is further approximated via a TT or HT sub-network.

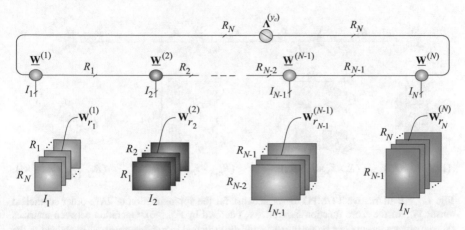

**Fig. 16** The Tensor Chain (TC) architecture for the representation of coefficient (weight) tensors $\underline{\mathbf{W}}_{y_c}$ of the score function $h_{y_c}(\mathbf{x}_1, \ldots, \mathbf{x}_N)$. Note that for $R_0 = R_N = 1$ the TC simplifies to the standard TT

## 6.3 Tensor Chain and TT/MPO Networks

Alternatively, the Tensor Chain (TC) network, called also TT/MPS with periodic bounded conditions, shown in Fig. 16 can be employed to represent individual hidden layers or output fully connected layers. This TC network is mathematically described through the following a recursive formulas as

$$\underline{\mathbf{W}}^{\leq 1} = \underline{\mathbf{W}}^{(1)} \in \mathbb{R}^{R_N \times I_1 \times R_1}$$

$$\underline{\mathbf{W}}^{\leq 2} = \sum_{r_1=1}^{R_1} \underline{\mathbf{W}}_{r_1}^{\leq 1} \circ \mathbf{W}_{r_1}^{(2)} \in \mathbb{R}^{R_N \times I_1 \times I_2 \times R_2}$$

$$\ldots$$

$$\underline{\mathbf{W}}^{\leq n} = \sum_{r_{n-1}=1}^{R_{n-1}} \underline{\mathbf{W}}_{r_{n-1}}^{\leq n-1} \circ \mathbf{W}_{r_{n-1}}^{(n)} \in \mathbb{R}^{R_N \times I_1 \times \cdots \times I_n \times R_n}$$

$$\ldots$$

$$\underline{\mathbf{W}}^{\leq N} = \sum_{r_{N-1}=1}^{R_{N-1}} \underline{\mathbf{W}}_{r_{N-1}}^{\leq N-1} \circ \mathbf{W}_{r_{N-1}}^{(N)} \in \mathbb{R}^{R_N \times I_1 \times \cdots \times I_N \times R_N}$$

$$\underline{\mathbf{W}}_{y_c} = \sum_{r_N=1}^{R_N} \lambda_{r_N}^{(y_c)} \underline{\mathbf{W}}_{r_N, r_N}^{\leq N} \in \mathbb{R}^{I_1 \times \cdots \times I_N}, \tag{44}$$

where $\underline{\mathbf{W}}_{r_{n-1}}^{(n)} = \underline{\mathbf{W}}^{(n)}(r_{n-1}, :, :) \in \mathbb{R}^{I_n \times R_n}$ are lateral slices of the core tensor $\underline{\mathbf{W}}^{(n)} \in \mathbb{R}^{R_{n-1} \times I_n \times R_n}$, $\underline{\mathbf{W}}_{r_n}^{\leq n} = \underline{\mathbf{W}}^{\leq n}(:, \ldots, :, r_n) \in \mathbb{R}^{R_N \times I_1 \times \cdots \times I_n}$ are sub-tensors of $\underline{\mathbf{W}}^{\leq n} \in$

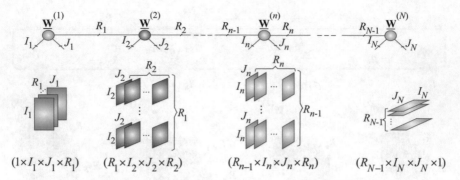

**Fig. 17** An alternative TT/MPO tensor network for the approximation of $2N$th-order coefficient tensor $\underline{\mathbf{W}}_{y_c}$ of the score function $h_c(\mathbf{x}_1, \ldots, \mathbf{x}_N)$ defined by Eq. (46). Operation between matrices (slices) of core tensors can be performed multi-linearly and in nonlinear way, as explained in the next section

$\mathbb{R}^{R_N \times I_1 \times \cdots \times I_n \times R_n}$, and $\underline{\mathbf{W}}^{\leq N}_{r_N, r_N} = \underline{\mathbf{W}}^{\leq N}(r_N, :, \ldots, :, r_N) = \mathrm{Tr}(\underline{\mathbf{W}}^{\leq N}) \in \mathbb{R}^{I_1 \times \cdots \times I_N}$ for $n = 1, \ldots, N$ and $R_0 = R_N \neq 1$.

The above TC network can be written in a compact form (due to the commutativity and associativity of the outer products) as

$$\underline{\mathbf{W}}_{y_c} = \sum_{r_1=1}^{R_1} \cdots \sum_{r_N=1}^{R_N} \lambda_{r_N}^{(y_c)} (\mathbf{w}_{r_N, r_1}^{(1)} \circ \mathbf{w}_{r_1, r_2}^{(2)} \circ \cdots \circ \mathbf{w}_{r_{N-1}, r_N}^{(N)}), \tag{45}$$

where $\mathbf{w}_{r_{n-1}, r_n}^{(n)} = \underline{\mathbf{W}}^{(n)}(:, i_n, :) \in \mathbb{R}^{I_n}$ are tubes of a core tensor $\underline{\mathbf{W}}^{(n)} \in \mathbb{R}^{R_{n-1} \times I_n \times R_n}$.

The ConvAC can be alternatively modeled via TT/MPO networks, as illustrated in Fig. 17 for a more general score function defined as

$$h_{y_c} = \sum_{i_1=1}^{I_1} \cdots \sum_{i_N=1}^{I_N} \sum_{j_1=1}^{J_1} \cdots \sum_{j_N=1}^{J_N} \underline{\mathbf{W}}_{y_c}(i_1, \ldots, i_N, j_1, \ldots, j_N) \prod_{n=1}^{N} f_{\theta_{i_n j_n}}(\mathbf{x}_n), \tag{46}$$

where $\underline{\mathbf{W}}_{y_c}(i_1, \ldots, i_N, j_1, \ldots, j_N)$ represent entries of an $2N$th-order coefficient tensor.

Such tensor networks are well understood and efficient algorithms exist to perform their learning, that is, to estimate the core tensors on the basis of a relatively small number of measurements or a small set of available training data.

# 7 Deep Convolutional Rectifier Using Nonlinear Tensor Networks Decompositions

The convolutional arithmetic circuits (ConvACs) model employs the standard outer (tensor) products, which for two tensors, $\underline{\mathbf{A}} \in \mathbb{R}^{I_1 \times \cdots \times I_N}$ and $\underline{\mathbf{B}} \in \mathbb{R}^{J_1 \times \cdots \times J_M}$, are defined as

$$(\underline{\mathbf{A}} \circ \underline{\mathbf{B}})_{i_1,\ldots,i_N,j_1,\ldots,j_M} = a_{i_1,\ldots,i_N} \cdot b_{j_1,\ldots,j_M}.$$

However, in order to convert ConvAC tensor models to popular and widely used convolutional rectifier networks we need to employ the generalized (nonlinear) outer products, defined as [30]

$$(\underline{\mathbf{A}} \circ_\rho \underline{\mathbf{B}})_{i_1,\ldots,i_N,j_1,\ldots,j_M} = \rho(a_{i_1,\ldots,i_N}, b_{j_1,\ldots,j_M}), \tag{47}$$

where the operator

$$\rho = \rho_{\sigma,P}(a,b) = P[\sigma(a), \sigma(b))], \tag{48}$$

is referred to as the activation-pooling operator or function,[9] which meets the associativity and the commutativity requirements (i.e., the operator satisfies the following properties: $\rho(\rho(a,b),c) = \rho(a, \rho(b,c))$ and $\rho(a,b) = \rho(b,a), \quad \forall a,b,c \in \mathbb{R}$).

The activation–pooling operator can take various forms. In particular, for the convolutional rectifier network with max pooling, we can use the following activation-pooling operator

$$\rho_{\sigma,P}(a,b) = \max\{[a]_+, [b]_+)\} = \max\{a,b,0\}. \tag{49}$$

As an example, consider a generalized CP decomposition, which represents a shallow rectifier network in the form

$$\underline{\mathbf{W}}_{y_c} = \sum_{r=1}^{R} \lambda_r^{(y_c)} (\mathbf{w}_r^{(1)} \circ_\rho \mathbf{w}_r^{(2)} \circ_\rho \cdots \circ_\rho \mathbf{w}_r^{(N)}), \tag{50}$$

where the coefficients $\lambda_r^{(y_c)}$ represent weights of the output layer, vectors $\mathbf{w}_r^{(n)} \in \mathbb{R}^{I_n}$ are weights in the hidden layer, and $R$ denotes the number of channels (using the language of deep learning community).

It should be noted that if we employ the weight sharing, then all vectors $\mathbf{w}_r^{(n)} = \mathbf{w}_r, \forall n$, and consequently the coefficient tensor, $\underline{\mathbf{W}}_{y_c}$, must be a symmetric tensor which further limits the ability of this model to approximate a desired function.

---

[9]The symbols $\sigma(\cdot)$ and $P(\cdot)$ are respectively the activation and pooling functions of the network.

As a second example, let us consider a nonlinear HT tensor network which models a deep convolutional rectifier. The TN shown in Fig. 17 can be compactly described as follows (assuming the generalized outer products defined above):

$$\underline{\mathbf{W}}_{y_c} = \sum_{r_1=1}^{R_1} \cdots \sum_{r_{N-1}=1}^{R_{N-1}} \lambda_{r_{N-1}}^{(y_c)} (\mathbf{W}_{1,r_1}^{(1)} \circ_\rho \mathbf{W}_{r_1,r_2}^{(2)} \circ_\rho \cdots \circ_\rho \mathbf{W}_{r_{N-1,1}}^{(N-1)}), \tag{51}$$

where $\mathbf{W}_{r_{n-1},r_n}^{(n)} \in \mathbb{R}^{I_n \times J_n}$ are block matrices of core tensor $\underline{\mathbf{W}}^{(n)} \in \mathbb{R}^{R_{n-1} \times I_n \times J_n \times R_n}$ (for more detail see [13, 14]).

The TT and TC networks[10] provide some simplicity in comparison to HT, together with very deep TN structures, that is, $N$ hidden layers. Note that the HT model generates architectures of DCNNs with $L = \log_2(N)$ hidden layers, while TT/TC tenor network employs $N$ hidden layers. Taking into account the current trend in deep leaning to use a large number of hidden layers, it would be a quite attractive to employ so called quantized TT/TC QTT/QTC networks with a relatively large number of hidden layers: $L = N \cdot \log_2(I)$ [13].

To summarize, deep convolutional neural networks may be considered as a special case of hierarchical architectures, which can be indirectly simulated and optimized via relative simple and well understood tensor networks, especially HT/TT (i.e., using unbalanced or balanced binary trees and graphical models), however, more sophisticated tensor network diagrams with loops, discussed in the next section may provide potentially better performance and the ability to generate novel architectures of DCNNs.

# 8   MERA Tensor Networks for a Next Generation of DCNNs

The Multiscale Entanglement Renormalization Ansatz (MERA) tensor network was first introduced by Vidal [95], and numerical algorithms to minimize the energy or local Hamiltonian already exist [96].

The MERA is a relatively new tensor network, widely investigated in quantum physics based variational Ansatz, since it is capable of capturing many of the key complex physical properties of strongly correlated ground states [97]. The MERA also shares many relationships with the AdS/CFT (gauge-gravity) correspondence by realizing a complete holographic duality within the tensor networks framework. Furthermore, the MERA can be regarded as a TN realization of an orthogonal wavelets transform acting on the mode space of the physical fermionic degrees of freedom [98–100].

---

[10]It is important to note that TT/TC tensor networks described in this section do not necessary need to have weight sharing and do not need even to be convolutional.

For simplicity, we focus in this section on the 1D binary and ternary MERA tensor networks (see Fig. 18a for basic binary MERA). Instead of writing of complex mathematical formulas it is more convenient to describe MERA tensor networks graphically.

Using, the terminology from quantum physics, the standard binary MERA architecture contains three classes of core tensors: (i) Disentanglers—4th-order cores; (ii) isometries also called the coarse-grainer, typically 3rd order cores for binary MERA and 4th-order cores for ternary MERA; and (iii) one output core which is usually a matrix or a 4th-order core, as illustrated in Fig. 18a and c. Each MERA layer is constructed of a row of disentanglers and a row of coarse-grainers or isometries. Disentanglers remove the short-scale entanglement between the adjacent modes, while isometries renormalise each pair of modes to a single mode Each renormalisation layer performs these operations on a different length scale.

The coarse-grainers take inputs from two modes on a lower scale in the MERA, and give an output onto one mode which is on a higher layer in the tensor network, while the disentangler removes entanglement between two neighboring modes (sites) on the same level. From the perspective of a mapping, the nodes (core tensors) can be considered as processing units, that is, the 4th-order cores map matrices to other matrices, while the coarse-grainers take matrices and map then to vectors. The key idea here is to realize that the "compression" capability arises from the hierarchy and the entanglement. As a matter of fact, the MERA network embodies the mutual information chain rule. In other words, the main idea underlying MERA is that of disentangling the system at various length scales as one follows coarse graining Renormalization Group (RG) flow in the system. The MERA is particularly effective for (scale invariant) critical points of the physical systems.

The key features properties of MERA can be summarized as follows [97]:

- MERA can capture scale-invariance of inputs data;
- It reproduces polynomial decay of correlations between inputs, in contrast to HT or TT tensor networks which reproduce only exponential decay of correlations;
- MERA has ability to much better compress tensor data that TT/HT tensor networks;
- It reproduces a logarithmic correction to the area law, therefore MERA is a more powerful tensor network in comparison to HT/TTNS or TT/TC networks;
- MERA can be efficiently contracted due to unitary constraints imposed on core tensors.

Motivated by these features, we are currently investigating MERA tensor networks as powerful tools to model and analyze DCNNs. A key objective is to establish a precise connection between MERA tensor networks and extended model of DCNNs. This connection may provide exciting new insights about deep learning and may also allow for construction of improved families of DCNNs, with potential application to more efficient data/image classification, clustering and prediction. In other words, we conjecture that the MERA will lead to useful new results, potentially allowing not only better characterization of expressive power of DCNNs, but

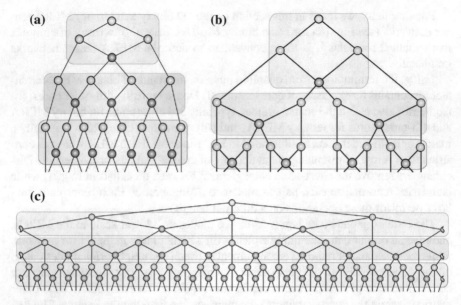

**Fig. 18** Various architectures of MERA tensor networks for the new generation of deep convolutional neural networks. **a** Basic binary MERA tensor network. Observe that the alternating layers of disentangling and coarse-graining cores. For the network shown in (**a**) the number of modes (tensor cores) after each such set of operations is approximately halved. **b** Improved (lower complexity) MERA network. **c** Ternary MERA in which coarse grainers are also 4th-order tensors, i.e., three sites (modes) are coarse-grained into one effective site (mode)

also new practical implementations. Going the other way, the links and relations between TNs and DCNNs could lead to useful advances in the design of novel deep neural networks.

The MERA tensor networks, shown in Fig. 18, may provide a much higher expressive power of deep learning in comparison to networks corresponding to HT/TT architectures, since this class of tensor networks can model more complex long term correlations between input instances. This follows form the facts that for HT and TT, TC tensor networks correlations between input variables decay exponentially and the entanglement entropy saturates to a constant, while the more sophisticated MERA tensor networks provide polynomially decaying correlations.

For future research directions, it would be very important to further explore the links between deep learning architectures, such as DCNN or deep Boltzmann machine, and TNs with hierarchical structures such as tree tensor network states (TTNS) and multi-scale entanglement renormalization ansatz (MERA), in order to better understand and improve the expressive power of deep feedforward neural networks. We deeply believe that the insights into the theory of tensor networks and quantum many-body physics can provide better theoretical understanding of deep learning, together with the guidance for optimized DNNs design.

To summarize, the tensor network methodology and architectures discussed in this section could be extended to allow analytic construction of new DCNNs. Moreover, systematic investigation of the correspondences between DNNs and wide spectrum of TNs can provide a very fruitful perspective including cashing the existing conjectures and claims about operational similarities and correspondences between DNNs and TNs into a more rigorous and constructive framework.

# 9 Conclusions and Discussions

The tensor networks (TNs) methodology is a promising paradigm for the analysis of extreme-scale multidimensional data. Due to their 'super' compression abilities and the distributed way in which they process data, TNs can be employed for a wide family of large-scale optimization problems, especially linear/multilinear dimensionality reduction tasks.

In this paper, we focused on two main challenges in huge-scale data analysis which are addressed by tensor networks: (i) an approximate representation of a specific cost (objective) function by a tensor network while maintaining the desired accuracy of approximation, and (ii) the extraction of physically meaningful latent variables from data in a sufficiently accurate and computationally affordable way. The benefits of multiway (tensor) analysis methods for large-scale data sets then include:

- Graphical representations of tensor networks allow us to express mathematical operations on tensors (e.g., tensor contractions and reshaping) in a simple and intuitive way, and without the explicit use of complex mathematical expressions;
- Simultaneous and flexible distributed representations of both the structurally rich data and complex optimization tasks;
- Efficient compressed formats of large multidimensional data achieved via tensorization and low-rank tensor decompositions into low-order factor matrices and/or core tensors;
- Ability to operate with noisy and missing data by virtue of numerical stability and robustness to noise of low-rank tensor/matrix approximation algorithms;
- A flexible framework which naturally incorporates various diversities and constraints, thus seamlessly extending the standard, flat view, Component Analysis (2-way CA) methods to multiway component analysis;
- Possibility to analyze linked (coupled) blocks of large-scale matrices and tensors in order to separate common/correlated from independent/uncorrelated components in the observed raw data.

In that sense, this paper both reviews current research in this area and complements optimisation methods, such as the Alternating Direction Method of Multipliers (ADMM) [101].

Tensor decompositions (TDs) have been already adopted in widely diverse disciplines, including psychometrics, chemometrics, biometric, quantum physics quantum chemistry, signal and image processing, machine learning, and brain science [12–14, 26, 29, 63, 102–105]. This is largely due to their advantages in the analysis of data that exhibit not only large volume but also very high variety (see Fig. 1), as in the case in bio- and neuroinformatics and in computational neuroscience, where various forms of data collection include sparse tabular structures and graphs or hypergraphs.

Moreover, tensor networks have the ability to efficiently parameterize, through structured compact representations, very general high-dimensional spaces which arise in modern applications [11, 14, 72, 106–110]. Tensor networks also naturally account for intrinsic multidimensional and distributed patterns present in data, and thus provide the opportunity to develop very sophisticated models for capturing multiple interactions and couplings in data—these are more physically insightful and interpretable than standard pair-wise interactions.

**Acknowledgements** This work has been partially supported by Misnistry of Education and Science of the Russian Federation (grant 14.756,0001).

# References

1. Zurada, J.: Introduction to Artificial Neural Systems, vol. 8. West St, Paul (1992)
2. LeCun, Y., Bengio, Y.: Convolutional networks for images, speech, and time series. In: The Handbook of Brain Theory and Neural Networks, MIT Press, pp. 255–258 (1998)
3. Hinton, G., Sejnowski, T.: Learning and relearning in boltzmann machines. In: Parallel Distributed Processing, MIT Press, pp. 282–317 (1986)
4. Cichocki, A., Kasprzak, W., Amari, S.: Multi-layer neural networks with a local adaptive learning rule for blind separation of source signals. In: Proceedings of the International Symposium Nonlinear Theory and Applications (NOLTA), Las Vegas, NV, Citeseer, pp. 61–65 (1995)
5. LeCun, Y., Bengio, Y., Hinton, G.: Deep learning. Nature **521**(7553), 436–444 (2015)
6. Goodfellow, I., Bengio, Y., Courville, A.: Deep Learning. MIT Press (2016). http://www.deeplearningbook.org
7. Cichocki, A., Zdunek, R.: Multilayer nonnegative matrix factorisation. Electron. Lett. **42**(16), 1 (2006)
8. Cichocki, A., Zdunek, R.: Regularized alternating least squares algorithms for non-negative matrix/tensor factorization. In: International Symposium on Neural Networks, pp. 793–802. Springer (2007)
9. Cichocki, A.: Tensor decompositions: new concepts in brain data analysis? J. Soc. Instr. Control Eng. **50**(7), 507–516. arXiv:1305.0395 (2011)
10. Cichocki, A.: Era of big data processing: a new approach via tensor networks and tensor decompositions, (invited). In: Proceedings of the International Workshop on Smart Info-Media Systems in Asia (SISA2013). arXiv:1403.2048 (September 2013)
11. Cichocki, A.: Tensor networks for big data analytics and large-scale optimization problems. arXiv:1407.3124 (2014)

12. Cichocki, A., Mandic, D., Caiafa, C., Phan, A., Zhou, G., Zhao, Q., Lathauwer, L.D.: Tensor decompositions for signal processing applications: from two-way to multiway component analysis. IEEE Signal Process. Mag. **32**(2), 145–163 (2015)
13. Cichocki, A., Lee, N., Oseledets, I., Phan, A.H., Zhao, Q., Mandic, D.: Tensor networks for dimensionality reduction and large-scale optimization: part 1 low-rank tensor decompositions. Found. Trends Mach. Learn. **9**(4–5), 249–429 (2016)
14. Cichocki, A., Phan, A.H., Zhao, Q., Lee, N., Oseledets, I., Sugiyama, M., Mandic, D.: Tensor networks for dimensionality reduction and large-scale optimization: part 2 applications and future perspectives. Found. Trends Mach. Learn. **9**(6), 431–673 (2017)
15. Oseledets, I., Tyrtyshnikov, E.: Breaking the curse of dimensionality, or how to use SVD in many dimensions. SIAM J. Sci. Comput. **31**(5), 3744–3759 (2009)
16. Dolgov, S., Khoromskij, B.: Two-level QTT-Tucker format for optimized tensor calculus. SIAM J. Matrix Anal. Appl. **34**(2), 593–623 (2013)
17. Kazeev, V., Khoromskij, B., Tyrtyshnikov, E.: Multilevel Toeplitz matrices generated by tensor-structured vectors and convolution with logarithmic complexity. SIAM J. Sci. Comput. **35**(3), A1511–A1536 (2013)
18. Kazeev, V., Khammash, M., Nip, M., Schwab, C.: Direct solution of the chemical master equation using quantized tensor trains. PLoS Comput. Biol. **10**(3), e1003359 (2014)
19. Kressner, D., Steinlechner, M., Uschmajew, A.: Low-rank tensor methods with subspace correction for symmetric eigenvalue problems. SIAM J. Sci. Comput. **36**(5), A2346–A2368 (2014)
20. Vervliet, N., Debals, O., Sorber, L., De Lathauwer, L.: Breaking the curse of dimensionality using decompositions of incomplete tensors: Tensor-based scientific computing in big data analysis. IEEE Signal Process. Mag. **31**(5), 71–79 (2014)
21. Dolgov, S., Khoromskij, B.: Simultaneous state-time approximation of the chemical master equation using tensor product formats. Numer. Linear Algebra Appl. **22**(2), 197–219 (2015)
22. Liao, S., Vejchodský, T., Erban, R.: Tensor methods for parameter estimation and bifurcation analysis of stochastic reaction networks. J. R. Soc. Interface **12**(108), 20150233 (2015)
23. Bolten, M., Kahl, K., Sokolović, S.: Multigrid methods for tensor structured Markov chains with low rank approximation. SIAM J. Sci. Comput. **38**(2), A649–A667 (2016)
24. Lee, N., Cichocki, A.: Estimating a few extreme singular values and vectors for large-scale matrices in Tensor Train format. SIAM J. Matrix Anal. Appl. **36**(3), 994–1014 (2015)
25. Lee, N., Cichocki, A.: Regularized computation of approximate pseudoinverse of large matrices using low-rank tensor train decompositions. SIAM J. Matrix Anal. Appl. **37**(2), 598–623 (2016)
26. Kolda, T., Bader, B.: Tensor decompositions and applications. SIAM Rev. **51**(3), 455–500 (2009)
27. Orús, R.: A practical introduction to tensor networks: matrix product states and projected entangled pair states. Ann. Phys. **349**, 117–158 (2014)
28. Dolgov, S., Savostyanov, D.: Alternating minimal energy methods for linear systems in higher dimensions. SIAM J. Sci. Comput. **36**(5), A2248–A2271 (2014)
29. Cichocki, A., Zdunek, R., Phan, A.H., Amari, S.: Nonnegative Matrix and Tensor Factorizations: Applications to Exploratory Multi-way Data Analysis and Blind Source Separation. Wiley, Chichester (2009)
30. Cohen, N., Shashua, A.: Convolutional rectifier networks as generalized tensor decompositions. In: Proceedings of The 33rd International Conference on Machine Learning, pp. 955–963 (2016)
31. Li, J., Battaglino, C., Perros, I., Sun, J., Vuduc, R.: An input-adaptive and in-place approach to dense tensor-times-matrix multiply. In: Proceedings of the International Conference for High Performance Computing, Networking, Storage and Analysis, p. 76. ACM (2015)

32. Ballard, G., Benson, A., Druinsky, A., Lipshitz, B., Schwartz, O.: Improving the numerical stability of fast matrix multiplication algorithms. arXiv:1507.00687 (2015)
33. Ballard, G., Druinsky, A., Knight, N., Schwartz, O.: Brief announcement: Hypergraph partitioning for parallel sparse matrix-matrix multiplication. In: Proceedings of the 27th ACM on Symposium on Parallelism in Algorithms and Architectures, pp. 86–88. ACM (2015)
34. Tucker, L.: The extension of factor analysis to three-dimensional matrices. In: Gulliksen, H., Frederiksen, N. (eds.) Contributions to Mathematical Psychology, pp. 110–127. Holt, Rinehart and Winston, New York (1964)
35. Tucker, L.: Some mathematical notes on three-mode factor analysis. Psychometrika **31**(3), 279–311 (1966)
36. Sun, J., Tao, D., Faloutsos, C.: Beyond streams and graphs: dynamic tensor analysis. In: Proceedings of the 12th ACM SIGKDD international conference on Knowledge Discovery and Data Mining, pp. 374–383. ACM (2006)
37. Drineas, P., Mahoney, M.: A randomized algorithm for a tensor-based generalization of the singular value decomposition. Linear Algebra Appl. **420**(2), 553–571 (2007)
38. Lu, H., Plataniotis, K., Venetsanopoulos, A.: A survey of multilinear subspace learning for tensor data. Pattern Recogn. **44**(7), 1540–1551 (2011)
39. Li, M., Monga, V.: Robust video hashing via multilinear subspace projections. IEEE Trans. Image Process. **21**(10), 4397–4409 (2012)
40. Pham, N., Pagh, R.: Fast and scalable polynomial kernels via explicit feature maps. In: Proceedings of the 19th ACM SIGKDD international conference on Knowledge discovery and data mining, pp. 239–247. ACM (2013)
41. Wang, Y., Tung, H.Y., Smola, A., Anandkumar, A.: Fast and guaranteed tensor decomposition via sketching. In: Advances in Neural Information Processing Systems, pp. 991–999 (2015)
42. Kuleshov, V., Chaganty, A., Liang, P.: Tensor factorization via matrix factorization. In: Proceedings of the Eighteenth International Conference on Artificial Intelligence and Statistics, pp. 507–516 (2015)
43. Sorber, L., Domanov, I., Van Barel, M., De Lathauwer, L.: Exact line and plane search for tensor optimization. Comput. Optim. Appl. **63**(1), 121–142 (2016)
44. Lubasch, M., Cirac, J., Banuls, M.C.: Unifying projected entangled pair state contractions. New J. Phys. **16**(3), 033014 (2014)
45. Di Napoli, E., Fabregat-Traver, D., Quintana-Ortí, G., Bientinesi, P.: Towards an efficient use of the BLAS library for multilinear tensor contractions. Appl. Math. Comput. **235**, 454–468 (2014)
46. Pfeifer, R., Evenbly, G., Singh, S., Vidal, G.: NCON: A tensor network contractor for MATLAB. arXiv:1402.0939 (2014)
47. Kao, Y.J., Hsieh, Y.D., Chen, P.: Uni10: An open-source library for tensor network algorithms. J. Phys. Conf. Ser. **640**, 012040 (2015). IOP Publishing
48. Grasedyck, L., Kessner, D., Tobler, C.: A literature survey of low-rank tensor approximation techniques. GAMM-Mitteilungen **36**, 53–78 (2013)
49. Comon, P.: Tensors: A brief introduction. IEEE Signal Process. Mag. **31**(3), 44–53 (2014)
50. Sidiropoulos, N., De Lathauwer, L., Fu, X., Huang, K., Papalexakis, E., Faloutsos, C.: Tensor decomposition for signal processing and machine learning. arXiv:1607.01668 (2016)
51. Zhou, G., Cichocki, A.: Fast and unique Tucker decompositions via multiway blind source separation. Bull. Pol. Acad. Sci. **60**(3), 389–407 (2012)
52. Phan, A., Cichocki, A., Tichavský, P., Zdunek, R., Lehky, S.: From basis components to complex structural patterns. In: Proceedings of the IEEE International Conference on Acoustics, Speech and Signal Processing, ICASSP 2013, Vancouver, BC, Canada, May 26–31, 2013, pp. 3228–3232
53. Phan, A., Tichavský, P., Cichocki, A.: Low rank tensor deconvolution. In: Proceedings of the IEEE International Conference on Acoustics Speech and Signal Processing, ICASSP, April 2015, pp. 2169–2173
54. Lee, N., Cichocki, A.: Fundamental tensor operations for large-scale data analysis using tensor network formats. Multidimension. Syst. Signal Process, pp 1–40, Springer (March 2017)

55. Bellman, R.: Adaptive Control Processes. Princeton University Press, Princeton, NJ (1961)
56. Austin, W., Ballard, G., Kolda, T.: Parallel tensor compression for large-scale scientific data. arXiv:1510.06689 (2015)
57. Jeon, I., Papalexakis, E., Faloutsos, C., Sael, L., Kang, U.: Mining billion-scale tensors: algorithms and discoveries. VLDB J. 1–26 (2016)
58. Phan, A., Cichocki, A.: PARAFAC algorithms for large-scale problems. Neurocomputing **74**(11), 1970–1984 (2011)
59. Klus, S., Schütte, C.: Towards tensor-based methods for the numerical approximation of the Perron-Frobenius and Koopman operator. arXiv:1512.06527 (December 2015)
60. Bader, B., Kolda, T.: MATLAB tensor toolbox version. **2**, 6 (2015)
61. Garcke, J., Griebel, M., Thess, M.: Data mining with sparse grids. Computing **67**(3), 225–253 (2001)
62. Bungartz, H.J., Griebel, M.: Sparse grids. Acta Numerica **13**, 147–269 (2004)
63. Hackbusch, W.: Tensor spaces and numerical tensor calculus. Springer Series in Computational Mathematics, vol. 42. Springer, Heidelberg (2012)
64. Bebendorf, M.: Adaptive cross-approximation of multivariate functions. Constr. Approx. **34**(2), 149–179 (2011)
65. Dolgov, S.: Tensor product methods in numerical simulation of high-dimensional dynamical problems. Ph.D. thesis, Faculty of Mathematics and Informatics, University Leipzig, Germany, Leipzig, Germany (2014)
66. Cho, H., Venturi, D., Karniadakis, G.: Numerical methods for high-dimensional probability density function equations. J. Comput. Phys. **305**, 817–837 (2016)
67. Trefethen, L.: Cubature, approximation, and isotropy in the hypercube. SIAM Rev. (to appear) (2017)
68. Oseledets, I., Dolgov, S., Kazeev, V., Savostyanov, D., Lebedeva, O., Zhlobich, P., Mach, T., Song, L.: TT-Toolbox (2012)
69. Oseledets, I.: Tensor-train decomposition. SIAM J. Sci. Comput. **33**(5), 2295–2317 (2011)
70. Khoromskij, B.: Tensors-structured numerical methods in scientific computing: Survey on recent advances. Chemometr. Intell. Lab. Syst. **110**(1), 1–19 (2011)
71. Oseledets, I., Tyrtyshnikov, E.: TT cross-approximation for multidimensional arrays. Linear Algebra Appl. **432**(1), 70–88 (2010)
72. Khoromskij, B., Veit, A.: Efficient computation of highly oscillatory integrals by using QTT tensor approximation. Comput. Methods Appl. Math. **16**(1), 145–159 (2016)
73. Schmidhuber, J.: Deep learning in neural networks: an overview. Neural Netw. **61**, 85–117 (2015)
74. Schneider, D.: Deeper and cheaper machine learning [top tech 2017]. IEEE Spectr. **54**(1), 42–43 (2017)
75. Lebedev, V., Lempitsky, V.: Fast convolutional neural networks using group-wise brain damage. arXiv:1506.02515 (2015)
76. Novikov, A., Podoprikhin, D., Osokin, A., Vetrov, D.: Tensorizing neural networks. In: Advances in Neural Information Processing Systems (NIPS), pp. 442–450 (2015)
77. Poggio, T., Mhaskar, H., Rosasco, L., Miranda, B., Liao, Q.: Why and when can deep–but not shallow–networks avoid the curse of dimensionality: a review. arXiv:1611.00740 (2016)
78. Yang, Y., Hospedales, T.: Deep multi-task representation learning: a tensor factorisation approach. arXiv:1605.06391 (2016)
79. Cohen, N., Sharir, O., Shashua, A.: On the expressive power of deep learning: a tensor analysis. In: 29th Annual Conference on Learning Theory, pp. 698–728 (2016)
80. Chen, J., Cheng, S., Xie, H., Wang, L., Xiang, T.: On the equivalence of restricted Boltzmann machines and tensor network states. arXiv e-prints (2017)
81. Cohen, N., Shashua, A.: Inductive bias of deep convolutional networks through pooling geometry. CoRR (2016). arXiv:1605.06743
82. Sharir, O., Tamari, R., Cohen, N., Shashua, A.: Tensorial mixture models. CoRR (2016). arXiv:1610.04167

83. Lin, H.W., Tegmark, M.: Why does deep and cheap learning work so well? arXiv e-prints (2016)
84. Zwanziger, D.: Fundamental modular region, Boltzmann factor and area law in lattice theory. Nucl. Phys. B **412**(3), 657–730 (1994)
85. Eisert, J., Cramer, M., Plenio, M.: Colloquium: Area laws for the entanglement entropy. Rev. Modern Phys. **82**(1), 277 (2010)
86. Calabrese, P., Cardy, J.: Entanglement entropy and quantum field theory. J. Stat. Mech. Theory Exp. **2004**(06), P06002 (2004)
87. Anselmi, F., Rosasco, L., Tan, C., Poggio, T.: Deep convolutional networks are hierarchical kernel machines. arXiv:1508.01084 (2015)
88. Mhaskar, H., Poggio, T.: Deep vs. shallow networks: an approximation theory perspective. Anal. Appl. **14**(06), 829–848 (2016)
89. White, S.: Density-matrix algorithms for quantum renormalization groups. Phys. Rev. B **48**(14), 10345 (1993)
90. Vidal, G.: Efficient classical simulation of slightly entangled quantum computations. Phys. Rev. Lett. **91**(14), 147902 (2003)
91. Perez-Garcia, D., Verstraete, F., Wolf, M., Cirac, J.: Matrix product state representations. Quantum Inf. Comput. **7**(5), 401–430 (2007)
92. Verstraete, F., Murg, V., Cirac, I.: Matrix product states, projected entangled pair states, and variational renormalization group methods for quantum spin systems. Adv. Phys. **57**(2), 143–224 (2008)
93. Schollwöck, U.: Matrix product state algorithms: DMRG, TEBD and relatives. In: Strongly Correlated Systems, pp. 67–98. Springer (2013)
94. Huckle, T., Waldherr, K., Schulte-Herbriggen, T.: Computations in quantum tensor networks. Linear Algebra Appl. **438**(2), 750–781 (2013)
95. Vidal, G.: Class of quantum many-body states that can be efficiently simulated. Phys. Rev. Lett. **101**(11), 110501 (2008)
96. Evenbly, G., Vidal, G.: Algorithms for entanglement renormalization. Phys. Rev. B **79**(14), 144108 (2009)
97. Evenbly, G., Vidal, G.: Tensor network renormalization yields the multiscale entanglement renormalization Ansatz. Phys. Rev. Lett. **115**(20), 200401 (2015)
98. Evenbly, G., White, S.R.: Entanglement renormalization and wavelets. Phys. Rev. Lett. **116**(14), 140403 (2016)
99. Evenbly, G., White, S.R.: Representation and design of wavelets using unitary circuits. arXiv e-prints (2016)
100. Matsueda, H.: Analytic optimization of a MERA network and its relevance to quantum integrability and wavelet. arXiv:1608.02205 (2016)
101. Boyd, S., Parikh, N., Chu, E., Peleato, B., Eckstein, J.: Distributed optimization and statistical learning via the alternating direction method of multipliers. Found. Trends Mach. Learn. **3**(1), 1–122 (2011)
102. Smilde, A., Bro, R., Geladi, P.: Multi-way Analysis: Applications in the Chemical Sciences. Wiley, New York (2004)
103. Tao, D., Li, X., Wu, X., Maybank, S.: General tensor discriminant analysis and Gabor features for gait recognition. IEEE Trans. Pattern Anal. Mach. Intell. **29**(10), 1700–1715 (2007)
104. Kroonenberg, P.: Applied Multiway Data Analysis. Wiley, New York (2008)
105. Favier, G., de Almeida, A.: Overview of constrained PARAFAC models. EURASIP J. Adv. Signal Process. **2014**(1), 1–25 (2014)
106. Kressner, D., Steinlechner, M., Vandereycken, B.: Low-rank tensor completion by Riemannian optimization. BIT Numer. Math. **54**(2), 447–468 (2014)
107. Zhang, Z., Yang, X., Oseledets, I., Karniadakis, G., Daniel, L.: Enabling high-dimensional hierarchical uncertainty quantification by ANOVA and tensor-train decomposition. IEEE Trans. Comput.-Aided Des. Integr. Circ. Syst. **34**(1), 63–76 (2015)
108. Corona, E., Rahimian, A., Zorin, D.: A tensor-train accelerated solver for integral equations in complex geometries. arXiv:1511.06029 (2015)

109. Litsarev, M., Oseledets, I.: A low-rank approach to the computation of path integrals. J. Comput. Phys. **305**, 557–574 (2016)
110. Benner, P., Khoromskaia, V., Khoromskij, B.: A reduced basis approach for calculation of the Bethe-Salpeter excitation energies by using low-rank tensor factorisations. Mol. Phys. **114**(7–8), 1148–1161 (2016)

# Local Data Characteristics in Learning Classifiers from Imbalanced Data

Jerzy Błaszczyński and Jerzy Stefanowski

**Abstract**  Learning classifiers from imbalanced data is still one of challenging tasks in machine learning and data mining. Data difficulty factors referring to internal and local characteristics of class distributions deteriorate performance of standard classifiers. Many of these factors may be approximated by analyzing the neighbourhood of the learning examples and identifying different types of examples from the minority class. In this paper, we follow recent research on developing such methods for assessing the types of examples which exploit either $k$-nearest neighbours or kernels. We discuss the approaches to tune the size of both kinds of neighborhoods depending on the data set characteristics and evaluate their usefulness in series of experiments with real-world and synthetic data sets. Furthermore, we claim that the proper analysis of these neighborhoods could be the basis for developing new specialized algorithms for imbalanced data. To illustrate it, we study generalizations of over-sampling in pre-processing methods and neighbourhood based ensembles.

## 1 Introduction

Supervised classification is one of the well studied tasks of machine learning, data mining and statistical data analysis. Its aim is to learn the relationship between values of attributes describing examples and a target class of interest. Since many problems can be represented in the attribute value form it has a wide spectrum of possible applications [1]. The classification relationships learned from labeled examples can be used as a classifier to predict class labels for new, unclassified examples. Numerous approaches, based on different principles, have been already introduced to learn

J. Błaszczyński · J. Stefanowski (✉)
Institute of Computing Science, Poznań University of Technology,
Piotrowo 2, 60-965 Poznań, Poland
e-mail: jerzy.stefanowski@cs.put.poznan.pl

J. Błaszczyński
e-mail: jerzy.blaszczynski@cs.put.poznan.pl

© Springer International Publishing AG 2018
A.E. Gawęda et al. (eds.), *Advances in Data Analysis with Computational
Intelligence Methods*, Studies in Computational Intelligence 738,
https://doi.org/10.1007/978-3-319-67946-4_2

classifiers. Nevertheless they may be insufficient when dealing with complexities affecting the data representation.

One of these complexities is *class imbalanced data*, where at least one of the target classes contains a much smaller number of examples than the other classes. This class is usually called the *minority class*, while the remaining classes are denoted as *majority class(es)*. Imbalanced data often occur in practical problems, such as, medical data analysis, fraud detection, technical diagnostics or image recognition, see, e.g., [8, 20, 60]. In all these problems correct recognition of the minority class is of key importance. Nevertheless, the standard learning algorithms usually do not work properly for these problems since they are biased toward better recognition of the majority classes and they met difficulties, or even are unable, to classify correctly new objects from the minority class [61].

Although the difficulty while learning classifiers from imbalanced data has been known in practical applications for decades, this problem received a particular, growing research interest in the beginning of the current century and several specialized methods have been proposed (for their review see, e.g., [7, 20, 21, 56]). They are usually categorized as classifier-independent pre-processing techniques or modifications of algorithms for learning particular classifiers.

Researchers still treat learning from class imbalanced data as a research challenge and look for new more effective directions. One of these directions includes studying the nature of the imbalanced data, key properties of its underlying distribution and consequences they bring for learning better classifiers or for constructing specialized pre-processing methods.

While examining these properties, it has been noticed that the high, global imbalance ratio between cardinalities of minority and majority classes is not the only and not even the main reason of difficulties in learning classifiers. Other, as we call them, *data difficulty factors*, referring to internal characteristics of class distributions, are also influential. They include: decomposition of the minority class into many rare sub-concepts playing a role of small disjuncts [25, 26], the effect overlapping between the classes [15, 46] or presence of many minority class examples inside the majority class region [39]. When these data difficulty factors occur *together* with class imbalance, they may seriously hinder the recognition of the minority class, see e.g., experimental studies [36, 40, 42, 48].

Please note that aforementioned data factors correspond to *local data characteristics*, occurring in some sub-regions of the minority class distribution rather than at the global level of the entire data set. Furthermore, the development of several informed pre-processing methods, such as [9, 31], is strongly based on exploiting information about example distribution in the neighborhood of considered minority examples.

In the previous research Napierala and Stefanowski have linked data difficulty factors to *different types of examples* forming the minority class distribution [39, 40, 52, 55]. It has led the authors to a differentiation between safe and unsafe examples for recognizing the minority class. These types of examples were identified by analyzing class labels distribution among examples' *neighbours* [40]. Two ways of modeling the neighbourhood have been proposed, either by considering, *k-nearest neighbours* or *kernel functions* [38, 40]. These approaches can be applied to several crucial issues for learning classifiers from imbalanced data:

- to analyze internal characteristics of real-world data sets and establish their difficulty for recognizing minority classes [38, 40];
- to support comparisons of algorithms for learning classifiers as well as preprocessing methods [42];
- to construct new, specialized algorithms for improving classifiers [5].

Nevertheless, in these studies the size of neighborhood was chosen in the simplest way and usually with the same value of the crucial hyper-parameter for all considered data sets. Although it has proven to be sufficiently effective in previous works, a more systematic tuning of this parameter with respect to data set characteristics is still an open research problem and requires more studies.

Therefore, the main aims of this paper are the following:

1. To introduce a new approach to tune the size of the neighborhood depending on the data characteristics. Unlike the previous works [40, 42], we pay more attention to using kernels in this analysis.
2. To experimentally study usefulness of kernels for an analysis of imbalanced data—also for identifying more types of examples than proposed in [40].
3. To discuss the applicability of this special tuned neighborhood for constructing dynamic pre-processing methods as well as to learning neighbourhood based ensembles dedicated to, imbalanced data.

The paper is organized as follows. The next section summarizes related works on data difficulty factors and using local information in pre-processing methods. The previous approach to an identification of types of minority examples is discussed in Sect. 3. The new proposal of tuning its parameters is introduced in Sect. 4 and validated in the experiments in Sect. 5. The following section discusses its use to construct new pre-processing techniques. Similarly, its applicability for the Nearest Neighbourhood Ensemble is presented in Sect. 7. Other possible extensions of the presented neighborhood analysis are discussed in Sect. 8. The final section draws conclusions.

## 2   Related Research on Imbalanced Data Characteristics

In this section we will briefly discuss the issues most related to studying local characteristics of class imbalanced data. We do not intend to provide here a comprehensive review of methods for dealing with these data. For such a review, the reader is referred to the monograph [20] covering the most representative issues, as well as to systematic surveys, such as [7, 8, 21, 56].

## 2.1 Nature of the Class Imbalance Problem

Recall that a data set is considered class imbalanced when it is characterized by an unequal distribution of objects in classes. Japkowicz names it a *between-class imbalance* [24]. It may be quantified by a *class imbalance ratio*—which represents a global point of view at data characteristics.

Generally speaking, any data set with unequal distribution of examples between class could be considered as imbalanced. However, there is no common agreement with regard to a precise threshold defined for the global imbalance ratio that would allow to distinguish imbalanced data sets [21]. Here we also do not define a precise threshold value but share an opinion saying that the class imbalance problem is associated with lack of data (called also absolute rarity [60]), which hinder the accurate recognition of minority classes [53].

In this study we consider a two class (minority class vs. majority class) formulation of class imbalance problem. It is justified by semantic importance of the rare class versus other classes, which can be considered as the two class problem. Moreover, this formulation of the imbalance problem is mostly studied in the current literature. Even if the original definition of the classification problem includes more classes, they are aggregated into one majority class. Note, however, that in some applications it may be reasonable to consider multi-class data sets, where imbalances may exist between various classes and it is required to improve classifier performance with respect to more than one minority class. We will come back to these issues in Sect. 8.

The class imbalance observed in a data set can be either *intrinsic* (in the sense that it is a direct result of the nature of the data space) or *extrinsic* (caused by reasons external to the data space). Extrinsic imbalance can be caused by high costs of acquiring the examples from the minority class, e.g., due to economic or privacy reasons or it comes from technical, time or storage limitations [60].

## 2.2 Data Complexity and Difficulty Factors

Although many authors have experimentally shown that standard classifiers meet difficulties while recognizing the minority class, it has also been observed that in some problems characterized by high imbalance between classes (expressed by the value of the global imbalanced data) standard classifiers are still sufficiently accurate [2]. For instance, Napierala reports several experimental studies which conclude that when there is a clear separation between classes, the minority class can be sufficiently recognized regardless of the high imbalance ratio [38].

These and other studies prove that the *global class imbalance ratio* is not necessarily the only, or even the main, problem causing the decrease of classification performance and focusing only on the global ratio may be insufficient for improving classification performance. *Data complexity*, understood here as the distribution of examples from both classes in the attribute space, has a crucial impact on learning. It is not particularly surprising, since data complexity affects learning also in standard, balanced domains. However, when data complexity occurs *together* with the class imbalance, the deterioration of classification performance is amplified and it affects mostly (or even only) the minority class.

In the context of learning from imbalanced data the term "data complexity" may comprise different data distribution patterns, such as: overlapping, small disjuncts, outliers or noise. Several authors call them as *data difficulty factors*. We describe them briefly below.

### Within Class Decomposition and Small Disjuncts

The experimental studies with several data sets have shown that minority class usually does not form a homogeneous, compact distribution of the target concept but it is often scattered into smaller sub-parts representing separate sub-concepts. Japkowicz named it *within-class imbalance* [26]. This is closely related to the problem of *small disjuncts* which are harder to learn and cause more classification errors than larger sub-concepts.

Although the problem of within-class imbalance may occur in both minority and majority classes, small disjuncts are more characteristic and more critical for a minority class. In the majority class, the sub-concepts will be most often represented by a sufficient number of examples forming larger disjuncts, while in the minority class, in which the examples are already rare, their further decomposition into several sub-concepts will produce small disjuncts, represented by a too small number of examples to be correctly learned. Such fragmentation of the minority class into five smaller sub-parts is illustrated in Fig. 1. Additionally each sub-part of the minority

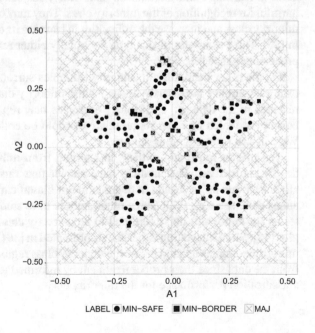

**Fig. 1** Visualization of sub-concepts of the minority class additionally affecting by class over-lapping (here represented by borderline examples) in flower data

LABEL ● MIN–SAFE  ■ MIN–BORDER  ⊠ MAJ

has a small overlapping with the neighbours from the majority class (which constitute an additional difficulty).

According to [25, 26] the higher deterioration of classification performance results from an increased decomposition of the minority class into many sub-parts containing too few examples rather than by changing the global imbalance ratio.

**Overlapping Between the Classes**

In the boundary regions between classes, the examples from different classes may overlap—which hinders learning classifiers even in a standard, balanced case. As the minority class is underrepresented in the data set, it may be underrepresented also in the overlapping region. Most learning algorithms tend to shift the decision boundary too close to the minority class, treating the whole overlapping area as belonging to the majority class. Indeed, the experiments on mainly artificial data with different degrees of overlapping have shown that overlapping deteriorated the classifier performance, especially when the minority class was concerned [46]. Furthermore, according to research of [15] the imbalance ratio calculated locally inside the overlapping regions is more influential for the minority class than the global ratio concerning the complete data. In other experiments a combination of increased overlapping between the classes with decomposition of the minority class influenced results more than changing the class imbalance ratio [39].

**Dealing with Noisy or Outlier Examples**

Single examples from one class, located far from the decision boundary inside the other class, are usually called noisy examples. Handling noise is often considered in standard machine learning problems, however it becomes even more important issue in learning from imbalanced data. Noisy majority examples are particularly harmful for recognition of the minority class. They may cause a fragmentation of the minority class and increase the difficulties in learning its definition—see a discussion in [38]. Thus, examples of this type are usually either removed/relabeled in the pre-processing phase [48, 55].

On the other hand, distant minority examples surrounded by the majority class examples are not necessarily noisy. As the minority class examples are underrepresented in the data set, such lonely examples may represent a rare but valid sub-concept of which no other representatives could be collected for training [38, 40]. We will call such examples *outliers*.

The role of noise and outliers in learning from imbalanced data has not been deeply studied yet. Few authors have shown that randomly introduced class or attribute value noise results in degradation of classification performance on imbalanced data, see e.g., [38]. Some other authors have studied the role of iterative filtering (or removing) noisy (difficult to be correctly classified) minority case examples [48]. More interesting experiments presented in [39] have also shown that single minority examples located inside the majority class regions cannot be simply deleted from the data since their proper treatment by informed pre-processing may improve classification performance for the minority class.

To summarize the discussion of the aforementioned data complexity factors we would like to stress that their identification in real world data sets is not a trivial task. The discussion of this issue and references to known methods are presented in [38, 53].

## 2.3   Local Data Characteristics in Informed Pre-processing

Recall that the pre-processing methods are classifier independent and they are designed to modify the imbalanced data set in a way that transforms the class distribution to a more appropriate one for learning classifiers. Many of these methods generate a more balanced distribution of examples into classes. In general, changing the class distribution towards a more balanced one improves the performance for most data sets and classifiers [21].

The simplest pre-processing methods arc random *over-sampling* which replicates examples from the minority class, and random *under-sampling* which randomly eliminates examples from the majority classes until a required degree of balance between class cardinalities is reached. Therefore these methods exploit global information about the data set: the current and expected imbalance ratios.

Since simple random pre-processing methods are often not effective, *focused* (also called *informed*) methods have been introduced; see their comprehensive reviews in [7, 21]. Many of these methods attempt to take into account internal characteristics of data regions around minority class examples. Historically, the first such method resulted from Kubat and Matwin's proposal of the *one-side-sampling* method (OSS) [29]. These authors observed that characteristics of mutual positions of examples from different classes is a source of difficulty. Thus, OSS is based on distinguishing different types of learning examples: safe examples (located inside the regions occupied by examples from the given class), borderline (located near the decision boundary) and, so called, noisy examples (these authors understood them as examples from the given class localed inside safe regions of the other classes). According to the OSS filtering approach, borderline and noisy examples are removed from the majority classes, while the minority class is kept unchanged (even for noisy minority examples).

Many other filtering (mainly under-sampling) methods exploits the paradigm of edited nearest classifiers. For instance, the *Nearest Cleaning Rule* (NCR) [31] applies it to removal of "difficult" examples from the majority classes. Briefly speaking, NCR first looks for a specific number $k$ of *nearest neighbours* ($k = 3$ is recommended in [31]) of the "seed" example. Then, it re-classifies seed example according to most frequent class label among neighbours. Finally, it removes from majority class these examples, which cause the wrong re-classification.

The analysis of class labels among $k$ nearest neighbors is also exploited in a hybrid method SPIDER that selectively filters out the majority examples which may lead to incorrect re-classification of the minority ones [55]. In the first stage it applies the edited nearest rule to distinguish between safe and unsafe examples (which is depending how strongly $k$ neighbours may correctly—or incorrectly—re-classify the given "seed" example). For the majority class, the neighbours which misclassify the seed minority example are either removed or relabeled. Then, in the next stage, the reclassification analysis is repeated and the remaining unsafe minority examples are additionally replicated depending on the number of majority neighbours.

The best known method of informative over-sampling is called Synthetic Minority Over-sampling Technique (SMOTE) [9]. It is also based on the $k$ nearest neighbourhood and exploits it to selectively over-sample the minority class by creating *new synthetic examples* with respect to the global parameter, called *over-sampling* ratio. SMOTE has been further extended in different ways—see reviews in [7, 21]. Quite often these extensions exploit different local information about the learning examples. For instance, the authors of BORDERLINE SMOTE do not treat all minority examples in the same way and focus oversampling around examples from borderline region between classes [19].

## 3 Analyzing Neighbourhoods of Minority Class Examples

### 3.1 Motivations

Following the critical analysis of earlier works on using local data characteristics in informed pre-processing and studies on the complexity of imbalanced data Napierala and Stefanowski have decided to link data difficulty factors to *different types of examples* forming the minority class distribution. They proposed to differentiate between safe and unsafe examples in learning from imbalanced data [40], however in a different way than earlier proposed, e.g. by [29]. Below we present this categorization following their definitions from [38, 40, 42].

*Safe examples* are ones located in the homogeneous regions populated by examples from one class only. Other examples are *unsafe* and more difficult for learning. Unsafe examples are categorized into *borderline* (placed close to the decision boundary between classes), *rare cases* (isolated groups of few examples located deeper inside the opposite class), or *outliers*. As the minority class can be highly underrepresented in the data, it is claimed that the rare examples or outliers, could represent a very small but valid sub-concepts of which no other representatives could be collected for training [38]. Therefore, they cannot be considered as noise examples which typically are then removed or re-labeled. In Fig. 2 all these four types of examples from the minority class are illustrated in the 2-dimensional distribution of the two class data set called paw.

Recall experimental studies from [38, 40], where the graphical visualizations techniques based on multi-dimensional scaling and non-linear t-SNE projection have confirmed the occurrence of this categorization of example types in several real-world imbalanced data sets. However, such an analysis cannot be directly applied to

**Fig. 2** Visualization of four types of minority class examples in paw data

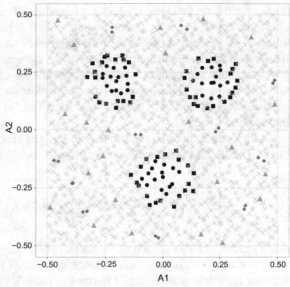

LABEL ● MIN-SAFE ■ MIN-BORDER ◆ MIN-RARE ▲ MIN-OUTLIER ⨯ MAJ

larger data. Napierala and Stefanowski have looked for new simple techniques which should more directly identify these types of examples.

Their method origins from the hypotheses [40] on role of the mutual positions of the learning examples in the attribute space and the idea of assessing the type of an example by analyzing class labels of the other examples in its *local neighbourhood*.

Following the proposal of [38, 40]—a term local refers to studying characteristics of the nearest examples due to the possible sparse decomposition of the minority class into rather rare sub-concepts with non-linear decision boundaries. Considering a larger size of the neighbourhood may not reflect the underlying distribution of the minority class.

Such a neighbourhood of an example could be modeled in different ways. In the previous research Napierala and Stefanowski proposed to construct it with:

- $k$-nearest neighbours,
- or kernel functions.

The analysis of class labels of examples in the $k$-nearest approach concerns a fixed number of nearest examples (without taking into account their distances to the seed examples) while in the kernel approach all examples within a given radius (the kernel bandwidth) are taken into account together with their distances. We will come back to the problem of tuning their proper values in Sect. 4. An analysis of the class label distribution of examples inside the neighborhood of the given example allow us to assess its level of difficulty and as a result its type (safe vs. unsafe to be learned).

Note, however, that constructing both types of the neighbourhood involves decisions on choosing the *distance function*. In previous considerations Napierala and

Stefanowski have followed results of analyzing different distance metrics [32] and chose the HVDM metric (*Heterogeneous Value Difference Metric*) [63]. Its main advantage for mixed attributes is that it aggregates normalized distances for qualitative and quantitative attributes. In particular, comparing to other metrics, HVDM provides more appropriate handling of qualitative attributes as instead of simple value matching, as it makes use of the class information to compute attribute value conditional probabilities by using a Stanfil and Valtz value difference metric for nominal attributes [63].

More precisely, let $x$ be a seed example and $y$ be another example (potential neighbour). The HVDM is defined over $m$ attributes as

$$D(x,y) = \sqrt{\sum_{i=1}^{m} d_i(x_i, y_i)^2}$$

All distances for single attributes are normalized in range 0 to 1. If one of the attribute values of $x_i, y_i$ is unknown, the distance $d_i$ is equal to 1. The partial distance for numeric attributes is defined as a normalized metric $(y_i - x_i)$. Then, the partial distance for nominal attributes is defined as:

$$d_i(x_i, y_i) = \begin{cases} 0 & \text{if } x_i = y_i \\ svdm & \text{if } x_i \neq x_i \end{cases}$$

Value difference metric *svdm* is defined as [10]:

$$svdm = \sum_{l=1}^{k} \left| \frac{N(x_i, K_l)}{N(x_i)} - \frac{N(y_i, K_l)}{N(y_i)} \right|$$

where $k$ is the number of classes, $N(x_i)$ and $N(y_i)$ are the numbers of examples for which the value on i-th attribute is equal to $x_i$ and $y_i$ respectively, $N(x_i, K_l)$ and $N(y_i, K_l)$ are the numbers of examples from the decision class $K_l$, which belong to $N(x_i)$ and $N(y_i)$, respectively.

In the next two sub-sections we will discuss more precisely previous proposals of modeling these two kinds of the neighbourhood (with $k$-nearest neighbours or kernel functions) and establishing types of minority class examples [38, 40].

In both cases, deciding about the type of minority examples is based on analyzing class labels of examples in its neighbourhood.

## 3.2 Modeling k-Neighbourhood

The $k$-nearest neighbourhood has been mainly exploited in the previous studies [38, 40, 42] and some applications of this approach to pre-processing [43, 62] or specialized ensembles [5]. These authors have aimed at distinguishing whether an example is safe, borderline, rare or outlier depending on the numbers of examples from minority vs. majority classes in the considered neighbourhood. As we will also discuss in the next section, the size neighbourhood $k$ should not be smaller than 5 as it may poorly distinguish between four types of examples.

In [40] the following rule has been introduced to identify the type of the given example. If all, or nearly all, its neighbours belong the same (usually minority) class, this example is treated as the safe example, otherwise it is one of unsafe types. If the number of both classes inside the $k$-neighbourhood are quite similar, the example is treating as borderline one. For an extreme situation—all neighbours belong to the opposite class it is clearly an outlier. Finally, the examples with one or sometimes two (for larger sized of the $k$) neighbours from its class was identified as a rare case.

For the most used size of neighbourhood $k = 5$, the proportion of neighbours from the same class against neighbours from the opposite class can range from 5:0 (all neighbours are from the same class as the analyzed example) to 0:5 (all neighbours belong to the opposite class). Depending on this proportion, Napierala and Stefanowski have proposed to assign the labels to the examples in the following way:

- 5:0 or 4:1—an example is labelled as a safe example.
- 3:2 or 2:3—a borderline example; Note that although the examples with the proportion 3:2 are still correctly re-classified by its neighbours, the number of neighbours from both classes is approximately the same, so it was assumed that this example could be located too close to the decision boundary between the classes.
- 1:4—labelled as a rare example.
- 0:5—an example is labelled as an outlier.

Similar interpretations has been extended for larger values of $k$. For instance, in case of $k = 7$ and the neighbourhood distribution 7:0 or 6:1 or 5:2—a safe example; 4:3 or 3:4—a borderline example; again the number of neighbours from both classes are approximately the same; 2:5 or 1:6—a rare example; and 0:7—an outlier [38].

Besides using such thresholding, these authors also considered defining the one coefficient expressing a safe level of the given example $x$—being an estimator of conditional probability of its assignment to the minority class as $p(C_{min}|x) = \frac{k_{min}}{k}$, where $C_{min}$ is a minority class, $k$ is the number of neighbours and $k_{min}$ is the number of minority class neighbours [42].

## 3.3 Modeling Kernel Neighborhood

An alternative approach to fixing the number of neighbours is to fix the local area around the example as it done in kernel approaches—which was discussed in [38] and studied in [42]. Note that due to the form of the kernel function, different weights (probabilities) could be assigned to the neighbours, based on their distance from the analyzed minority example $x$. Moreover, unlike having always the same number of examples in the $k$-neighbourhood modeling, each kernel may cover different number of examples within a fixed radius which rises wider interpretation of local density (see our further experimental analysis in Sect. 5.2).

Several kernel functions could be considered—besides the most popular Gaussian kernel, other triangular or Epanechnikov functions are among common choices. In this study we have decided to apply Epanechnikov function which is defined as:

$$K(u) = \frac{3}{4}(1 - u^2)\mathbf{1}_{|u|\leq 1},$$

where $u = \frac{d_i}{h}$, $d_i$ is the distance of $i$-th example $(x_i)$ to the considered example $x$, and $h$ is bandwidth of the kernel. Epanechnikov kernel is suitable for our purposes since it takes values 0 when $d_i > h$. In this sense, it resembles limits of $k$-neighbourhood. Moreover, this property will be very useful inside the procedure for tuning the neighborhood size discussed in Sect. 5.2. The distance $d_i$ between examples is calculated according to HVDM metric (see motivations presented in the earlier Sect. 3.1). Given the definition of the kernel function we estimate a weighted sum of all minority neighbours, where weights depend on the distance from the analyzed example. Comparing it to the weighted sum calculated for the majority class neighbours we can estimate the probability that the analyzed example $x$ may belong to the minority class $p(C_{min}|x)$.

To assess the type of a minority example, we need to discretize the range of this value into subintervals. Inspired by earlier research [38], in this paper we proposed the following rule: if $1 \geq p(C_{min}|x) > 0.7$ then label $x$ as safe; if $0.7 \geq p(C_{min}|x) > 0.4$ then label $x$ as borderline; if $0.4 \geq p(C_{min}|x) > 0.2$ then label $x$ as rare; if $0.2 \geq p(C_{min}|x) > 0$ then label $x$ as outlier (we keep this type similarly to earlier name); if $p(C_{min}|x) = 0$ then label $x$ as a new type called *zero*. Finally, if there is no other example inside the neighbourhood of $x$ (even from the opposite majority class), then label $x$ as a singleton in an empty sub-region (further called simply *empty*).

Note that this rule is different than the one proposed in [38, 42] as it introduces two new labels, which allow to better understand types of the kernel neighbourhood discovered in data.

## 3.4  Experiences with Analyzing Types of Minority Examples

The previous experiments with modeling $k$-nearest neighbourhood applied to UCI imbalanced data sets are described in [38, 42]. They have clearly demonstrated that most of these real-world data do not include many safe minority examples. They rather contain all types of examples, but in different proportions. Depending on the dominating type of identified minority examples, the considered data sets could be labeled as: safe, border, rare or outlier—which show the level of their potential difficulty. Moreover, the thesis [38] has shown that the classifier performance could be related to the category of data. First, for the safe data nearly compared single classifiers (SVM, RBF, $k$-NN, decision trees or rules) have achieved good, comparable prediction results. The larger differentiation among these classifiers has been noticed for more unsafe data sets (e.g. SVM is worse than $k$-NN and trees for data with higher number of rare cases and outliers). The similar analysis has been carried out for the most representative pre-processing approaches, showing that the competence area of each method depends on the data difficulty level, based on the types of minority class examples. For more details see [38, 42].

## 4 Tuning the Neighbourhood Size

In this paper we focus our interest on tuning the size of the neighborhood with respect to characteristics of each data set.

### 4.1 Tuning k Value

In the previous studies Napierala and Stefanowski [38, 40, 42] exploited mainly $k$ nearest neighbourhood and they showed that values smaller than 5, e.g., $k = 1$ and $k = 3$, may poorly distinguish the type of examples, especially if one wants to assign them to four types. Too high values, on the other hand, would be inconsistent with the assumption of the locality of the method (see [42] for more details of the discussion why the locality is important for analyzing complex minority class distributions in imbalanced data).

They proposed to set $k = 5$ as the default value. To check whether this parameter $k$ could strongly influence the results of labelling minority examples, a special sensitivity analysis over 26 different data sets was carried out in [42]. Its results have shown that proportions of identified types of examples are quite stable while changing $k$ values (between 5 and 13—globally defined for all of these data sets). The recommendation of the smallest value of $k$ has come from the paradigm of the most local analysis of the complex decision boundaries of the minority class and its sparsity. Furthermore, the authors pointed out that the parameter $k = 5$ was recommended for many related, informed pre-processing methods (see e.g. [9, 31, 55]).

Nevertheless, the idea of tuning of $k$ parameter, for each imbalanced data set individually, has not been considered so far. Studying the literature one may find some positions that consider changing size of neighbourhoods in a standard $k$-NN classifier for class balanced data. In these works choosing value $k$ is made with respect to the data set or class cardinality. Refer, e.g., to [17] which recommends approximating $k \approx \sqrt{n}$, where $n$ is the total number of learning examples. However, we hypothesize that in case of imbalanced data $n$ should be rather the size of the minority class. Other researches have proposed some slightly different approximations. Enas and Chai [12] postulated to take

$$k = n^{2/8} \text{ or } k = n^{3/8}.$$

See also [16] for a more detailed presentation of similar proposals. Since these formulas have been designed with typical problems and $k$-NN classifier in mind, Napierala and Stefanowski have expressed their doubts whether they can be directly transferred into a different context of modeling neighborhoods for class imbalanced data [42].

Here, we share this point of view and we propose a method of tuning $k$ value in a cross-validation procedure. The important question concerns the choice of optimization criterion for the tuning method. If one refers to the idea of recognizing the minority class examples as good as possible (which is a key issue in learning from imbalanced data)—such a criterion may reflect abilities of $k$ neighborhood to correctly re-classify examples. This idea is consistent with some earlier proposals of using cross-validation to choose $k$ value which minimize the classification error of a standard $k$-NN classifier, as it was argued by Dasarathy [11]. We will describe it in more detail in Sect. 4.3.

## 4.2 Tuning Kernel Bandwidth

Modeling neighbourhood with kernels was preliminary discussed in [38, 42] as an alternative to using $k$ neighbours analysis of imbalanced data. The authors postulated that the Epanechnikov function should be equal to the average distance to the 5th neighbour of each minority example in the data set, as they wanted to keep the link to their basic $k$ neighbourhood method. Furthermore, in [42] they presented an comparative experiment of labelling the minority class examples in 26 popular imbalanced data sets and demonstrated that using the kernel method does not change the results of $k$ neighbourhood more than by 5–10%.

In this paper we want to consider new approaches for tuning the size of kernel neighbourhood with respect to each data set. Firstly, note that the kernel analysis is often related to *kernel density estimation*, i.e., non-parametric approach to estimation of probability density function, which is one of the most fundamental issues in statistics [33, 50, 51]. Although there are important differences between the density estimation and our problem, one can still notice some similarities while calculating probabilities in considered points of the example space. Recall that exploiting class probabilities inside the kernel neighbourhood of the seed example $x$ may be equivalent to operating on contribution of neighbours with respect to their kernel distance to $x$. It may be also interpreted in the context of the kernel density estimator

$$\hat{f}_h(x) = \frac{1}{n} \sum_{i=1}^{n} K_h(x - x_i),$$

where $n$ is a number of neighbours $x_i$ (or more generally considered data points), $K_h$ a kernel function with a bandwidth size $h$.

It is also known that the kernel bandwidth is this parameter which strongly influences the resulting probability estimate. Its tuning has been already intensively studied in statistics. The most of approaches attempt to optimize a criterion referring to the expected $L_2$ risk, which is a kind of the mean integrated squared error between $\hat{f}_h(x) - f(x)$. Although basic formulations involve unknown density function $f$ many automatic, data-based methods have been developed for selecting the bandwidth $h$; for some reviews refer, e.g., to [27].

If Gaussian basis kernel functions are used to approximate univariate data, and the underlying density being estimated is assumed to be Gaussian, the choice for $h$ (that is, the bandwidth that minimizes the mean integrated squared error) is often estimated as

$$h = \left( \frac{4\hat{\sigma}^5}{3n} \right)^{\frac{1}{5}} \approx 1.066\hat{\sigma}n^{-1/5}.$$

where $\hat{\sigma}$ is the standard deviation of the examples in the data. This approximation is known as Silverman's rule of thumb [51] and quite often implemented in statistical software. Other bandwidth selection methods were also proposed, for instance Terrell and Scott proposed oversmoothed density estimates which in case of the standard Gaussian kernel leads to the oversmoothed bandwidth $h = 1.144\hat{\sigma}n^{-1/5}$. These considerations could be generalized for the multi-dimensional kernel with $H$—a symmetric positive bandwidth matrix [33]. For instance the aforementioned rules of thumbs are generalized to

$$h_i = \hat{\sigma}_i \left( \frac{4}{(d+2)/n} \right)^{\frac{1}{d+4}}.$$

Nevertheless, the above tuning methods concern a typical estimation of density function in the unsupervised setting. Although they are sometimes applied as a kind of pre-processing inside the supervised classifiers—in particular Bayesian classifiers, see e.g., [34], in our opinion these methods cannot be transferred directly to our problem of supervised neighbourhood analysis for imbalanced data. However, due to some similarities, we acknowledge inspiration in specialized density estimation methods, which are based on cross-validation optimization of Least Squares forms representing the integrated squared error (ISE) of density functions or, so called, biased versions [50].

## 4.3 A New Tuning Method Based on Cross-Validation

Following the critical analysis of tuning $k$ parameter (see Sect. 4.1), and kernel bandwidth in density estimation (in Sect. 4.2), we propose a simple cross-validation method to tune both of these parameters. Our motivation is to make use of ability of neighbourhoods of an example to correctly recognize its class label. Recall that in learning classifiers from imbalanced data one attempts to improve recognition of the minority class, so studying the neighborhood from the re-classification perspective may be connected with this aim.

The tuning method is based on the optimization procedure which scans a value of neighbourhood parameter ($k$ for $k$ nearest neighbourhood and bandwidth $h$ for kernel neighbourhood) from a pre-defined set of possible values. In our further experiments, for the kernel version we will refer these values to the average distances between minority class examples calculated for a given data set (see Sect. 5.2). However, in general, they could be other appropriate values. In case of $k$ nearest neighbourhood we will enumerate $k$ values starting from the smallest possible value.

As the optimization criterion we should choose a measure reflecting ability of the neighborhoods built on the training examples to recognize the type of the testing example. In further experiment we have decided to apply popular G-mean measure as it aggregates re-classifications of examples from both classes.

For a given value of an analyzed parameter (bandwidth $h$ or $k$) the data set is split into training and testing parts following the stratified version of cross validation technique. For each split the following schema is carried out:

- For each example from the training part its neighborhood is constructed and tuned with respect to the given parameter value—its size.
- Each example from the testing part is classified by the tuned neighborhood (of the same size as the optimized parameter).
- The classification by the neighbourhood is performed according to highest probability $p(C_i|x)$ that example $x$, from the test set may belong to class $C_i$ (for problems considered in this paper $i = \{1, 2\}$, since we have only minority class $C_{min}$, and majority class $C_{maj}$), estimated according to distribution of classes of examples in the neighbourhood constructed in the training set.
- The value of the optimization criterion is calculated on the basis of how many examples from a test set are correctly classified by the tuned neighbourhood.

The final value of the optimization criterion comes from averaging over several folds inside the cross-validation. The cross-validation may be repeated several times to reduce variance of optimization criterion. The value of the finally chosen neighbourhood parameter that corresponds to the best average optimization criterion is the result of this tuning method.

## 5    Experimental Analysis of Data Characteristics

### 5.1    Experimental Setup

In this section we will carry out two kinds of experiments. Firstly, we will show how to tune the kernel neighbourhood and $k$-neighbourhood sizes, i.e., bandwidth $h$ and $k$, over different benchmark real-world data sets and synthetic data sets. It should

**Table 1**  Characteristics of real-world data

| Data set | # examples | # attributes | Minority class | IR |
|---|---|---|---|---|
| abalone | 4177 | 8 | 0–4, 16–29 | 11.47 |
| breast-cancer | 286 | 9 | Recurrence-events | 2.36 |
| car | 1728 | 6 | Good | 24.04 |
| cleveland | 303 | 13 | 3 | 7.66 |
| cmc | 1473 | 9 | 2 | 3.42 |
| ecoli | 336 | 7 | imU | 8.60 |
| haberman | 306 | 4 | 2 | 2.78 |
| hepatitis | 155 | 19 | 1 | 3.84 |
| scrotal-pain | 201 | 13 | Positive | 2.41 |
| solar-flare | 1066 | 12 | F | 23.79 |
| transfusion | 748 | 4 | 1 | 3.20 |
| vehicle | 846 | 18 | Van | 3.25 |
| yeast | 1484 | 8 | ME2 | 28.10 |

illustrate the usefulness of the method presented in Sect. 4. Secondly, given the tuned sizes of neighbourhood, we will analyze the internal characteristics of imbalanced data sets and establish the level of their difficulty (with respect to different types of minority examples). This part of experiment should show the applicability of the neighbourhood analysis to recognize the different categories of imbalanced data sets.

Similarly to the related study [42] we will focus our experiments on 13 benchmark real-world imbalanced data sets. Their characteristics is presented in Table 1. We have chosen the data sets which have been often studied in many experimental studies with imbalanced data. They represent different sizes, imbalance ratios (denoted by IR), domains and have both continuous and nominal attributes. Following the most related results [42] some of these data sets should be easier to learn standard classifiers while most of them constitute different degrees of difficulties.

Nearly all of benchmark real-world data sets come from the UCI repository.[1] One data set is medical data set which was used in the earlier works of Stefanowski et al. on class imbalance.[2] In data sets with more than one majority class, they are aggregated into one class to have only binary problems, which is also typically done in the literature.

Furthermore, we have decided to study few synthetic data sets with known data distribution. We apply a specialized generator for imbalanced data [64] and produced two different types of data sets. The examples of both minority classes are generated randomly inside predefined spheres and the majority class examples are

---

[1] http://www.ics.uci.edu/mlearn/MLRepository.html.
[2] We are grateful to Prof. W. Michalowski and the MET Research Group from the University of Ottawa for providing us an access to scrotal-pain data set.

uniformly distributed in an area surrounding them. We consider two configurations of these minority class spheres: called paw and flower—see their 2-D illustrations at Figs. 1 and 2. In both data sets the global imbalanced ratio *IR* is equal to 7, and the total cardinality of examples are 1200 for paw and 1500 for flower always with three attributes. The minority class is decomposed into 3 sub-parts or 5 sub-parts. Moreover, each of this data sets has been generated with different numbers of unsafe examples—which is denoted by four numbers inside the name of data. For instance flower5-3d-30-40-15-15 means that the generated minority class should contain approximately 30% of safe examples, 30% inside the class overlapping, 15% rare and 15% outliers.

## 5.2    Tuning Kernel Bandwidth and k-Neighbourhood

In this experiment we used the method presented in Sect. 4 to tune the best size of kernels' bandwidth *h* and the best value of parameter *k* representing the number of nearest neighbours. The results of the tuning on benchmark real-world data are presented in Table 2, while the results of tuning on synthetic data are presented in Table 3. The results presented in these tables come from stratified 10-fold cross-validation averaged 5 times to improve reproducibility and reduce possible variance of the optimization criterion (here G-mean).

Note that the considered bandwidth *h* sizes refer to the average distance to *k*-th nearest neighbour in the minority class of the given data set. This setting allows us to obtain more comparable results and make the bandwidth size dependent on the char-

**Table 2**    Bandwidth *h* and *k* tuned on real-world data

| Data set | Kernel | | | *k*-NN | |
|---|---|---|---|---|---|
| | Avg. *k* | *h* | G-mean | *k* | G-mean |
| abalone | 6.5 | 0.074 | 36.679 | 5 | 45.547 |
| breast-cancer | 8 | 0.087 | 52.480 | 7 | 57.324 |
| car | 8 | ≃0 | 77.265 | 5 | 87.627 |
| cleveland | 1 | 0.523 | 22.190 | 5 | 41.997 |
| cmc | 1 | 0.059 | 47.963 | 5 | 58.233 |
| ecoli | 7 | 0.332 | 76.739 | 9 | 80.300 |
| haberman | 9 | 0.328 | 43.624 | 5 | 56.552 |
| hepatitis | 6 | 0.812 | 65.695 | 7 | 71.893 |
| scrotal-pain | 8.5 | 0.408 | 55.955 | 9 | 77.244 |
| solar-flare | 1 | 0.038 | 27.095 | 5 | 50.609 |
| transfusion | 3 | 0.128 | 53.976 | 7 | 60.710 |
| vehicle | 8.5 | 0.516 | 88.682 | 5 | 93.883 |
| yeast | 2.5 | 0.430 | 34.391 | 5 | 60.018 |

**Table 3**  Bandwidth $h$ and $k$ tuned on synthetic data

| Data set | Kernel | | | $k$-NN | |
|---|---|---|---|---|---|
| | Avg. $k$ | $h$ | G-mean | $k$ | G-mean |
| flower5-3d-10-20-35-35 | 0.5 | 0.058 | 43.199 | 7 | 52.549 |
| flower5-3d-100-0-0-0 | 9 | 0.077 | 91.906 | 9 | 96.407 |
| flower5-3d-30-40-15-15 | 2.5 | 0.103 | 79.623 | 9 | 80.998 |
| flower5-3d-30-70-0-0 | 9 | 0.076 | 89.802 | 9 | 96.082 |
| flower5-3d-50-50-0-0 | 9 | 0.077 | 92.757 | 8 | 96.506 |
| paw3-3d-10-20-35-35 | 0.5 | 0.066 | 44.088 | 7 | 49.319 |
| paw3-3d-100-0-0-0 | 8.5 | 0.099 | 95.425 | 9 | 97.067 |
| paw3-3d-30-40-15-15 | 2 | 0.113 | 78.178 | 7 | 79.186 |
| paw3-3d-30-70-0-0 | 9 | 0.100 | 90.252 | 7 | 93.189 |
| paw3-3d-50-50-0-0 | 8.5 | 0.098 | 92.458 | 9 | 95.090 |

acteristics of each data set that was analyzed. Please note that value of $k$-neighbour according to the average distance in the minority class relates to some extend to the value of $k$ in the other approach based on nearest neighbours. Technically, we considered values of the kernel bandwidth corresponding to average distance to $k$-th neighbour, with $k$ from interval [5, 9] with a basic step 0.5.

We have chosen these values as we wanted to check smaller neighbourhoods, which was already well motivated in the previous research presented in [42]. In case of the other approach based on nearest neighbours, we considered only $k = \{5, 6, 7, 8, 9\}$ for the same reasons. The choice of $k \geq 5$ is motivated here by the fact that neighbourhoods smaller than 5 do not allow to perform sensible labelling of example types that we presented in Sect. 5.3. This argument is not viable for average $k$ values related to the bandwidth size. In Tables 4 and 5, we present an average number of examples inside the kernel for bandwidths tuned in experiments on real-world and synthetic data sets, respectively.

Note that average numbers of nearest neighbours in kernels of real-world data sets, presented in Table 4, are always higher than 5. For synthetic data sets, presented in Table 5, one can observe that the average number of examples inside kernels is smaller than 3 in case of the most difficult to learn distributions of examples (data sets: flower5-3d-10-20-35-35, paw3-3d-10-20-35-35). In case of these two data sets, rare and outlier examples are the most numerous in the minority class. This result can be explained when we take a look at results from the Table 3. For these data sets the value of average $k$ is the smallest possible, which means that it was better to keep the neighbourhood (and the bandwidth) as small as possible to obtain the best optimization result of G-mean.

A comparison of results obtained with tuning kernels and nearest neighbours variants, reported in Tables 2, and 3, shows that kernel neighbourhoods works differently than $k$ nearest neighbourhoods. This observation comes mainly from the comparison of G-mean values obtained in the tuning process. Regardless whether we compare on

**Table 4** Average $k$ (for tuned bandwidth) and average number of examples inside a kernel for real-world data

| Data set | Avg. $k$ | Avg. n |
|---|---|---|
| abalone | 6.5 | 115.04 |
| breast-cancer | 8 | 41.12 |
| car | 8 | 14.39 |
| cleveland | 1 | 18.74 |
| cmc | 1 | 6.96 |
| ecoli | 7 | 25.37 |
| haberman | 9 | 54.25 |
| hepatitis | 6 | 36.69 |
| scrotal-pain | 8.5 | 58.46 |
| solar-flare | 1 | 273.93 |
| transfusion | 3 | 38.55 |
| vehicle | 8.5 | 22.33 |
| yeast | 2.5 | 62.24 |

**Table 5** Average $k$ (for tuned bandwidth) and average number of examples inside a kernel for synthetic data

| Data set | Avg. $k$ | Avg. n |
|---|---|---|
| flower5-3d-10-20-35-35 | 0.5 | 3.10 |
| flower5-3d-100-0-0-0 | 9 | 12.56 |
| flower5-3d-30-40-15-15 | 2.5 | 18.16 |
| flower5-3d-30-70-0-0 | 9 | 12.96 |
| flower5-3d-50-50-0-0 | 9 | 12.55 |
| paw3-3d-10-20-35-35 | 0.5 | 2.88 |
| paw3-3d-100-0-0-0 | 8.5 | 12.28 |
| paw3-3d-30-40-15-15 | 2 | 15.82 |
| paw3-3d-30-70-0-0 | 9 | 14.81 |
| paw3-3d-50-50-0-0 | 8.5 | 12.94 |

real-world or synthetic data sets, $k$-neighbourhood achieves higher G-mean values than kernel neighbourhood.

However, one should be careful with drawing conclusions from comparing average $k$ related to the tuned kernel bandwidth with $k$ tuned directly for nearest neighbours as the kernel approach uses other ranges. Nevertheless, it is visible that higher values of bandwidths in kernels relate always to higher values of $k$ in nearest neighbours. We can also notice that larger neighbourhoods are selected for easier data sets.

The size of the kernel bandwidth (the distance values) presented in Tables 2, and 3 is not easy to interpret since it is a value of HVDM metric (please see Sect. 3). Note, however, that values of the bandwidth on real-world data sets have higher variance than these observed for synthetic data sets. It seems natural that real-world data sets should present more variability than synthetic ones.

## 5.3 Analyzing Types of Minority Examples

In this part experiment, we used the previously tuned bandwidths of kernels and $k$-neighbourhoods to label different types of minority class examples in real-world and synthetic data sets (it is somehow inspired by the earlier analysis in [40]). The results obtained for benchmark real-world data sets with kernel neighbourhood are presented in Table 6, and the ones obtained with $k$-neighbourhood are presented in Table 7.

Let us first explain differences in the number of example types identified by the two approaches to model neighbourhoods. Recall that differently than in [42], we have not applied the same labelling rule and the tuned values of $k$ are different and vary depending on the given data set (see values of $k$ for $k$-NN in Table 2 for details). Instead we used analogous rules, which are formulated according to estimated values of probability of minority class, for both kernels and $k$-neighbourhood (please see Sect. 4 for details).

**Table 6** Labelling of minority class examples in real-word data for the tuned bandwidth

| Data set | Safe [%] | Borderline [%] | Rare [%] | Outlier [%] | Zero [%] | Empty [%] |
|---|---|---|---|---|---|---|
| abalone | 4.78 | 10.15 | 8.66 | 70.75 | 3.58 | 2.09 |
| breast-cancer | 17.65 | 18.82 | 31.76 | 29.41 | 1.18 | 1.18 |
| car | 0.00 | 47.83 | 43.48 | 8.70 | 0.00 | 0.00 |
| cleveland | 2.86 | 2.86 | 25.71 | 42.86 | 17.14 | 8.57 |
| cmc | 13.81 | 21.32 | 24.02 | 13.21 | 20.42 | 7.21 |
| ecoli | 5.71 | 68.57 | 14.29 | 5.71 | 5.71 | 0.00 |
| haberman | 1.23 | 25.93 | 39.51 | 29.63 | 2.47 | 1.23 |
| hepatitis | 28.12 | 21.88 | 3.12 | 34.38 | 6.25 | 6.25 |
| scrotal-pain | 15.25 | 20.34 | 28.81 | 22.03 | 1.69 | 11.86 |
| solar-flare | 4.65 | 6.98 | 16.28 | 65.12 | 4.65 | 2.33 |
| transfusion | 5.06 | 38.76 | 27.53 | 16.85 | 6.74 | 5.06 |
| vehicle | 55.78 | 35.68 | 5.53 | 0.00 | 0.50 | 2.51 |
| yeast | 7.84 | 11.76 | 27.45 | 39.22 | 9.80 | 3.92 |

**Table 7** Labelling of minority class examples in real-word data for tuned $k$

| Data set | Safe [%] | Borderline [%] | Rare [%] | Outlier [%] |
|---|---|---|---|---|
| abalone | 11.04 | 8.36 | 23.58 | 57.01 |
| breast-cancer | 29.41 | 28.24 | 29.41 | 12.94 |
| car | 60.87 | 21.74 | 13.04 | 4.35 |
| cleveland | 0.00 | 22.86 | 17.14 | 60.00 |
| cmc | 23.72 | 18.32 | 31.23 | 26.73 |
| ecoli | 28.57 | 48.57 | 14.29 | 8.57 |
| haberman | 14.81 | 29.63 | 38.27 | 17.28 |
| hepatitis | 43.75 | 28.12 | 12.50 | 15.62 |
| scrotal-pain | 38.98 | 42.37 | 15.25 | 3.39 |
| solar-flare | 0.00 | 18.60 | 32.56 | 48.84 |
| transfusion | 26.97 | 33.71 | 15.17 | 24.16 |
| vehicle | 78.89 | 13.57 | 6.03 | 1.51 |
| yeast | 15.69 | 19.61 | 21.57 | 43.14 |

The next important difference comes from the new assumption that the kernel approach allows us to identify more types of examples. It is clearly visible for the real-world data sets (see Table 6) which contain minority examples of all six different types. A similar observation is valid for the same data sets analyzed with $k$-neighbourhood (in Table 7), although here we distinguish four types. Let us also note that the results presented in Table 7 correspond well with the previous ones presented in [42]. Nevertheless, some differences in proportions are visible mostly for more difficult data sets (e.g., abalone, solar-flare, yeast).

Even though numbers of examples into different types labelled by kernel neighbourhood and $k$-neighbourhood are not exactly the same, the characteristics of the particular data sets (i.e. their categorization with respect to dominating types of minority examples) are generally quite similar. In particular, the highest number of outliers is discovered for the same data sets: yeast, solar-flare, abalone, cleveland. The highest number of rare type examples is also discovered for the same data sets: cmc, breast-cancer (although $k$-neighbourhood discovers the same number of safe examples), haberman. The same applies to borderline and safe examples. The highest number of borderline examples is discovered for data sets: transfusion, and ecoli. The highest number of safe examples is discovered by both kernel and $k$ neighbourhood for vehicle. Limited differences in labeling are observed for few data sets only: hepatitis, scrotal-pain, and car.

One can notice that new types of examples discovered by the kernel neighbourhood are present in almost all data sets. There are two exceptions: zero type examples are not discovered in car; then empty type examples are not found in car, and ecoli. These type of examples are not dominant in any data set. Since they reflect poor performance of kernel neighbourhood at estimating probability of minority class, one should not expect to find a lot of them. Still, relatively high numbers

**Table 8** Labelling of minority class examples in synthetic data for tuned bandwidth

| Data set | Safe [%] | Borderline [%] | Rare [%] | Outlier [%] | Zero [%] | Empty [%] |
|---|---|---|---|---|---|---|
| flower5-3d-10-20-35-35 | 20.21 | 22.87 | 21.28 | 0.00 | 35.11 | 0.53 |
| flower5-3d-100-0-0-0 | 84.57 | 14.89 | 0.53 | 0.00 | 0.00 | 0.00 |
| flower5-3d-30-40-15-15 | 35.64 | 34.04 | 3.19 | 14.36 | 12.77 | 0.00 |
| flower5-3d-30-70-0-0 | 76.60 | 23.40 | 0.00 | 0.00 | 0.00 | 0.00 |
| flower5-3d-50-50-0-0 | 77.13 | 22.34 | 0.53 | 0.00 | 0.00 | 0.00 |
| paw3-3d-10-20-35-35 | 14.67 | 20.67 | 24.67 | 0.67 | 36.00 | 3.33 |
| paw3-3d-100-0-0-0 | 65.33 | 34.67 | 0.00 | 0.00 | 0.00 | 0.00 |
| paw3-3d-30-40-15-15 | 26.00 | 42.67 | 4.67 | 11.33 | 15.33 | 0.00 |
| paw3-3d-30-70-0-0 | 44.67 | 52.00 | 3.33 | 0.00 | 0.00 | 0.00 |
| paw3-3d-50-50-0-0 | 57.33 | 40.67 | 2.00 | 0.00 | 0.00 | 0.00 |

**Table 9** Labelling of minority class examples in synthetic data for tuned $k$

| Data set | Safe [%] | Borderline [%] | Rare [%] | Outlier [%] |
|---|---|---|---|---|
| flower5-3d-10-20-35-35 | 25.00 | 5.32 | 36.17 | 33.51 |
| flower5-3d-100-0-0-0 | 87.77 | 12.23 | 0.00 | 0.00 |
| flower5-3d-30-40-15-15 | 52.66 | 17.55 | 17.02 | 12.77 |
| flower5-3d-30-70-0-0 | 77.13 | 22.87 | 0.00 | 0.00 |
| flower5-3d-50-50-0-0 | 90.43 | 9.57 | 0.00 | 0.00 |
| paw3-3d-10-20-35-35 | 18.00 | 12.00 | 34.67 | 35.33 |
| paw3-3d-100-0-0-0 | 70.67 | 29.33 | 0.00 | 0.00 |
| paw3-3d-30-40-15-15 | 54.00 | 16.00 | 14.67 | 15.33 |
| paw3-3d-30-70-0-0 | 76.00 | 23.33 | 0.67 | 0.00 |
| paw3-3d-50-50-0-0 | 66.00 | 34.00 | 0.00 | 0.00 |

of zeros and empty type examples is found in data sets: cleveland and cmc. Relatively high number of zero examples only is found in yeast. Furthermore, a relatively high number of empty type examples is found in scrotal-pain. Some relations between the numbers of discovered zero and empty type examples and the predictive performance of kernel neighbourhood (in Table 2) can be also observed.

The labeling results obtained for synthetic data sets with kernel neighbourhood and $k$-neighbourhood are presented in Table 8 and in Table 9, respectively.

We can conclude that the types of examples injected to synthetic data sets are rather well discovered by both kernel neighbourhood and $k$-neighbourhood. Safer distributions of examples in data sets (without rare and outlier type examples) are recognized in the best way. There is a tendency to mislabel some of safe examples as borderline (which could explained for examples located very closed to the decision boundaries that they are too dominated by neighbors from the opposite class), however, the reverse tendency (to mislabel borderline as safe) is also observable (especially for $k$-neighborhood). Rare and outlier types of examples are much better recognized by $k$-neighborhood than kernel neighborhood. We can hypothesize that the kernel neighborhood expresses a worrying tendency to discover outliers as zero type (and also sometimes empty type) examples. This result can be linked to choosing too small bandwidth by the tuning procedure for difficult distributions of examples.

To sum up, this kind of labeling analysis shows the usefulness of modeling the neighborhood to identify the level of difficulty of the studied data set. Generally speaking, the less safe examples, the more difficult could be the data set. It is also interesting to notice that most of studied data sets do not contain too many safe examples. The percentage of rare, outlier or even empty example is quite high for some of data sets. In particular the kernel analysis may provide more information than $k$ neighborhood approach due to new types of examples.

## 6  Improving Pre-processing Techniques with the Neighbourhood Analysis

One can ask whether the estimation of probability of minority class examples, which is behind the labelling of minority class, may be useful to improve pre-processing of imbalanced data sets. Therefore, we compare performance of a standard unprunned J48 classifier trained on data sets pre-processed according to the neighbourhood analysis with kernel and $k$-neighbourhoods against the same classifier trained on not-processed and randomly over-sampled data sets. The choice of over-sampling is motivated by its' ease of implementing as compared to under-sampling.

The proposed over-sampling technique uses probability of the minority class estimated for each of minority class example according to the frequency of examples in tuned kernel neighbourhood and $k$ neighbourhood (we use the same tuning as comes from the analysis carried out in Sect. 5.3). The estimated probability is used as a weight of example in the sampling procedure. The difference with respect to the neighbourhood analysis is that, since we apply over-sampling, we want difficult examples (thus, having low value of the probability) to be more represented in the over-sampled data set than safe ones. To achieve this result we simply use inverse of the probability as the weight and replicate them proportionally to this value. In general, we want to achieve approximately balanced classes, so we estimate the global number of need copies and divide this number among all minority examples with respect to their weights.

Classification performance of J4.8 with pre-processing technique is measured by standard measures such as G-mean and sensitivity. G-mean results are presented in Tables 10, and 11, for real-world, and synthetic data sets, respectively.

G-mean classification results on real-world data sets show rather limited influence of the proposed pre-processing on predictive performance. In general, one

**Table 10** G-mean [%] for unprunned J48 learned on base (original) and over-sampled real-world data

| Data set | Base | Random | Kernel | k-NN |
|---|---|---|---|---|
| abalone | 53.790 | 60.198 | 60.802 | 60.481 |
| breast-cancer | 56.495 | 68.139 | 68.764 | 68.791 |
| car | 89.851 | 90.356 | 90.157 | 89.681 |
| cleveland | 48.984 | 56.570 | 50.365 | 51.716 |
| cmc | 56.706 | 64.142 | 64.541 | 64.494 |
| ecoli | 70.489 | 74.011 | 74.080 | 74.401 |
| haberman | 56.060 | 54.559 | 57.394 | 56.492 |
| hepatitis | 63.136 | 72.058 | 66.507 | 68.809 |
| scrotal-pain | 69.563 | 70.570 | 70.313 | 71.781 |
| solar-flare | 44.249 | 44.522 | 42.867 | 44.110 |
| transfusion | 60.018 | 56.071 | 56.456 | 56.564 |
| vehicle | 91.929 | 94.405 | 93.912 | 92.609 |
| yeast | 54.564 | 53.735 | 55.535 | 57.219 |

**Table 11** G-mean [%] for unprunned J48 learned on base and over-sampled synthetic data

| Data set | Base | Random | Kernel | k-NN |
|---|---|---|---|---|
| flower5-3d-10-20-35-35 | 0.000 | 39.627 | 38.835 | 38.426 |
| flower5-3d-100-0-0-0 | 89.410 | 88.692 | 87.245 | 88.190 |
| flower5-3d-30-40-15-15 | 72.924 | 72.281 | 70.576 | 73.215 |
| flower5-3d-30-70-0-0 | 87.205 | 87.496 | 86.000 | 85.125 |
| flower5-3d-50-50-0-0 | 90.530 | 89.306 | 89.834 | 88.442 |
| paw3-3d-10-20-35-35 | 0.000 | 33.252 | 34.634 | 33.474 |
| paw3-3d-100-0-0-0 | 88.205 | 89.231 | 89.894 | 88.192 |
| paw3-3d-30-40-15-15 | 71.320 | 73.613 | 74.417 | 74.074 |
| paw3-3d-30-70-0-0 | 88.491 | 85.650 | 86.153 | 84.993 |
| paw3-3d-50-50-0-0 | 89.499 | 87.421 | 86.449 | 86.088 |

can observe improvements for several difficult data sets: yeast, haberman, then smaller improvements are also noted for: abalone, breast-cancer, and ecoli. For safer data sets like: vehicle, car one may expect that no over-sampling (base) or random over-sampling may be sufficient solutions (i.e., they may perform better). Then, we acknowledge that no oversampling is best performing on transfusion. Moreover, random over-sampling works best on two data sets: solar-flare, and cleveland.

The results on synthetic data sets also show no significant improvement when kernel and k-neighborhood over-sampling is applied. Better performance in comparison to random over-sampling and no over-sampling (base) can be observed on some more difficult distributions. Sensitivity results confirm the observations made with respect to G-mean. Thus, we do not include tables with these results due to the page limits.

More encouraging results have been obtained for modifications of SMOTE, in particular the recent proposal called Local Neighbourhood extension of SMOTE (briefly LN-SMOTE) which is inspired by the analyzing local data characteristics of the minority examples [37]. Its comparative study against basic SMOTE and two other related generalizations applied with 3 different classifiers (J48, Naive Bayes and $k$-NN) showed that it improved G-mean and F-measure on several of real world data sets. Yet another modifications of SMOTE with respect to individual difficulty weights of examples has been also considered in [43].

# 7  Neighbourhood Based Ensembles

Ensembles are another kind of methods which could be improved by the neighbourhood analysis. The current proposals of ensembles dedicated to class imbalanced data are mainly extensions of known strategies as bagging, boosting or random trees. They usually either employ pre-processing methods before learning component classifiers or embed the cost-sensitive framework in the ensemble learning process; see their review in [14]. Previous comparative studies, such as [4, 14], have showed that extensions of bagging ensembles are quite promising. The most popular extensions pre-process bootstrap samples by under-sampling the majority class or over-sampling the minority class to obtain a balance of class cardinalities in each bootstrap sample. Roughly Balanced Bagging (RBBag), which is a kind of specialized under-sampling approach leads to best improvements [30, 54].

In this section we want to show that using the neighbourhood based approach to change distributions of minority class examples in bootstrap samples may improve performance of bagging ensemble classifiers and result in solutions being competitive to Roughly Balanced Bagging.

We focus on $k$-neighbourhoods in bagging ensembles, since they proved to better render the distribution of minority class examples in Sect. 5.2. Moreover, they have been already successfully integrated in the Neighbourhood Balanced Bagging (NBBag), which we have proposed [5].

Neighbourhood Balanced Bagging is based on a different principle than all known bagging extensions for class imbalance. First, instead of integrating bagging with pre-processing, it keeps the standard bagging idea. What changes are probabilities of sampling examples to bootstraps. The chance of drawing minority examples is, sometimes radically, amplified (which is controlled by a special hyper-parameter $\psi$). Furthermore, the amplification depends on the type of difficulty of minority example identified according to its $k$-neighbourhood.

We have already shown that NBBag works in both types of bagging generalizations: over-sampling and under-sampling [5]. In first type of generalization, it is similar to over-sampling minority class examples into bootstraps, however, at the same time, the probabilities of drawing majority class examples are decreased. The size of bootstrap is kept the same as the size of the original learning set. The second type is inspired by under-sampling generalizations, which predicts better than over-sampling generalizations [5]. The probabilities of drawing minority class examples are increased, while probabilities of drawing majority class examples are decreased.

Most of the extensions of bagging for imbalanced data are non-parametric [6]. They do not introduce any new parameters, which need to be adjusted during construction of an ensemble of classifiers. On the one hand, one can argue that bagging itself is a parametric method since the adequate size of the ensemble for a given problem is not known a priori. The size of the ensemble is a parameter, which may influence the performance of each of the considered extensions. On the other hand, fixing this parameter enables comparison of ensembles of the same size, which should allow to distinguish ones which perform better than the others under the same conditions.

Different types of parameters are introduced in NBBag [5] to control the characteristics of neighbourhood: size of neighbourhood $k$, and amplification factor $\psi$. In the experiments comparing NBBag to other bagging extensions presented in [5] these two parameters were carefully selected to provide the best average performance. The previous tuning of these parameters was made post-hoc, i.e., first results were obtained for a number of promising pairs of parameter values and then the best values were chosen. On the other hand, we need to look for more appropriate approaches to tune these parameter inside learning an ensemble rather than in a post-hoc way.

Tuning of such model parameters is a known problem in machine learning [18]. However, to our best knowledge, this problem has drawn rather limited attention in the context of learning ensembles from imbalanced data. Class imbalance may limit using some more advanced parameter tuning techniques. To put it simply, minority class examples are to valuable to spare them for tuning purposes only, while majority class examples are not. Following this observation, we investigate a basic technique taken from tree learning. In the same way as reduced-error pruning uses training data [47], we divide training data set into two stratified samples. The first sample is used for training NBBag models and the second one to validate the trained models. After the best parameters are selected, NBBag classifier is constructed on the whole training set. Contrary to what was presented in [5], this technique does not allow to distinguish best values of parameters for all data sets nor even for one data set when learning of a classifier is repeated, as e.g., in cross-validation. Tuning of parameters is performed independently for each constructed component classifier.

In the following, we present performance of two variants of Neighbourhood Balanced Bagging: under-sampling (uNBBag) and over-sampling (oNBBag) with tuning of $k$ and $\psi$ parameters among a limited set of values (small $k$, and limited amplification of examples weight represented by $\psi$—please consult [6] for details). Tuning of best parameter values is performed on $2/3$ of the training set. The remaining $1/3$ of training set is used for the validation.

Now we experimentally compare classification performance of uNBBag and oNBBag to Exactly Balanced Bagging (EBBag) [23], Over-Bagging (OverBag) [58],

**Table 12** G-mean [%] of NBBag and other bagging ensembles on real-world data

| Data set | EBBag | OverBag | uNBBag | oNBBag | RBBag |
|---|---|---|---|---|---|
| abalone | 78.845 | 69.230 | 79.517 | 78.706 | 79.035 |
| breast-cancer | 58.175 | 60.718 | 58.465 | 58.795 | 60.091 |
| car | 96.668 | 96.959 | 96.356 | 96.851 | 96.568 |
| cleveland | 73.628 | 51.629 | 73.260 | 66.754 | 71.130 |
| cmc | 64.191 | 61.036 | 65.051 | 63.787 | 65.350 |
| ecoli | 88.178 | 83.896 | 88.435 | 85.380 | 88.430 |
| haberman | 64.144 | 63.329 | 63.742 | 61.779 | 63.533 |
| hepatitis | 79.137 | 75.816 | 78.035 | 74.762 | 79.457 |
| scrotal-pain | 73.679 | 74.038 | 72.923 | 71.997 | 75.618 |
| solar-flare | 83.710 | 64.649 | 83.149 | 79.994 | 83.421 |
| transfusion | 66.607 | 67.748 | 66.449 | 66.476 | 67.143 |
| vehicle | 95.038 | 94.934 | 95.440 | 95.115 | 95.417 |
| yeast | 84.018 | 63.167 | 84.475 | 79.557 | 85.016 |

and Roughly Balanced Bagging (RBBag) [22]. The size of ensembles is fixed to 50 components, J48 with exactly the same parameters as in Sect. 6 is used as component classifier. We restrict our comparison to real-world data sets only.

The results of G-mean and sensitivity are presented in Tables 12 and 13, respectively. These results were estimated by a stratified 10-fold cross-validation repeated ten times to reduce the variance of measures.

Looking at both Tables 12 and 13, one can notice that uNBBag and RBBag stand out as the best performing classifiers. Another observation is that over-sampling extensions of bagging, represented by OverBag and oNBBag, provide worse performance that under-sampling extensions. When we compare G-mean performance of ensemble classifiers to performance of over-sampled single classifiers (see Table 10) it is clear that ensembles provide better performance except for breast-cancer, where ensembles are only better than single classifier trained on not pre-processed data (i.e., base). A more detailed comparison on G-mean shows that RBBag and uNBBag does not perform best only in case of some relatively safe data sets like: car (both classifiers), scrotal-pain (uNBBag) or more difficult breast-cancer (uNBBag), and cleveland (RBBag).

With respect to values of sensitivity (Table 13) uNBBag and EBBag are clearly the best performing classifiers. uNBBag provides the best recognition of the minority class in case of almost all of considered real-world data sets.

This analysis of classification performance of bagging extensions leads to conclusions, which are concordant with the ones presented in [5] and in [6]. RBBag and uNBBag are identified as two outstanding alternatives. Moreover, an exploitation of a relatively simple parameter tuning technique, including a dynamic adaptation of the neighborhood size, allowed us to obtain quite satisfactory predictive performance of NBBag.

**Table 13** Sensitivity [%] of NBBag and other bagging ensembles on real-world data

| Data set | EBBag | OverBag | uNBBag | oNBBag | RBBag |
|---|---|---|---|---|---|
| abalone | 80.925 | 51.224 | 80.776 | 75.851 | 77.045 |
| breast-cancer | 63.412 | 54 | 65.176 | 59.059 | 58.471 |
| car | 100 | 95.652 | 100 | 95.942 | 100 |
| cleveland | 80.286 | 30.571 | 79.143 | 63.429 | 69.143 |
| cmc | 70.240 | 50.721 | 68.739 | 63.423 | 64.685 |
| ecoli | 92 | 76 | 92 | 84 | 90.571 |
| haberman | 56.914 | 59.136 | 63.827 | 66.543 | 55.802 |
| hepatitis | 83.438 | 67.188 | 79.062 | 69.688 | 77.500 |
| scrotal-pain | 76.271 | 70.169 | 76.441 | 73.051 | 75.763 |
| solar-flare | 88.140 | 46.977 | 86.744 | 81.395 | 85.581 |
| transfusion | 66.517 | 61.236 | 72.697 | 67.753 | 65.674 |
| vehicle | 97.236 | 94.523 | 97.286 | 95.477 | 96.935 |
| yeast | 91.765 | 40.980 | 90.392 | 73.529 | 88.431 |

# 8 Extensions of the Neighbourhood Analysis

In this section we briefly point out potential extensions of the neighbourhood approaches which may be useful for some applications—although they are not studied in this paper. We focus our attention on the following three issues:

**Identification of Class Decomposition into Sub-concepts**
The discussed neighbourhood analysis may approximate some data difficulty factors only. In particular, it does not directly identify a decomposition of the minority class into sub-concepts. As it was discussed in the Sect. 2.1 research of Japkowicz and her collaborators on *within-class imbalance* showed that increasing the number of the sub-concepts decreased classification performance more than increasing the global imbalance ratio *between class imbalance* [24, 26]. The comprehensive summary of other studies on the role of such class decomposition is presented in [53].

The open question is how to automatically identify such sub-concepts in real-world data sets. In cluster-oversampling proposal, Japkowicz applied k-means clustering algorithm to examples from each class separately [44]. However, it is necessary to estimate the unknown number of expected clusters or to choose an optimization criterion (the most popular criteria are not defined for the context of imbalanced data). Moreover, these kinds of algorithms are not appropriate for dealing with complex decision boundaries or outlier examples. In our opinion there is a need for developing a new kind of a semi-supervised algorithm (where it is necessary to deal with presence of minority vs. majority examples inside clusters).

**Highly-Dimensional Data Sets**
The presented approach uses HVDM metric to calculate distances between examples. Similarly to using Euclidean metric in most of pre-processing methods it is

more suitable for problems with relatively small or medium number of attributes. On the other hand, highly dimensional data sets may occur in image analysis, biomedical data analysis, genetics or other fields. The use of such dissimilarity measures and $k$-nearest neighbor principle on such data sets may suffer from the curse of dimensionality as it has been recently showed by Tomasev's research on, so called, *hubness*-aware shared neighbor distances for high-dimensional $k$-nearest neighbor classification [57].

Recall that this problem is also a challenge for standard learning of classifiers as it increases risks of over-fitting as well as spurious findings. However, considering it with class-imbalanced predictions presents an additional source of difficulties, as it biases classification towards majority class for most classifiers (see, e.g., experimental analyses from [3, 30]). In standard balanced classification feature selection or projections techniques, such as: SVD or PCA, are often applied to enhance predictive performance. Even though these methods have been extensively studied, they mey be too biased toward majority class. Although, some new class imbalance techniques have been recently introduced [35], we postulate still more research also in the context of an identification of types of examples.

**Multiple Imbalanced Classes**
A binary classification task is mostly studied in case of imbalanced data. This formulation is justified by focus an interest on the most important class and real-world semantics, like in medical diagnosis (distinguishing sick vs. healthy patients). On the other hand, in some situations it may be reasonable to distinguish more classes with low cardinalities [59].

Considering multiple minority classes makes the learning task more difficult as relations between particular classes become more complex [59]. Internal data distributions or decision boundaries will be different than in case when some classes are aggregated. Techniques developed for binary imbalanced problems are usually not directly applicable to multi-class problems. Quite often they lose performance on one class while trying to gain it on another. A brief review of current specialized techniques is available in [49].

We could ask a question on possible generalizations of the neighbourhood analysis for more than one minority class. Although it has not been studied yet, two directions could be considered. Either one can decompose the multi-class imbalanced data set to a set of binary problems—one minority class vs. all other classes; consider them independently and somehow aggregate results. According to [28] it is a dominating strategy in specialized ensembles, see e.g., [13].

However, in such decomposition of the multiple imbalanced classes, pairwise relations between two classes may be too strongly over-simplified and they do not reflect more complex relations/interactions between several of classes, as one class influences several neighboring classes at the same time. Therefore, it may be more interesting to consider interaction of examples from various minority classes while defining types of examples or exploiting other information from the neighbourhood analysis—however, it is still a topic for further research.

# 9  Final Remarks

In this paper we follow earlier research on studying the internal characteristics of class imbalanced data and its consequences for difficulties while learning classifiers. We share opinions of researches [15, 25, 26, 36] who showed that the high imbalance ratio between the minority and majority classes (measured on the global level of the data) is not the only and not even the main reason of these difficulties. Other data difficulty factors, such as decomposition of the minority class into many rare sub-concepts, the effect of too strong overlapping between the classes or a presence of too many minority examples inside the majority class region, referring to more local characteristics of class distributions, are more influential.

Our current study on these local data characteristics and difficulties goes along research lines introduced by Napierala and Stefanowski in [40, 42]. They have proposed to capture the aforementioned data difficulty factors by considering the local characteristics of learning examples from the minority class and by an identification of four basic types of examples: safe, borderline, rare case and outlier. It has been achieved by analyzing the class distribution of examples from different classes inside a *local neighborhood* of the considered example which could be modeled either by means of $k$-neighbours or kernels.

As the tuning the size of these two kinds of neighbourhoods with respect to characteristics of given data sets have not been sufficiently studied yet, the first contribution of this paper is discussing tuning methods. In our opinion simple rules of thumb are simply not suitable. We have rather promoted tuning bandwidth of a kernel neighbourhood or number $k$ of nearest neighbours using the adapted version of cross validation optimization methods.

Results of many experiments presented in Sect. 5 have confirmed usefulness of these tuning methods. Moreover, they were sufficiently consistent with earlier results of establishing categories of data set difficulty with respect to dominating types of minority class examples [40, 42]. However, unlike the earlier studies, in this paper we have managed to find an individual size of neighbourhood for each data sets. A general observation is that this size is larger for easier imbalanced data while it becomes smaller for data sets treated as more difficult to be learned.

The other contribution of the current paper is to promote incorporating the results of analyzing this neighbourhood of minority class examples in construction of new methods for learning classifiers from imbalanced data. We have "implemented" this postulate by considering two main categories of methods specialized for imbalanced data: (1) the most popular over-sampling and (2) the generalization of bagging ensembles which incorporates the results of an analyzing the local neighbourhood to re-sample examples into bootstrap samples.

The experiments presented in Sect. 7 have demonstrated that Nearest Balanced Bagging in the version of under-sampling with local tuning the size of neighbourhoods and the level of re-sampling achieved the best predictive results. Furthermore, experiments presented in Sects. 5.2, and 6 have shown that the $k$ nearest neighbours variant has led to better predictions than the kernel neighbourhood. On the other hand, the kernel analysis allows to identify new types of minority class examples: singletons in empty sub-regions (which is an extreme rarity situation being different to single examples surrounded by $k$-neighbours from opposite classes—this extension may be valuable in studying medical complex data with many untypical cases of disease, see [45]).

Issues of dealing with the local characteristics of imbalanced data may still open several lines of future research. Besides already mentioned semi-supervised clustering for detecting small disjuncts, re-considering the neighbourhood based methods in highly dimensional spaces or multi-class imbalanced problems one could look for other tasks such as:

- Other, more sophisticated proposals of dynamic re-sampling (also under-sampling) of both classes with respect to identified different, local characteristics of sub-regions of imbalanced data.
- Considering a new type of cost-sensitive re-sampling where costs of misclassification between classes will be taken into account while defining types of the minority examples; Then, the cost post-posterior probability should be joined together with an estimation of different density of examples in various sub-regions.
- Studying differences between outliers and real noise in imbalanced data; detecting them, developing a new method for dealing with such noisy examples.
- Exploiting information about types of examples in modifications of other algorithms, see e.g., promising results of the rule induction algorithm, called BRACID [41].
- Studying imbalanced data streams affected by concept drifts, i.e., changes in definitions of target classes over time [65]; In particular, recent studies have shown needs for developing new kinds of ensembles for the imbalanced and evolving data streams.

**Acknowledgements** The research was funded by the the the Polish National Science Center, grant no. DEC-2013/11/B/ST6/00963. Close co-operation with Krystyna Napierala in research on modeling types of examples and with Mateusz Lango in research on ensemble models is also acknowledged.

# References

1. Aggarwal, C.C. (Ed.): Data Classification: Algorithms and Applications. Chapman & Hall/CRC (2015)
2. Batista, G., Prati, R., Monard, M.: A study of the behavior of several methods for balancing machine learning training data. ACM SIGKDD Explor. Newslett. **6**(1), 20–29 (2004)
3. Blagus, R., Lusa, L.: Class prediction for high-dimensional class-imbalanced data. BMC Bioinf. **11**, 523 (2010)
4. Błaszczyński, J., Stefanowski, J., Idkowiak, L.: Extending bagging for imbalanced data. In: Proceedings of the 8th CORES 2013. Springer Series on Advances in Intelligent Systems and Computing, vol. 226, pp. 269–278 (2013)
5. Błaszczyński, J., Stefanowski, J.: Neighbourhood sampling in bagging for imbalanced data. Neurocomputing **150 A**, 184–203 (2015)

6. Błaszczyński, J., Lango, M.: Diversity analysis on imbalanced data using neighbourhood and roughly balanced bagging ensembles. In: Proceedings of ICAISC 2016. Lecture Notes in Computer Science, vol. 9692, pp. 552–562 (2016)
7. Branco, P., Torgo, L., Ribeiro, R.: A survey of predictive modeling under imbalanced distributions. ACM Comput. Surv. (CSUR) **49**(2), 31:1–31:50 (2016)
8. Chawla, N.: Data mining for imbalanced datasets: an overview. In: Maimon O., Rokach L. (eds.) The Data Mining and Knowledge Discovery Handbook, pp. 853–867. Springer (2005)
9. Chawla, N., Bowyer, K., Hall, L., Kegelmeyer, W.: SMOTE: synthetic minority over-sampling technique. J. Artif. Intell. Res. **16**, 341–378 (2002)
10. Cost, S., Salzberg, S.: A weighted nearest neighbor algorithm for learning with symbolic features. Mach. Learn. J. **10**(1), 1213–1228 (1993)
11. Dasarathy, B.V.: NN concepts and techniques: an introductory survey. In: Nearest Neighbor Norms, NN Pattern Classification Techniques, pp. 1–30. IEEE Press (1991)
12. Enas, G., Chai, S.: Choice of the smoothing parameter and efficiency of the k-nearest neighbour classification. Comput. Math. Appl. **12**, 308–317 (1986)
13. Fernandez, A., Lopez, V., Galar, M., Jesus, M., Herrera, F.: Analysis the classification of imbalanced data sets with multiple classes, binarization techniques and ad-hoc approaches. Knowl. Based Syst. **42**, 97–110 (2013)
14. Galar, M., Fernandez, A., Barrenechea, E., Bustince, H., Herrera, F.: A review on ensembles for the class imbalance problem: bagging-, boosting-, and hybrid-based approaches. IEEE Trans. Syst. Man Cybern. Part C: Appl. Rev. **99**, 1–22 (2011)
15. Garcia, V., Sanchez, J.S., Mollineda, R.A.: An empirical study of the behaviour of classifiers on imbalanced and overlapped data sets. In: Proceedings of Progress in Pattern Recognition, Image Analysis and Applications 2007. LNCS, vol. 4756, pp. 397–406. Springer (2007)
16. Gatnar, E.: Multimodel Approach to Discrimination and Regression Issues. PWN Warszawa (2008) (in Polish)
17. Goldstein, M.: $K_n$-nearest neighbour classification. IEEE Trans. Inf. Theory 627–630 (1972)
18. Guyon, I., Saffari, A., Dror, G., Cawley, G.: Model selection: beyond the Bayesian/frequentist divide. J. Mach. Learn. Res. **11**, 61–87 (2010)
19. Han, H., Wang, W., Mao, B.: Borderline-SMOTE: a new over-sampling method in imbalanced data sets learning. In: Proceedings of ICIC. LNCS, vol. 3644, pp. 878–887. Springer (2005)
20. He, H., Yungian, M. (eds): Imbalanced Learning. Foundations, Algorithms and Applications. IEEE, Wiley (2013)
21. He, H., Garcia, E.: Learning from imbalanced data. IEEE Trans. Data Knowl. Eng. **21**(9), 1263–1284 (2009)
22. Hido S., Kashima H.: Roughly balanced bagging for imbalance data. In: Proceedings of the SIAM International Conference on Data Mining, pp. 143–152 (2008). An Extended Version in Statistical Analysis and Data Mining, vol. 2, no. 5–6, pp. 412–426 (2009)
23. Hoens, T., Chawla, N.: Generating diverse ensembles to counter the problem of class imbalance. Proc. PAKDD **2010**, 488–499 (2010)
24. Japkowicz, N.: Concept-learning in the presence of between-class and within-class imbalances. In: Proceedings of Canadian Conference on AI, vol. 2001, pp. 67–77 (2001)
25. Japkowicz, N., Stephen, S.: Class imbalance problem: a systematic study. Intell. Data Anal. J. **6**(5), 429–450 (2002)
26. Jo, T., Japkowicz, N.: Class imbalances versus small disjuncts. ACM SIGKDD Explor. Newslett. **6**(1), 40–49 (2004)
27. Jones, M.C., Marron, J.S., Sheather, S.J.: A brief survey of bandwidth selection for density estimation. J. Am. Stat. Assoc. **91**(433), 401–407 (1996)
28. Krawczyk, B.: Learning from imbalanced data: open challenges and future directions. Prog. Artif. Intell. **5**(4), 221–232 (2016)
29. Kubat, M., Matwin, S.: Addressing the curse of imbalanced training sets: one-side selection. In: Proceedings of the 14th International Conference on Machine Learning ICML-97, pp. 179–186 (1997)

30. Lango, M., Stefanowski, J.: The usefulness of roughly balanced bagging for complex and high-dimensional imbalanced data. In: Proceedings of International ECML PKDD Workshop on New Frontiers in Mining Complex Patterns NFmCP 2015. LNAI 9607, pp. 93–107. Springer (2015)

31. Laurikkala, J.: Improving identification of difficult small classes by balancing class distribution. Tech. Report A-2001-2, University of Tampere (2001)

32. Lumijarvi, J., Laurikkala, J., Juhola, M.: A comparison of different heterogeneous proximity functions and Euclidean distance. Stud. Health Technol. Inform. 107(Part 2), 1362–1366 (2004)

33. Ledl, T.: Kernel density estimation: theory and application in discriminant analysis. Austrian J. Stat. 33(3), 267–279 (2004)

34. Liu, B., Yang, Y., Webb, GT., Boughton, J.: A comparative study of bandwidth choice in kernel density estimation for Naive Bayesian classiffication. In: Proceedings of the 13th Pacific-Asia Conference on Advances in Knowledge Discovery and Data Mining, PAKDD '09. LNCS, vol. 5476, pp. 302–313. Springer (2009)

35. Lin, W., Chen, J.: Class-imbalanced classifiers for high-dimensional data. Brief. Bioinform. 14(1), 13–26 (2013)

36. Lopez, V., Fernandez, A., Garcia, S., Palade, V., Herrera, F.: An insight into classification with imbalanced data: empirical results and current trends on using data intrinsic characteristics. Inf. Sci. 257, 113–141 (2014)

37. Maciejewski, T., Stefanowski, J.: Local neighbourhood extension of SMOTE for mining imbalanced data. In: Proceedings of IEEE Symposium on Computational Intelligence and Data Mining, pp. 104–111 (2011)

38. Napierala, K.: Improving rule classifiers for imbalanced data. Ph.D. Thesis. Poznan University of Technology (2013)

39. Napierala, K., Stefanowski, J., Wilk, Sz.: Learning from imbalanced data in presence of noisy and borderline examples. In: Proceedings of 7th International Conference on RSCTC 2010. LNAI, vol. 6086, pp. 158–167. Springer (2010)

40. Napierala, K., Stefanowski, J.: The influence of minority class distribution on learning from imbalance data. In: Proceedings of 7th Conference on HAIS 2012. LNAI, vol. 7209, pp. 139–150. Springer (2012)

41. Napierala, K., Stefanowski, J.: BRACID: a comprehensive approach to learning rules from imbalanced data. J. Intell. Inf. Syst. 39(2), 335–373 (2012)

42. Napierala, K., Stefanowski, J.: Types of minority class examples and their influence on learning classifiers from imbalanced data. J. Intell. Inf. Syst. 46(3), 563–597 (2016)

43. Napierala, K., Stefanowski, J., Trzcielinska, M.: Local characteristics of minority examples in pre-processing of imbalanced data. In: Andreasen, T., et al. (eds.) Proceedings of ISMIS 2014. LNAI, vol. 8502, pp. 123–132. Springer (2014)

44. Nickerson, A., Japkowicz, N., Milios, E.: Using unsupervised learning to guide re-sampling in imbalanced data sets. In: Proceedings of the 8th International Workshop on Artificial Intelligence and Statistics, pp. 261–265 (2001)

45. Niemann, U., Spiliopoulou, M., Volzke, H., Kuhn, J.P.: Subpopulation discovery in epidemiological data with subspace clustering. Found. Comput. Decis. Sci. 39(4), 271–300 (2014)

46. Prati, R., Batista, G., Monard, M.: Class imbalance versus class overlapping: an analysis of a learning system behavior. In: Proceedings of 3rd Mexican International Conference on Artificial Intelligence, pp. 312–321 (2004)

47. Quinlan, R.: C4.5: Programs for Machine Learning. Morgan Kaufmann Publishers, San Mateo, CA (1993)

48. Saez, J., Luengo, J., Stefanowski, J., Herrera, F.: Addressing the noisy and borderline examples problem in classification with imbalanced datasets via a class noise filtering method-based re-sampling technique. Inf. Sci. 291, 184–203 (2015)

49. Seaz, J., Krawczyk, B., Wozniak, M.: Analyzing the oversampling of different classes and types in multi-class imbalanced data. Pattern Recogn. 57, 164–178 (2016). doi:10.1016/j.atcog.2016.03.012

50. Sheather, S.J.: Density estimation. Stat. Sci. **19**(4), 588–597 (2004)
51. Silverman, B.W.: Density Estimation for Statistics and Data Analysis. Chapman and Hall/CRC (1986)
52. Stefanowski, J.: Overlapping, rare examples and class decomposition in learning classifiers from imbalanced data. In: Ramanna, S., Jain, L.C., Howlett, R.J. (eds.) Emerging Paradigms in Machine Learning, pp. 277–306 (2013)
53. Stefanowski, J.: Dealing with data difficulty factors while learning from imbalanced data. In: Mielniczuk, J., Matwin, S. (eds.) Challenges in Computational Statistics and Data Mining, pp. 333–363. Springer (2016)
54. Stefanowski, J.: On properties of under-sampling bagging and its extensions for imbalanced data. In: Proceedings of the 9th International Conference on Computer Recognition Systems CORES 2015, pp. 407–417. Springer (2016)
55. Stefanowski, J., Wilk, Sz.: Selective pre-processing of imbalanced data for improving classification performance. In: Proceedings of the 10th International Conference on DaWaK 2008. LNCS, vol. 5182, pp. 283–292. Springer (2008)
56. Sun, Y., Wong, A., Kamel, M.: Classification of imbalanced data: a review. Int. J. Pattern Recogn. Artif. Intell. **23**(4), 687–719 (2009)
57. Tomasev, N., Mladenic, D.: Class imbalance and the curse of minority hubs. Knowl.-Based Syst. **53**, 157–172 (2013)
58. Wang, S., Yao, T.: Diversity analysis on imbalanced data sets by using ensemble models. In: Proceedings of IEEE Symposium on Computational Intelligence and Data Mining, pp. 324–331 (2009)
59. Wang, S., Yao, X.: Mutliclass imbalance problems: analysis and potential solutions. IEEE Trans. Syst. Man Cybern. Part B **42**(4), 1119–1130 (2012)
60. Weiss, G.M.: Mining with rarity: a unifying framework. ACM SIGKDD Explor. Newslett. **6**(1), 7–19 (2004)
61. Weiss, G.M., Provost, F.: Learning when training data are costly: the effect of class distribution on tree induction. J. Artif. Intell. Res. **19**, 315–354 (2003)
62. Wilk, S., Stefanowski, J., Wojciechowski, S., Farion, K.J, Michalowski, W.: Application of preprocessing methods to imbalanced clinical data: an experimental study. In: Pietka E. (ed.) Information Technologies in Medicine, pp. 503–515. Springer (2016)
63. Wilson, D.R., Martinez, T.R.: Improved heterogeneous distance functions. J. Artif. Intell. Res. **6**, 1–34 (1997)
64. Wojciechowski, S., Wilk, Sz.: Difficulty Factors and Preprocessing in Imbalanced Data Sets: An Experimental Study on Artificial Data. Found. Comput. Decis. Sci. **42**(2), 149–176 (2017)
65. Zliobaite, I., Pechenizkiy, M., Gama, J.: An overview of concept drift applications. In: Japkowicz, N., Stefanowski, J. (eds.) Big Data Analysis: New Algorithms for a New Society. Springer Studies in Big Data Series, pp. 91–11 (2016)

# Dimensions of Semantic Similarity

Paweł Szmeja, Maria Ganzha, Marcin Paprzycki
and Wiesław Pawłowski

**Abstract** Semantic similarity is a broad term used to describe many tools, models and methods applied in knowledge bases, semantic graphs, text disambiguation, ontology matching and more. Because of such broad scope it is, in a "general" case, difficult to properly capture and formalize. So far, many models and algorithms have been proposed that, albeit often very different in design and implementation, produce a single score (a number) each. These scores come under the single term of *semantic similarity*. Whether one is comparing documents, ontologies, entities, or terms, existing methods often propose a *universal* score—a single number that "captures all aspects of similarity". In opposition to this approach, we claim that there are many ways, in which semantic entities can be similar. We propose a division of knowledge (and, consequently, similarity) into categories (*dimensions*) of semantic relationships. Each *dimension* represents a different "type" of similarity and its implementation is guided by an interpretation of the *meaning* (semantics) of that similarity score in a particular dimension. Our proposal allows to add extra information to the similarity score, and to highlight differences and similarities between results of existing methods.

P. Szmeja · M. Ganzha · M. Paprzycki · W. Pawłowski
Systems Research Institute, Polish Academy of Sciences, Warsaw, Poland
e-mail: Pawel.Szmeja@ibspan.waw.pl

M. Ganzha (✉)
Warsaw University of Technology, Warsaw, Poland
e-mail: Maria.Ganzha@ibspan.waw.pl

M. Paprzycki
Warsaw Management Academy, Warsaw, Poland
e-mail: Marcin.Paprzycki@ibspan.waw.pl

W. Pawłowski
Faculty of Mathematics, Physics, and Informatics,
University of Gdańsk, Gdańsk, Poland
e-mail: Wieslaw.Pawlowski@inf.ug.edu.pl

© Springer International Publishing AG 2018
A.E. Gawęda et al. (eds.), *Advances in Data Analysis with Computational Intelligence Methods*, Studies in Computational Intelligence 738,
https://doi.org/10.1007/978-3-319-67946-4_3

# 1  Introduction

Semantic similarity, understood broadly, has been applied in very different fields such as psychology, linguistics, biology, knowledge modeling, artificial intelligence, and others. Even though, in our work, we focus on computer science (and mathematics), the understanding of similarity in any domain is influenced by other domains. Within the scope of computer science there are many areas of interest for similarity scoring, such as graphics (e.g. face recognition), information retrieval, machine learning, etc. In this context, *semantic* similarity algorithms are focused mostly on computational linguistics and semantic reasoning, each with multiple applications. The most popular direct areas of application are ontology matching and document (i.e. text) similarity scoring.

To introduce some order into our considerations, let us declare specific *objects* that we consider most relevant (in the scope of this work). Those are *documents* (natural language texts organized in corpora), *terms* (e.g. atomic parts of a text), *ontologies* (a representation of a knowledge base, e.g. a semantic graph) and *entities* (atomic parts of a knowledge base or an ontology). We also look at *entity descriptions* contained in ontologies. We mostly consider pairwise comparisons between any two objects of the same type (e.g. two documents, and not one document and one ontology). However, to avoid gratuitous verbosity, the focus is on comparison of entities and their descriptions. Nevertheless, this paper presents a theoretical approach, or a "meta-model" of semantic similarity and the presented ideas can be applied in different fields where similarity is relevant and to objects other than entities.

To formulate specific examples we use OWL [1] (the most popular ontology description language) and, occasionally, *description logic* formulas. This is done in order to illustrate practical applicability of our ideas.

This paper is an extension and continuation of our previous work [2], where we have briefly introduced and justified the idea of semantic similarity dimensions.

We proceed as follows. Section 2 contains a short introduction to relevant concepts from the description logic. In following Sects. 3 and 4 we summarize the existing approaches (both theoretical and practical) to calculating semantic similarity. We also briefly explore the general truths about similarity. Semantic *similarity dimensions* are introduced in Sect. 5. Its subsections present archetypes of *dimensions*, along with general information and examples. Section 6 illustrates practical application of the *dimensional* similarity method in the field of ontologies, while Sect. 7 confronts the *dimensional* similarity score and results of other similarity methods. Properties of *similarity dimensions* are examined in Sect. 8, while Sect. 9 outlines more general use cases. Finally Sect. 10 presents a summarized case for *semantic similarity dimensions*.

## 2 Description Logic

Let us start with a brief introduction to description logic (DL). Information presented here is needed in order to understand some examples given in later sections. This is because OWL is based on DL, and OWL axioms can be written in the form of mathematical formulas expressed in DL. This representation gives a useful perspective on ontological entities.

In description logic, knowledge is stored in *knowledge bases* that contain *axioms* (also called *facts*). An ontology is a collection that contains knowledge and, within the scope of this paper, is considered to be an equivalent of a knowledge base. More formally, a knowledge base is part of a mathematical model and ontology document (specifically in computer science) is a presentation of this model.

Each knowledge base (KB) (and each ontology) can be partitioned into *TBox*, *ABox* and *RBox*. Each of those boxes contains different kinds of axioms. This division extends to different *entities* and *entity descriptions*.

The TBox contains *concepts* (also called *classes*), i.e. the declarations and descriptions of concepts. A *declaration* is simply a statement about the kind of an *entity* (in this case—a class), and each *description* of a concept is constructed from *concept names*, *role names*, constants and a set of DL *constructors*. The *description* is said to provide the explanation of a semantic meaning of a class.

A set of classes, organized into a hierarchy, is called a *taxonomy*. In taxonomy classes may have descendants (specializations) and ancestors (*subsumers* or generalizations). A relation of subsumption in many semantic graphs has the name *IS-A* e.g. "computer IS-A machine". The IS-A relations occur very often and form directed sub-graphs in semantic graphs. In DL every taxonomy contains a special class—*Thing* ($\top$), that is the "top" of the hierarchy, i.e. it has no ancestors. Moreover each concept is necessarily a descendant of $\top$, which means that every concept in an ontology is a part of the same hierarchy. In other words every concept is of type *Thing* and, in semantic graphs, there exists at least one path along the IS-A relation between any two concepts. *Thing* is considered to be the root of a taxonomy tree. In this paper, concepts with rich *descriptions* are called *complex*, as opposed to simple concepts, descriptions of which consist only of class names and describe only the concepts' position within the taxonomy. Historically, simple concept descriptions are important, because many old ontologies were formulated exclusively in terms of taxonomies, and, consequently, some similarity algorithms consider only the taxonomic part of ontologies. Taxonomies and classes are still considered central to a lot of ontologies.

The ABox contains *declarations* and *descriptions* of *individuals*. *Individuals* are *instances* of classes, i.e. each instance is of at least one *type* and, necessarily, of type *Thing*. A description of an *individual* is comprised of *types* and *assertions* about *properties* built from *concept* names, *role* names and constants. Property assertions are parts of a description that are specific to each individual and together with individual's types describe the meaning of the entity. Individuals usually do not form a hierarchy. They are, nevertheless, strongly tied (via the *types*) to the taxonomy. As a consequence of the meaning of subsumption, an individual that is explicitly of type *A* is also of all types that are ancestors of *A* (including ⊤).

The RBox contains *declarations* and *descriptions* of *roles* (also called *properties*). A role description defines the role's domain, range and characteristics (such as symmetry, transitivity and others), which likens roles to mathematical binary relations. In OWL DL [1], an OWL variant, two types of roles are distinguished: *object* properties and *data* properties. The range of an object property is a class, while the range of a data property is a literal (e.g. a numerical value or a string). For both types of roles the domain is a concept. The RBox is often considered to be a part of ABox, as opposed to a "box" of its own. In any case the main line of division of a knowledge base lies between ABox and TBox.

Following the division of KB into "boxes", within the scope of this paper the term (ontological) "*entities*" refers to *concepts*, *individuals* or *properties*. The most relevant similarity calculation is between entities from the same "box", although comparison between, for instance, a class and an individual is also, theoretically, possible (we touch upon this later).

Different varieties of DL (sometimes called *profiles*) determine what *constructors* are available when formulating axioms, as well as what syntactic variant is allowed (i.e. what symbol sequences are allowed). In order to clearly present our ideas, in this paper, most examples of DL expressions are given in a simple DL formalism $\mathcal{EL}$ (see Table 1), unless otherwise noted. We also discuss how the ideas may be extended to more expressive description logics.

In what follows, *concept* names are denoted by capital letters $A, B, C, \ldots$, *individual* names by lowercase letters $a, b, \ldots, o$, and *role* names by lowercase letters $p, r, \ldots, z$. Each name may have an optional index, e.g. $C_1$. An expression $C(a)$ means that individual *a* is of type *C*. Expression $r(a, b)$ is a role assertion and denotes that *a* is related to *b* by role *r*, *a* being the realization of the roles domain and *b* its range. Table 1 summarizes relevant DL constructors. Constructors in $\mathcal{EL}$ are (by definition) limited to: top concept, concept conjunction and existential quantification (restriction). For more details about DL, its constructors, semantics, and varieties refer to [3].

Some relations defined in DL are of special importance to similarity scoring. Those are primarily subsumption (⊒), inclusion (⊑) and equivalence (≡). The ⊑ cor-

**Table 1** DL constructors

| $\mathcal{EL}$ | Name | Syntax |
|---|---|---|
| * | Top concept | ⊤ |
| | Bottom concept | ⊥ |
| | Concept (class) | $C$ |
| | Concept negation | $\neg C$ |
| * | Concept conjunction | $C_1 \sqcap C_2$ |
| | Concept disjunction | $C_1 \sqcup C_2$ |
| * | Existential restriction | $\exists R.C$ |
| | Universal restriction | $\forall R.C$ |

responds directly to the IS-A relation, and the $\sqsupseteq$ combined with the $\sqsubseteq$ means the same as the $\equiv$. While subsumption is usually reserved for classes, the equivalence relation can be applied to any entity and has special interpretation of "maximal similarity". Unsurprisingly, when two entities are equivalent, they should be treated as one and the same entity, and their similarity score should be maximal. In similar fashion, a negated ($\neg$) entity should be maximally different from the entity it negates, but most often negation is neither part of the DL profile of an ontology, nor is considered when calculating similarity.

There are two definitions that are highly relevant to calculation of similarity, that can be expressed in DL—the *least common subsumer* [4] and the *most specific concept* [5].

The *least common subsumer* (LCS) of two entities $X$ and $Y$ is the most specific (i.e. farthest from the root) entity that is an ancestor to both $X$ and $Y$. In a taxonomy, it is a concept that shares the most types with compared concepts, and itself is a type (generalization) of compared concepts. In other words, the LCS is a class that is a superclass to both $X$ and $Y$. Since there may be multiple such superclasses, we choose the one that is least general, i.e. as far from the *Thing* and as close to $X$ and $Y$ as possible. Since LCS represents, in a sense, the most information the entities have in common, it is sometimes treated as the central part of similarity algorithms.

The *most specific concept* (MSC) of an individual is a concept whose description is built from assertions about the individual in a way that includes every such assertion. In other words it is a class that is built specifically to contain an individual and in its construction is guided by the description of the individual. It is often so specific that it only contains the one individual it was constructed for. The process of construction of the MSC utilizes standard semantic deduction [6] and is described in detail in [7]. The MSC is, in general case, not unique and, because of that, its usefulness is put into question. The details are beyond the scope of this work and we refer interested readers to [7]. Even more information about both LCS and MSC can be found in [3–5]. In further sections we give specific examples for LCSs.

# 3 Similarity

There is a multitude of works dedicated to similarity across multiple domains, including psychology and sociology [8, 9], as well as more technical fields, such as mathematics, computer science, or engineering (i.e. similitude [10]). To keep the text coherent, and not to stray too far from the core ideas, in this section we present some general observations about similarity, relevant to the main content of this work. Since computer science is the main focus, articles relevant to this field of science are referenced throughout the text. Let us now start our general considerations.

*Features of Similarity*

There are many properties that may apply to a similarity measure. Most measures reflect the following two general observations about similarity:

- It grows with the commonality of objects
- It decreases with the difference between objects

Those two terse and laconic statements are ones of very few that, for all practical purposes, can be applied to vast majority of formalized similarity measures. Whether we compare documents, terms, ontologies or other objects, the factors that increase similarity can be summarized as the *commonality* between objects. What contributes to decrease of similarity is the *disparity*, often included in the score implicitly, as opposite to the *commonality*.

Here, it should be stressed that, although other observations can be made about similarity, there is no strong/community-wide consensus, based on which, one could construct a general definition. Furthermore, there are a number of "features" of similarity, both formal and informal, that have support and opposition. A notable informal feature of similarity is that, in human judgment, common features carry more weight than disparate ones [11]. This is exemplified in measures that explicitly consider only *commonality* and discard differences, i.e. work under the supposition that the initial similarity is minimal (zero) and each common feature increases it. A different approach relies on a "balancing" between commonalities and disparities, where each of those increase and decrease similarity (respectively) that, initially, is set at a middle score (e.g. 0.5 on a scale from 0 to 1).

### Attentiveness

Another concept relevant to similarity is classification that stems from computer graphics, and divides similarities into "Pre-attentive" and "attentive" [12, 13]. Pre-attentive similarity is measured before "interpretation" of entities (or "stimuli", in graphics processing terminology), while attentive methods are used after entities have been interpreted, classified and put into context. The "interpretation" process, while specific to computer graphics, can be extended to semantic similarity in general. At first glance, semantic similarity falls squarely into the attentive category, because semantic descriptions or features are already an interpretation of entities. On the other hand, similarity algorithms commonly do not distinguish between possible different interpretations of the same entity. This is highly relevant in graphic databases, where, for instance, searching for images most similar (to a reference image) requires different queries, depending on whether we are interested in shape or color palette similarity. Other, advanced features of an image may be calculated, such as painting style, and they all require attentive methods. Extending this idea into general semantic similarity, we may add similarity scoring with respect to provenance (e.g. authorship) information for the same images. The different "views" on similarity are, from the point of view of attentive methods, different interpretations of the same entities. This idea is expanded upon, although with different terminology, in later sections, where we describe semantic similarity dimensions.

### Reference Similarities

Another notable point is that similarity often depends on the *context*, where each feature has a weight based on *subjective* evaluation of importance (i.e. opinion) and circumstances surrounding the comparison. In human judgment, knowledge of the person performing the comparison is very important [14] and is de facto the implicit "knowledge base" that we score against. However, it is not directly relevant in case of comparison of DL entities, because, the knowledge used for such comparison is explicitly defined and available in KB. It has, however, heavy bearing on the quality of similarity methods, because their results are often judged against human (expert) opinions. In this way, the evaluation of a similarity algorithm is dependent not only on the expert doing the evaluation, but also on the quality of the ontology.

There are reference sets of similarities, such as the Miller's benchmark [15]. The benchmark itself (building on previous work [14]) produced a set of 30 pairs of generic terms along with a similarity score averaged from judgments of 38 students. It was conceived as a way to gain insight into how humans score similarity. The terms were chosen to be purposefully ambiguous, which adds another layer of (dis-ambiguation) challenge for the similarity algorithms. Moreover, such benchmarks are not a good reference point for big ontologies, since those are usually detailed and contain expert knowledge of a certain domain, which means that there are no ambiguous terms in them. Later, the benchmark was reproduced [16] with consid-erably different results, which further questions its validity as a similarity algorithm evaluation tool. Nevertheless, many works refer to this specific benchmark as a ref-erence and proof of good (i.e. correct) results [17–21].

There are organized efforts to counteract the difficulties in evaluating (semi-) automated methods of similarity scoring, such as OAEI (Ontology Alignment Evalu-ation Initiative) [22]. This initiative is dedicated specifically to evaluating only ontol-ogy alignment tools and its evaluations are organized into yearly editions, each with many "tracks". In each track there are some reference ontologies and alignments pre-pared. By ensuring that both ontologies and reference alignments are of good qual-ity, OAEI is able to, at least in principle, provide more meaningful evaluation results, than simple comparison to a benchmark. Another approach was adopted by Reuters, which publishes corpora of documents for text categorization, e.g. Reuters-21578 or RCV1 (Reuters Corpus Volume 1). Instead of declaring an authoritative reference categorization, the corpora may be used to compare results of different methods, or to test improvement of a single method over its previous iteration. Other similar cor-pora exist, e.g. Ohsumed [23] for medical documents and 20NewsGroups [24] for newsgroup documents.

### *Granularity*

It is worth noting that granularity of information has an influence on similarity. Infor-mally, the more details we include in our comparison, the less impact on similar-ity each of them will have. Consequently, a general feature of observation about an object has a bigger weight than a detailed description of the same feature. In other words, an expert (or an expert ontology) has a different (more detailed) view than a layman (a general ontology). In case of human judgment, but also algorithms mod-eled after it, taking into account that *commonality* carries more weight, the similarity of a feature is likely to decrease, when we take a closer look at it. For example, a sim-ple property, such as age of people, might be exactly the same, when we look only at the birth year, but will decrease with the increase of accuracy to days or even minutes. For a layman, creatures such as monkeys and chimpanzees would be much more similar, than in the eyes of an expert. A more general observation is that the amount of information we have about an entity greatly influences similarity, which is most pronounced in probabilistic methods, such as in [16], where similarity can change, even if we add knowledge seemingly unrelated to the compared objects. In particular, granularity of information may strongly influence human performed quality evaluation of similarity measures (cf. [15]).

### Distance and Closeness

Similarity is often considered in the context of distance between entities. The idea comes from psychology [8] and states that entities may be put into a multi-dimensional space, where each dimension is a separate characteristic. In such theoretical space, the distance between entities is the evidence of dissimilarity, which is, informally, an inverse of similarity. While in specific applications it may be possible to construct a finite set of dimensions [13], in general, the sheer number of possible characteristics in semantic descriptions makes construction of general algorithms based on this idea difficult [25]. The distance-based similarity is also applicable to graph structures (more on this in later sections).

### Similarity Ordering

It has been argued that just an ordering of concepts, with respect to similarity, is more useful than a number. In many applications one is primarily interested in finding an entity that is closest to a reference entity, the actual value of "closeness" being secondary. In case of three objects, one reference object and two compared objects, we might be satisfied just knowing, which of the comparison objects is more similar to the reference object, rather than learning the numerical score. This is clearly pronounced in the difference between regular search and search based on similarity measures. The first is a partition of the search space into entities that fit the search criteria (i.e. the query), and those that do not. Similarity search, essentially, responds with the complete search space, ordered by similarity. Note that, in practice, we are usually interested only in those entities that have an extreme value of similarity— either very high or very low. In both cases, the actual numerical scores, often are irrelevant, as long as they are above (or below) some threshold.

## 4 Similarity Calculation Methods

Let us now present a, non-exhaustive, list of selected algorithms and methods of calculation of similarity, that are both used in practice and relevant to presentation of dimensions of semantic similarity. In order to focus on presentation of our own ideas, rather than summarizing all existing similarity methods, we have chosen to describe only a few methods. For a richer list of similarity methods, see [26]. Note that the semantic measures library (SML) [27] contains implementations of a large number of methods described here, as well as many others.

## Edge Methods

Edge-based models work on assumption that edge distance in a graph is meaningful for similarity. Needless to say, edge-based methods require a graph structure, such as a taxonomy. Those methods view ontologies as directed graphs, where small distance along some edge type is the evidence of similarity, and long paths indicate dissimilarity. A common criticism of edge methods, when applied to ontologies, is that they work under the assumption, that each edge in a path has the same semantic distance. In practice, however, there is no formal evidence to back up this assumption, and some evidence that indicates the opposite [28, 29].

The simplest approach considers similarity to be equal to the length of the shortest path between a pair of entities (e.g. concepts) $S_{Rada}(X, Y) = min(paths(X, Y))$ [30], where $paths(X, Y)$ is the set of path lengths in an IS-A graph (see, Sect. 5). More sophisticated methods, such as e.g. $Sim_{Wu}(X, Y) = \frac{2*depth(LCS(X,Y))}{depth(X)+depth(Y)}$ [31], involve normalization and take into account depth of the compared entities, depth of their LCS, or length of path between the root, LCS and entities. Finally, [32] utilizes multiple relations (not just IS-A) in a graph (multigraph). Because edge methods regularly use simple mathematical rations, they are often applicable to the dimensional approach to similarity, where the same formulas are used on distances along dimensions, instead of path lengths.

## Feature Methods

A big class of similarity algorithms are *feature* methods. In these methods, each entity is represented by a set of semantic *features* and the entity similarity is equivalent to similarity of feature sets. What the features are, and how to identify and construct them, depends on the domain of application and, sometimes, on a particular implementation. For instance, in computer graphics, a set of features for each image depends on the software, and may include shape, size, texture, color, position, etc. In general case, however, the situation is complicated and there is no universal algorithm of representing entities as a set of features. Feature methods are applied in different domains (e.g. graphics [33], reasoning [34] and others) and "features" have slightly different meaning and are constructed (or extracted) in a different way, depending on the domain model. Usually, features are crisp (not fuzzy), i.e. a feature either belongs to a feature set, or not. This approach makes it difficult to evaluate similarity of features that have numerical representations, like the aforementioned color (e.g. in the RGB color space). When put into a set, a specific numerical value of a color is, in general, an entirely different feature than any other color value, no matter how close the color are. This property is another reason for the division of feature methods into specific implementations (e.g. in graphics) and general formulas (with crisp feature sets).

In ontologies there are many ways in which *feature* sets can be constructed from a description of an entity. In case of concepts, the features are usually considered to be the concept's ancestors, its roles, instances, or a set of all of those. Details of how a complex description is converted into a set of features depend on particular method and underlying logic (see, Sect. 5 and onward). A set of *features* of an individual may be constructed from its types and role assertions. A usual approach is to use only role assertions directly mentioned in the entity definition. In such approach, the semantic descriptions with color properties (e.g. *X hasColor red* and *Y hasColor light-red*) consider the property and its value as atomic, and do not go into detail about possible similarity of *red* and *light-red*. It is worth noting that, in some specific cases, the feature sets can be constructed in a very natural way. This is the case of WordNet [35], which explicitly defines synsets (sets of synonyms) that can be, with no additional effort, treated as feature sets. Another noteworthy property is that the feature sets in a taxonomy may be defined as a IS-A neighborhood of a class. In such case, the feature method is, conceptually, very close to an edge method, because they both use very similar information as input.

In *Tversky's ratio model* [11] similarity of two sets of features $X_F$ and $Y_F$ is given by the formula $S_{T_V}(X_F, Y_F) = \frac{\alpha f(X_F \cap Y_F)}{\alpha f(X_F \cap Y_F) + \beta f(X_F - Y_F) + \gamma f(Y_F - X_F)}$, where $X - Y$ is a set difference (relative complement) of $Y$ in $X$), $f$ is a monotonically increasing function (usually set cardinality), while $\alpha$, $\beta$ and $\gamma$ are positive coefficients. The coefficients control importance ("weight") of common features and features exclusive to either set. For different choices of values of the coefficients, Tversky's *ratio model* has different properties and produces different formulas. In particular, for $\alpha = \beta = \gamma = 1$ and $f = |\cdot|$ the model becomes the *Jaccard index* [36] $J(X_F, Y_F) = \frac{|X_F \cap Y_F|}{|X_F \cup Y_F|}$. Some methods [37, 38] include an ad-hoc weighting of coefficients for different features, instead of a fixed set of coefficients for all features. Tversky also proposed a *contrast model* represented by the formula $S_{T_VC}(X_F, Y_F) = \alpha f(X_F \cap Y_F) - \beta f(X_F - Y_F) - \gamma f(Y_F - X_F)$.

### Information-Theoretic Methods

Methods from an information-theoretic class approach similarity from the point of view of information theory [39] and assume that similarity is strictly related to the amount of information each of compared entities provides. This class is represented by *Information Content* (IC) model proposed by Resnik [16]. IC of an entity $e$ is computed from its *probability* $p(e)$: $IC(e) = -\log(p(e))$. When applied to a textual entity in a corpus, $p(e)$ is equal to the probability that this entity appears in a given document from the corpus. In the context of a taxonomy, probability of an entity is inversely proportional to the number of entities it subsumes. By this definition, IC is monotonically decreasing from leaves (most informative) to the root (least informative). Resnik's similarity is calculated from IC of the *most informative common ancestor* (MICA)—a common subsumer that has the maximum IC: $S_{Res}(X, Y) = IC(MICA(X, Y))$. MICA is closely related to LCS. Some works build on Resnik's approach by relaxing it's reliance on the LCS. For instance, Lin proposed a formula that involves the Information Content of the entities themselves, alongside their LCS: $S_{Lin}(X, Y) = \frac{2 \times S_{Res}(X,Y)}{IC(X) + IC(Y)}$. Other methods, such as the one proposed in [40], use only the number of immediate children as a measure of IC, where high number of children denotes low IC. Several other methods of calculating IC have been proposed [41–43], with a lot of them focusing specifically on WordNet.

### Geometric Methods

In the geometric approach to similarity, objects are represented as points in a multi-dimensional geometric space. Any feature of an object is converted to a number that serves as a coordinate. This approach directly corresponds to multi-dimensional similarity space described earlier. A set of coordinates represents the entire object in a space. The similarity is simply calculated as the shortest geometric (usually Euclidean) distance between two points. As such, just like any metric distance, it has the properties of minimality, and those of a mathematical metric. Unfortunately, many features are not easily subjected to conversion into a geometric dimension, as it requires them to be represented as a set of points on a continuous line. While some features have a natural representation in a geometric dimension, such as the RGB model of color, others do not, unless they are specifically designed to have that property (such as a brand of a product). A work on high-dimensional spaces [25] describes other problems that are relevant when dealing with a high number of geometric dimensions. In practice, only domain and problem specific implementations of geometric methods exist, such as the one proposed in [13] for an image database.

### Other Methods

Some methods do not fit neatly into the above categories and are considered "hybrid". Such methods (e.g. [38, 44]) use characteristics from multiple categories and combine, for instance, path length with depth in taxonomy, or taxonomic neighborhood. Often, such methods use weighted sum (with weights tuned to a specific data set) of separate results for each considered perspective, or "sub-method".

A decisively syntactic (i.e. non-semantic) class of methods that deserve mention, are the edit distance methods, most prominent of which is the Hamming distance [45]. The general idea is that high number of edits that need to be done in order to transform one entity into the other is indicative of dissimilarity. Different methods define "edits" in a different way, and for strings those usually include removing a character or adding one. Edit distance is usually not applied to feature sets, although some feature set methods compute similarity score in a way reminiscent to the edit distance. It is, however, relevant to ideas presented in Sect. 5.

Notice that each of the methods, presented in this section, makes some assumption as to the model of information. A feature method requires a set of features, an edge method needs a graph, etc. In order to apply each of those methods to a knowledge base we need to present it in a particular fashion—as a graph, a DL formula, a set of features and so on. In order to be applied to an ontology, each method requires a different *perspective*.

## 5  Semantic Similarity Dimensions

Let us now recall that, in their foundation, similarity and meaning (semantics) are inherently human concepts. From this point of view, a similarity score should have an explanation (or interpretation) that is understandable for a human. Let us consider a simple example of comparison of two physical objects. There are many *ways* in which they can be similar or dissimilar, two of them being shape and color. Those two kinds of features are independent with respect to similarity, e.g. objects can have similar color and different shape (and vice-versa). Canonical ways of automated calculation of similarity (described above) would produce a single score that would in some way combine similarity of shape and color. However, the two similarities, when treated separately, provide more information to a human, because they have a clear *interpretation*. Therefore we can assume that a person that knows this interpretation has a better understanding of how similar any two physical objects are. In this toy example shape and color contribute to two separate *dimensions* of similarity.

From this, it can be conjectured that similarity of semantic entities has many different aspects that are being grouped together, based on what part of available data (or knowledge) is used (regardless of the actual similarity method). Those groups represent different types (dimensions) of semantic relationships and, therefore, similarity. This idea draws on the concept of *knowledge dimensions* originating from [46], where authors also divided ontological knowledge into subsets (dimensions) and applied (in [47]) to calculation of similarity in WordNet. However, there the scores from each dimension were still *combined* into a final similarity score (a single number). Our idea also borrows from geometric and feature models of similarity, and is closely related to attentiveness and multi-dimensional similarity space, described in Sect. 3.

When approaching this from a different perspective, observe that approaches to either grouping attributes (features), or dividing the data into (what we call) dimensions were focused on results given by some predetermined method. In other words, *the starting point was a method* that provided foundation to interpret the result. Our approach starts from explaining the *nature / semantics* of the dimension that we are interested in, and then finds the method that would produce said result. In this way the recognized dimensions are *interpretation* driven, rather than method driven. In fact, the same method may be used in different dimensions, as exemplified in Sect. 6.

Note that the concept of different kinds of similarity has been present in the literature, in one form or another, for a long time. For instance [48] contains a summary of ontology matching methods and categorizes them by the kind of data they use. Sample "kinds" of methods use comparisons of entity labels, their "attributes" (in DL terms—assertions), instances of classes, position of entities in taxonomy and others. Categorization described in [48] complements a more general work on schema matching [49] that presents its own division of matching methods by type. Later work [50] reviews methods of ontology matching and distinguishes methods that use structure of the ontologies and those that utilize entities (called *structure-level* and *element-level* dimensions). The categorization goes deeper with dimensions such as *syntactic*, *semantic*, *external*, *terminological*, *extensional* and others, some of which overlap (for more detais, see [50]). The state of the art for ontology matching, described in [51], contains more detailed descriptions of different matching methods with specific examples of implementations. Later work [52], proposes a slightly different division into language-, linguistic-, string-, and structure-based approaches. Another example is [53], which mentions in its opening chapters that different simi-

larity measures have differing implicit assumptions, hinting at the existence of similarity dimensions. The Gene Ontology [54] defines two types of similarity measures,
namely pairwise and groupwise, which are akin to kinds of similarity, albeit specific
to that ontology only. Another work on semantic similarity [55], classifies existing
methods for biomedical ontologies with respect to scope (what entities are taken into
account), data source (edges, nodes or other) and metric (used algorithm). Authors
of [55] observe that methods that use different metric and data produce different
results, but all claim to produce a similarity score that is "universal". To the best of
our knowledge [46] was the first paper in which different kinds of similarity were
the explicit focus, and were given the name "dimensions". Finally, let us also note
that an implemented ontology matching system ASMOV [56] utilizes four (*lexical*,
*relational*, *internal* and *extensional*) dimensions that are weighted and summed to
obtain the final score.

It's important to note that, ideally, similarity dimensions should form an orthogonal partition of a "total" similarity. Since, as discussed previously, the distinction
lies in the kind of data used, the available knowledge should be partitioned into subsets, one for each dimension. Under this characteristic, the similarity scores for each
dimension would be independent. In practice, however, such clear division is not
always possible (see, following sections).

In this context, let us now introduce selected dimensions and formalization of a
general case of pairwise comparison of entities in an ontology based on description
logic.

## External and Internal Dimensions

From the point of view of the origin of data, similarity dimensions can be categorized
into *external* and *internal* ones.

**External Dimensions**. An *External* similarity dimension uses information from
outside of the main knowledge base or ontology. In case of entities, *external* methods
use a small (likely atomic) part of an entity description that serves as an identifier,
to find information about it in external sources. A good candidate for such an identifier is a label of an entity, because it is available for all named entities and may be
written in a natural language. A complex description is, in such case, simplified into
a single term, so that it can be easily identified and searched for in outside sources.
The assumption behind this operation is that the entity has a meaning outside of the
original knowledge base, and this meaning is relevant to calculation of similarity.

For a pairwise comparison of entities this means that the used methods are actually independent of the DL formalism used to describe the entities. The information
we use for scoring comes from an independent, external source (possibly having
its own formalization). For English words a method commonly used in ontology
alignment systems [56–58] is to utilize the English WordNet ontology and calculate similarity with a method specific to WordNet (e.g. feature method that works on
synsets).

External methods have two inherent weaknesses, as they rely on: existence of a
good term that describes each compared entity, and on the quality and relevance of
the external data source.

The *External* dimension gives a perspective on how similarity of entities is viewed outside of known and specific context. The lack of this specific context causes disambiguation problems. For that reason *external* similarity scoring is best suited for entities that are general and relatively insensitive to context. For instance it is useful when performing entity resolution for duplicate detection in data analysis, because often our data describes a broad range of items from multiple domains (e.g. items from a big online store). One of the simplest techniques of doing that is to convert entities into a canonical form, that serves as a representative and could be used as an *external* identifier.

**Internal Dimensions**. *Internal* dimensions are those that make use of information either explicitly provided in the knowledge base of compared entities, or inferred from that knowledge base. In this case we are not interested in any independent outside sources and assume that any knowledge we might use must come from what we already have in the knowledge base, or an ontology connected to it (e.g. via Linked Data [59]). Vast majority of similarity methods are *internal*; especially, ad-hoc ones that are restricted to a single ontology (e.g. [47] that works on WordNet only). Here, we assume that any discussed ontology can be expressed in description logic.

Usually, in description logic, entities compared *internally* are of the same type, i.e. both come from the same "box"—ABox, TBox or RBox (we compare concept with concept, role with role, etc.), although it is possible to compare "across boxes" (e.g. a concept with an individual), which is explained in what follows.

Note that some similarity dimensions have interpretations for both categories— *internal* and *external*, while others are exclusive to one category.

### Lexical Dimension

*Lexical* methods utilize dictionaries and lexical ontologies to asses similarity of entities (e.g., see [47]). In a general case of an ontology, *lexical* methods are *external*. Entities are considered in the context of a dictionary (where they are referenced to by an identifier) and not the original ontology. A pair of labels, or entity names, written in a natural language can be subject to the *lexical* dimension similarity methods. The methods themselves might be very complex and utilize big ontologies (such as WordNet).

*Lexical* dimension is most useful when entities have uniquely identifying labels. For that reason we can expect that, for example, comparing terms "dolphin" and "porpoise" will yield useful results. A simple *lexical* method for concepts could, for instance, extract the labels of entities and use WordNets *synsets* of the labels as features, in a feature-based method. On the other hand, this is not the case when labels are human names (e.g. Mary, Adam), because, even though technically being labels, those are a properties of an individual, rather than unique identifiers. In different ontologies these might refer to different people. Similar problem arises when identifying terms are words with multiple meanings (e.g. "seal"). Generally, any identifier that is sensitive to context of a knowledge base (like human names) is not a good candidate for a *lexical* similarity scoring. This is because in any external similarity dimension we lose the original context. In a *lexical* method the additional context we need to consider is the natural language itself (e.g. English, French etc.).

*Lexical* scores might differ between languages, because of varying sets of homonyms and many natural differences between languages. Despite this, as mentioned before, many ontology alignment methods use external *lexical* similarity as a way to find connections between ontologies that have no links defined between them. Sometimes the *lexical* scores are used as a bootstrap to discover other connections between ontological entities and improve the alignment.

Informally, the *lexical* dimension specifies similarity of names of entities in a dictionary. Unfortunately, it suffers from the problem common in dictionaries, i.e. ambiguity. So-called "word sense" disambiguation is a big issue in text processing [60] and semantic ontologies (e.g. applied to named entities [61]). Ambiguity of language negatively impacts accuracy of the *lexical* similarity score. Notice that, in the case of a well defined ontology, there is no ambiguity problem, because the entity descriptions are compared directly. When comparing terms we first need to find out what entity each term represents (what is the underlying entity) and then compare the entities. Miller's benchmark [15], often used to evaluate WordNet methods, does not, unfortunately, have explicit concept descriptions, and the word sense in each word pair needs to be decided solely on the two words in each pair.

In short, the interpretation of the *lexical* dimension is that entities *lexically* similar have names that are similar, according to a dictionary. In order for a *lexical* method to be semantic, it should not rely on any edit distance.

## Co-occurrence

Another group of methods dealing mostly with the *external* dimension are the *co-occurrence* methods. Like *lexical* methods, they also use a single term or label (identifier). Similarity is calculated based on a highly controversial assumption that entities that often appear together are similar. For instance, the web search co-occurrence methods measure the number of web pages that contain both identifiers (or terms). Methods in this dimension are often used for text similarity scoring, and work under the assumption that words that appear together in a high number of text corpora are similar. More advanced co-occurrence methods distinguish between different meanings of words [62]. Their authors, realized that, like in the case of *lexical* methods, disambiguation is an issue. In data analysis *co-occurrence* is used as an evidence of similarity (called "linkage pattern").

Co-occurrence methods usually do not take into account the reason for two entities appearing together. For instance, they do not take into account that co-occurrence might be a result of a single event, local culture, specific names (e.g. names of sports teams), or even a coincidence. In this way, *co-occurrence* is an evidence of relatedness, but not necessarily similarity. Overall, *co-occurrence* methods are known to give questionable results [28].

*Co-occurrence* dimension is *external*, because it uses many data sources (e.g. web pages, documents etc.). A commonly used sources are those that are publicly available, such as Wikipedia [17], or Freebase [21]. Although, in an ontology we might construct a co-occurrence method based on an assumption that entities that appear in a high number of axioms together are similar. Such methods would give results that would come under the same questions as ones from other co-occurrence methods [28]. Moreover, since the axioms contain detailed knowledge about semantic relationships between entities it is better to consider *why* the entities appear together, rather than disregarding that information. For this reason the type of axiom (e.g. RDF predicate, if available) should always play an important role in similarity scoring.

Interpretation of a *co-occurrence* similarity is, simply put, that entities often appear together and are referenced in the same contexts.

### *Taxonomic (Sort) Dimension*

Similarity in the *sort* dimension (also called *hierarchical* or *taxonomic*) describes how similar entities are, according to data from taxonomy and, therefore, uses mostly the TBox.

Theoretically *sort* dimension is most easily described in terms of *types* of concepts (i.e. subsumers or ancestors). For instance, in a hypothetical ontology of genetic ancestry two classes of creatures are similar, if both are reptiles (they share a type). Similarity increases with each type that the creatures have in common. At the same time it decreases with each disparate type (e.g. when one creature is a lizard, and the other a snake). This is in accordance with the general tenets of the concept of similarity (see, Sect. 3).

Practically, taxonomy is often visualized as a graph, where nodes are concepts and edges are IS-A relations. Because of the structure of a description logic taxonomy, each common type of compared entities lies on some path from the root ($\top$) to either of the entities. More precisely the commonality is defined by any path to the lowest common subsumer (LCS) of both entities. Any edge on such path is between two common types. Any edges from $\top$ to any of the entities that does not lie between $\top$ and LCS is an evidence of dissimilarity.

Many edge-counting methods (that use taxonomic ancestry of entities, [63]), some IC methods (like [64] or [65]) and feature methods (e.g. [66]) can be used in this dimension.

Formally *hierarchical* dimension includes information exactly about DL relations of subsumption ($\sqsupseteq$) and, consequently, inclusion ($\sqsubseteq$) and equivalence ($\equiv$). Recall that a concept is a specialization of all its types (classes), including the root, and a generalization of all its children (subclasses). The root ($\top$) is a generalization of any concept. Similarity measures that work on subsumptions usually take into account subsumers of measured classes, rather than descendants. For instance edge-counting methods "count" classes (types) that are on a path between the LCS and the root. Some IC methods make use of number of descendants (subsumed classes) to calculate "probability" of a node. *Taxonomic* similarity is also linked to distance between measured entities, either directly, or through IC of LCS.

Other than concepts, *sort* dimension can also be applied to roles or individuals. In some profiles of description logic roles have their own hierarchy (e.g. $\mathcal{H}$ subprofile of DL [67]) with a separate set of IS-A relations (whose domain and range are roles, not concepts) that also form a set of data for the *sort* dimension. In practice, however, the hierarchies of roles are almost never rich and deep enough to provide enough information for a useful hierarchical similarity score. Simply put, such score would not be useful. Individuals do not form their own hierarchy, however, there are ways to relate entities of that type to the taxonomy, e.g. the most specific class (MSC, [7]). Another method is to use only concept membership (asserted and inferred) for compared individuals as *taxonomic* knowledge. In this method we essentially compare sets of types of each compared individual, which is a good fit for a feature method, where each type would represent a feature. Since we can construct a set of ancestors for a concept and set of types for an individual (both ancestors and types are concepts themselves) it is possible to compare concepts with individuals using a feature method. Thus, in *taxonomic* dimension we can compare pairs of concepts, individuals and roles as well as a concept—individual pairs.

In $\mathcal{EL}$, concepts can be represented as an intersection of terms e.g. $C \equiv D_1 \sqcap D_2 \sqcap \exists p_1 \sqcap \exists p_2$ and $E \sqsubseteq D_1 \sqcap \exists p_3 \sqcap \exists p_4$. In this example, the knowledge that pertains to *sort* dimension is the part of the expressions that contains concept names, namely $C \equiv D_1 \sqcap D_2$ and $E \sqsubseteq D_1$. The role assertions are not considered a part of this dimension, so we do not take them into account. If we are interested in comparing the two example concepts $C$ and $E$ with respect to subsumption (note that $A \equiv B$ is equivalent to $A \sqsubseteq B$ and $B \sqsubseteq A$) we would use two expressions: $C \sqsubseteq D_1 \sqcap D_2$ and $E \sqsubseteq D_1$ that can easily be converted into sets (through itemization with respect to intersection) $[D_1, D_2]$ and $[D_1]$ respectively, and used in a feature method. For individuals $a, b$, assuming $C(a), E(b)$, *sort* similarity of $a$ and $b$ is equal to *sort* similarity of $C$ and $E$.

The general idea of "truncating" a complex description to one containing only symbols for concepts and constructors (to "extract" *sort* similarity) holds for more expressive DLs. For instance the expression $C_2 \equiv D_1 \sqcup \forall p_1.(D_2 \sqcap \exists p_1.D_2)$ does not seem to be easily subjected to "extraction" of *sort* terms. In practice, however, we can rely on semantic reasoners to build an inferred taxonomy that puts all named classes in order with respect to subsumption (and inclusion) while taking into account complex expressions [3]. New concepts, such as MSC, can also be put in a proper place in a taxonomy with the help of semantic reasoners. It is also common for the taxonomy tree to be explicitly created (asserted) by the author of the knowledge base. In *sort* dimension we are only interested in the existence of IS-A relation between entities and not the reasons for existence of such relation. Combining asserted and inferred hierarchies produces data that accurately represents *taxonomic* dimension.

In summary, *taxonomic* similarity of two entities is interpreted as the entities being of similar type or class, or sharing a number of types. While, in layman terms a "type" is a vague term, it has a very specific meaning in practical applications i.e. ontologies.

### Descriptive Dimension

From a theoretical point of view the *descriptive* dimension contains properties that an entity "has", as opposed to what it "is" (which is covered in *taxonomic* dimension). For animals, similarities in size, weight or age belong to the *descriptive* dimension. Generally speaking, *descriptive* dimension encapsulates attributes, characteristics, or properties of entities. Properties such as "having a child" are also included (and distinct from "being a child", which belongs to *taxonomic* dimension). Again, the more disparate attributes, the less similarity and vice-versa.

In certain ontologies, clearly distinguishing between *taxonomic* and *descriptive* data might be problematic, when it comes to entities that form a hierarchy. The difference between two dimensions, and whether they overlap or are entirely orthogonal, comes down to the way the hierarchy is constructed by the authoring ontology engineer. Let's consider a hierarchy of classes. A taxonomy might be created in an entirely expert-driven fashion, in which case it would not contain any explanation as to why any given class has the subsumers that it does. It would be simply an assertion of expert knowledge, stating that any instance of an example class *A is* of every type that subsumes *A*. On the other hand the taxonomy construction might be driven by roles of every class. Here, the reason for enclosing two classes in a subsumption relation is that they share a role restriction. In this case, a subsumption implies that *A has* a property that is shared among all its subsumers. In other words the basis for subsumption is inheritance of role restrictions. Informally, if an information about a role is "included" in a taxonomy (or used in its creation), then it overlaps with the *taxonomic* dimension, where it is included implicitly. The *descriptive* dimension considers all roles explicitly. More formally, orthogonality of *taxonomic* and *descriptive* dimensions depends on whether the ontology follows the principle of cognitive saliency [68]. Overall, this principle states that new concepts are created and subsumed only when there is a need to differentiate them, and put them in their own class. This principle is, often unknowingly followed in a lot of ontologies, and one can assume that the *taxonomic* and *descriptive* overlap is small or does not exist at all.

Practically, in ontologies that have both subsumption relations and role restrictions, the taxonomy includes results of both methods described above—expert assertion and inheritance. Specifically the inherited roles are the cause of partial overlap between *taxonomic* and *descriptive* dimensions. Notice that for any two concepts, the set of role restrictions that they have in common is at least the set of roles of their LCS, because both concepts inherit those roles from the LCS. In a very special case, where each class has only one non-inherited role restriction, each IS-A relation has a corresponding role restriction. Numerically, this means that number of contributing relations for both dimensions is exactly the same, so we can expect the results from both dimensions to be close. Such cases are very rare in practice, where some roles are the explicit reason for the shape of taxonomy, and some are independent of it. In an example biological ontology of creatures, properties such as type of reproductive system are the base of phylogenetic taxonomy. Other, such as diet or geo-spatial distribution are not considered in phylogeny. This is because they are not *inherited* genetically, which is the basic requirement for a phylogenetic subsumption.

Notice that, even when considering full set of roles of a concept (both inherited and not inherited), its cardinality might be different from the cardinality of the set of types (ancestors). For every item in the set of types we might have any number of role restrictions inherited for that type. In other words, every ancestor contributes only one piece of data (one superclass for *taxonomic* dimension), while roles inherited from the ancestor might contribute (to *descriptive* dimension) a different number

(0 or more). For two concepts with complex descriptions that are on the first level of taxonomy (i.e. direct children of T), if we apply an edge-counting method, their descriptions are essentially irrelevant for *taxon omic* similarity score (only distance to each other or T—their LCS—matters). In fact, this is the case in the *semantic sensor network ontology* (SSN) [69], where concepts close to T have a lot of roles. Those roles could have big impact on a similarity score, but are disregarded by *taxonomic* methods. This observation suggests that a clear way to distinguish the two discussed dimensions is to consider only non-inherited part of complex concept description, or, alternatively, not consider part of the description inherited from the LCS. The lack (in practice) of full overlap of dimensions suggests that *descriptive* similarity is useful along side of *sort* similarity and produces results with a different interpretation.

More formally, in *descriptive* dimension we are interested in relations that are roles and are not of type IS-A. Hierarchy constructed from any such relation is not taken into account. Instead, only existence of a relation and its value is considered. In DL terms, those are either role assertions (e.g. $r(a, b)$) in case of individuals, or quantified restrictions (e.g. $\exists p.C$, $\forall t.5$) in case of concept descriptions. *Descriptive* dimension fits naturally with feature methods, because we can treat each role assertion or restriction (a "descriptive" expression) as an item in a set of features, either for TBox or ABox. In $\mathcal{EL}$, extracting a set of such features from concept description is simple and very similar to the method described for *sort* dimension (it yields sets $[\exists p_1, \exists p_2]$ and $[\exists p_3, \exists p_4]$ for concepts $C$ and $E$, defined earlier, respectively).

Individuals do not form a hierarchy, so comparisons between this type of entities do not suffer from overlap with *taxonomic* dimension and as such are a good fit for *descriptive* dimension. For a set of statements (role assertions) $r(a, b)$, $r(b, b)$, $r(a, c)$, $p(a, c)$, $t(b, 5)$ about individuals $a$, $b$ and roles $r$, $p$ and $t$, the first two ($r(a, b)$, $r(b, b)$) contribute to similarity of $a$ and $b$, because the predicate (role) and object (individual) are the same for both $a$ and $b$. Expressions $r(a, c)$, $p(a, c)$ and $t(b, 5)$ contribute to dissimilarity of $a$ and $b$, because those assertions do not share both role and object for a subject of $a$ or $b$.

Comprehensive implementation of *descriptive* dimension in expressive DLs is highly problematic. While it's relatively easy to construct a transformation of a complex description to a normal form (e.g. conjunctive normal form) there is no universal way to compare complex restrictions. For instance there is no universally accepted method to calculate similarity of each pair of $\exists r.A$, $\forall r.A$ and $\exists r.B$ other than to treat those as entirely different (similarity score of 0), even though intuitively we might conclude that, since all 3 expressions pertain to the same role $r$, they are not *absolutely* different and the similarity score should not be zero, even if its close to it. Unfortunately, when it comes to roles in DL, the canonical approach is that they can be either identical or not, with no degrees of similarity in-between. The binary treatment of role restrictions or, widely speaking, features is a big weakness of many similarity methods. Moreover, comparison of complex descriptions, especially in expressive DLs, is a complicated problem, and is beyond the scope of this paper. For those interested, [70] proposes a method of comparison of complex descriptions.

It should be stressed that, unfortunately, existing methods usually do not distinguish between *taxonomic* and *descriptive* data, instead implicitly assuming that every role restriction contributes to a concepts place in a taxonomy and has no additional bearing on similarity. Consequently, there are no methods known to us that are purely *descriptive*. That being said, feature set methods are a natural fit for this dimension, because of the clear division between *descriptive* features, and other features.

*Descriptive* similarity is (informally) interpreted as standing for similarity of properties, attributes or characteristics, i.e. the items that describe what an entity "has".

## Other Dimensions

Up to this point we have presented four similarity dimensions that are very general and thus widely applicable. There are many other ways to divide knowledge, as was suggested in Sect. 5. Each of the relevant works [46–49, 53, 55] uses different kind of semantic relations and axioms. One could even argue that any partition of knowledge forms a set of semantic dimensions. The ones proposed in this paper were designed (on the basis of analysis of existing methods and ontologies) to be relatively simple in interpretation and generic enough to be available in almost any knowledge base. There are, however, other, more specific dimensions, that are worth mentioning.

Let us start from the the *membership* dimension. It can be used to measure similarity (only) between concepts by gathering and comparing sets of individuals that are of specific type. Compared to others dimensions, this one produces simple data even for expressive DLs, because the membership function is a binary predicate—an individual either is or is not of a given type. From simple statements $A(a)$, $A(b)$, $C(a)$ we know that concept $A$ has members $a$ and $b$, and $C$ has member $a$. This knowledge can be easily used to construct a feature method. The *membership* dimension is implicitly used in [7] where authors build feature sets composed of members and calculate similarity in a way very similar to Tversky's feature method. Because any individual of any type $A$ is also of all types that are ancestors of $A$, the *membership* dimension uses data that overlays, in part, with the *taxonomic* dimension, but still brings its own perspective on similarity.

Separately, the *descriptive* dimension contains knowledge about all of the properties without discrimination. One simple way to create a new similarity dimension is to isolate a set of types of roles from the *descriptive* dimension. The resulting set should have its own specific interpretation to be considered a separate dimension.

An example resulting from this method is the *compositional* dimension. It is comprised of roles that denote "being a part of," "having parts," "having ingredients," etc. It has a very clear interpretation and, as humans, we can often look at composition of any physical object. Formally, it is represented by roles such as *hasPart*, *isPartOf*, *isIngredient*, etc. In SSN [69] this kind of relations are represented by the *hasPart* role (inherited from *DUL* ontology). A similar role exists in WordNet [35] (also named *hasPart*) and in many other ontologies.

Another "sub-descriptive" dimension is the *physical* dimension. It contains all roles that describe any kind of physical characteristic. What roles are included specifically varies between ontologies. They might include size (e.g. height, width, area), mass, color, shape and others.

A practical problem with the subdivision of the *descriptive* dimension is that application of a dimension constructed by this method requires specific roles. Even guided by the interpretation, the specific dimensions might be represented by different roles in different knowledge bases. In one ontology the *physical* dimension would include a *hasWeight* and *hasHeight* roles, while in another by a *hasArea* role. A third ontology might not contain any roles relevant to the *physical* dimensions and, therefore, the *physical* similarity score would not be available. Any subdivision of the *descriptive* dimension generally means a loss of universality, i.e. one cannot apply our new dimension to every ontology. Another downside of this method is that the "sub-descriptive" knowledge in a very obvious way overlaps with the *descriptive* dimension. As a consequence, for instance, the *compositional* score and *descriptive* score are not independent (in fact, one is contained within the other), and the dimensions are not orthogonal. On the other hand, sub-descriptive dimensions are easy to implement in edge methods, such as [21]. What is required is simply to use only edges of a certain type, instead of all edges. One needs to be mindful that not all edge types appear often enough in a graph to form an interesting and useful dimension.

Let us recall that conversion of roles into a set of features is easy for simple DLs, but gets complicated for more expressive DLs. This is relevant for the *descriptive* dimension and its sub-dimensions, where we need to compare DL expressions (role assertions or restrictions). For a sub-dimension that contains only role $p$, a simple, single-term expression, such as $p.D_1$ is easy to parse and compare. A complicated expression, such as $p.((D_1 \sqcup (\forall p.D_2)) \sqcap (D_1 \sqcup \forall p.(\exists p.D_2)))$ is difficult to use in a comparison with others, because the class expression under the property restriction in the example is very complex. Moreover, it might have many equivalent forms, which are relevant in practical implementations of similarity algorithms. The simplest approach to solving this problem is to consider only the binary similarity of complex expressions.

Section 8 contains a description of interesting properties of semantic similarity dimensions that should be considered when designing new dimensions. Before that, let us present an example of application of dimensions introduced up to this point.

# 6 Example of Multi-dimensional Similarity

Let us consider an example of dimensional similarity scores in a mock-up biological ontology. The ontology in question (see Fig. 1) is an extract of a phylogenetic ontology with added roles. It compares three concepts—short-beaked common dolphin [71], silvertip shark [72] and lesser electric ray [73] denoted $D$, $S$ and $R$ respectively. Taxonomy describes the current understanding of the genetic ancestry of these creatures. It is complemented by roles selected to best aid in presentation of the idea of semantic similarity dimensions. The roles represent traits or features that are not genetically inherited and, therefore, in the example the *descriptive* dimension does

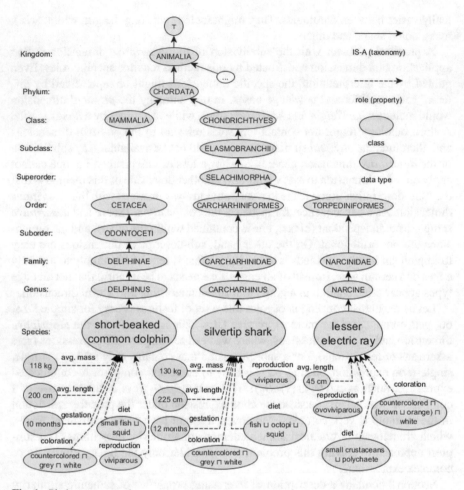

**Fig. 1** Phylogeny example

not overlap with the *taxonomic* one (see, Sect. 5). Note that this is in no way a complete set of information about these creatures. Data contained in these roles comes from [71–75] and was prepared with ease of understanding of the example in mind. Note that biology is not the main focus of this paper and accuracy of the data was not verified. This example is meant to demonstrate usage and indicate usefulness of similarity dimensions. Let us note that all used formulas are symmetric, normalized and have the properties of minimality and maximality.

Data used by *taxonomic* methods is the hierarchy of concepts and in this example there are 20 phylogenetic concepts (classes, including ⊤). Resnik's method [16] specifies similarity as the IC (information content) of the MICA (most informative common ancestor), which in the example is equivalent to the LCS (least common subsumer); $S_{Res}(X, Y) = IC(MICA(X, Y))$, $IC(e) = -\log(p(e))$. According to this method, similarity scores are as follows: $S_{Res}(D, S) = IC(CHORDATA)$ $= -\log(\frac{18}{20}) \approx 0.105$, $S_{Res}(S, R) = IC(SELACHIMORPHA) = -\log(\frac{9}{20}) \approx 0.799$, $S_{Res}(D, R) = IC(CHORDATA) \approx 0.105$. Note that this example contains only a fraction of available phylogenetic classes and in a full ontology Resnik's method would give a different score, because IC is sensitive to the total number of concepts, which is the basis of calculating the "probability" of a concept. Calculation of Jaccard index $J(A, B) = \frac{|A_f \cap B_f|}{|A_f \cup B_f|}$, where $A_f$ is a set of features of $A$ and assuming that each ancestor of a concept (including ⊤) is a feature, gives the following results: $J(D, S) = \frac{3}{16} \approx 0.188$, $J(S, R) = \frac{6}{14} \approx 0.429$, $J(D, R) = \frac{3}{16} \approx 0.188$.

In the *descriptive* dimension, the data we use are role restrictions. We can, again, use Jaccard index, this time using roles as features. This is an indication that we can use one method to calculate similarity in many different dimensions. In this example $D$ and $S$ have 6 roles each, while $R$ has 4 roles. As mentioned in Sect. 5, the simplest way to compare two values of a single role is to say that the similarity is binary (1 only if those values are identical and 0 otherwise); i.e. $Sim(r.5)$ and $Sim(r.4.99)$ is 0, despite their perceived numerical "closeness". Under this condition, *descriptive* Jaccard scores are as follows: $J(D, S) = \frac{2}{10} = 0.2$, $J(S, R) = \frac{0}{10} = 0$, $J(D, R) = \frac{0}{10} = 0$.

Final dimension considered here is the *physical* dimension that is meant to represent any physical feature, i.e. roles for mass, length and coloration. In order to better represent difference between numerical values, a simple ratio method is used for data values of the same role (assuming the same unit). This similarity is equal to the smaller value divided by the larger one $S_{val}(k^r, l^r) = \frac{\min(k^r, l^r)}{\max(k^r, l^r)}$, where $k^r$ and $l^r$ are values of role restrictions or assertions, about the same role $r$. For instance similarity of average weight between $D$ and $S$ is $\frac{118}{130} \approx 0.907$. Total similarity is this dimension is calculated by taking arithmetic average over similarity of each relevant role. The scores are: $Sim_{ph}^{avg}(D, S) = \frac{\frac{118}{130} + \frac{200}{225} + 1}{3} \approx 0.932$, $Sim_{ph}^{avg}(S, R) = \frac{\frac{45}{225} + 0 + 0}{3} \approx 0.067$, $Sim_{ph}^{avg}(D, R) = \frac{\frac{45}{200} + 0 + 0}{3} \approx 0.075$. Using the same method of simple arithmetic average the results for the entire *descriptive* dimension are as follows: $Sim_{desc}^{avg}(D, S) = \frac{\frac{118}{130} + \frac{200}{225} + \frac{10}{12} + \frac{1}{4} + 1 + 1}{6} \approx 0.813$, $Sim_{desc}^{avg}(S, R) = \frac{0 + \frac{45}{225} + 0 + 0 + \frac{2}{4} + 0}{6} \approx 0.117$, $Sim_{desc}^{avg}(D, R) = \frac{0 + \frac{45}{200} + 0 + 0 + \frac{2}{4} + 0}{6} \approx 0.121$.

### Analysis of Results

Obtained similarity scores are summarized in Table 2. Observe that *each method produces different similarity scores*, even in the same dimension. In the *taxonomic* dimension, Resnik's and Jaccard's methods produce different scores. This is, for instance, because of the assumption of Resnik that distance to the root in an ontology (a level) is significant. The levels of example concepts do not correspond with levels of phylogenetic classification, e.g. the dolphin does not have a biological *subclass* or *superorder*, so technically its *order* (CETACEA) is on the same ontological level as *subclass* of the shark (ELASMOBRANCHII), even though intuitively (and

**Table 2** Approximate similarity scores

|  | $Sim(D, S)$ | $Sim(S, R)$ | $Sim(D, R)$ |
|---|---|---|---|
| *Taxonomic* | | | |
| Resnik | 0.105 | 0.799 | 0.105 |
| Jaccard | 0.188 | 0.429 | 0.188 |
| *Descriptive* | | | |
| Jaccard | 0.2 | 0.0 | 0.0 |
| Arithmetic average | 0.813 | 0.117 | 0.121 |
| *Physical subdimension* | | | |
| Arithmetic average | 0.932 | 0.067 | 0.075 |

in accordance with biological research) an *order* should be more informative than a *subclass*. Such structure is a good example of a graph, in which edges do not uniformly represent the same value of difference in specificity (this was described in more detail in Sect. 4).

Differences between dimensions are very apparent in the results. In particular $S$ and $R$ have very small *descriptive* similarity (Jaccard gives a score of 0), while their *taxonomic* similarity is significant. Explanation of those results lays in the fact that *descriptive* features from the example were not used when constructing phylogeny. Features such as diet, type of reproduction, coloration, period of gestation, and others vary in the same *genus*, so *species* are not classified based on those characteristics. Purely *taxonomic* methods (such as Resnik's) do not take such features into account at all. Consequently, in this case, *descriptive* results are independent of taxonomy.

Another noteworthy observation is that the *physical* dimension score does not coincide with the *descriptive* Jaccard score, even though the former is, theoretically, a subdimension of the latter. This difference stems from difference in used methods. The *physical* arithmetic average method takes into account degree of difference between values corresponding to the same role, while the *descriptive* one does not, and only accepts identical values as similar. This, very simple, method works for this example, but cannot be applied universally (e.g. because of the division by zero problem). Unfortunately, disregarding custom ad-hoc methods (that work well, but cannot be easily applied outside of one specific ontology), there is no good and universal method that would compare complex descriptions in expressive DLs in an in-depth manner.

Notice that ordering of similarity changes between dimensions. *Taxonomicaly S* is closer to $R$ than to $D$, while *descriptively S* is closer to $D$. Taking into account interpretation of the used dimensions this suggests that short beaked dolphins and silvertip sharks look similar (high *physical* similarity), but their evolutionary ancestry is different (low or average *taxonomic* score). This statement is possible because separate dimensions of similarity have been independently evaluated and thus can be interpreted on the basis of their own semantics. It is impossible to infer such information from a single score.

One dimension that was not included in the example (for sake of brevity) is the *compositional* dimension. It would comprise of physical "components" of the animals with additional details, for instance, fins (e.g. small pointed dorsal fin), details of bone structure (e.g. serrated teeth), specific organs and functions (e.g. Ampullae of Lorenzini [76]). In this case the *hasPart* properties would refer to body parts. Note that this dimension has an interpretation in the context of phylogeny that fits the general interpretation and description from Sect. 5.

The final answer to the question of "how similar are two concepts?," for the dolphin and shark, according to Resnik's method is 0.105. According to the method of dimensional similarity $D$ and $S$ have *taxonomic* similarity of 0.188, *descriptive* similarity of 0.2 and *physical* similarity of 0.932. However, on the basis of the discussion presented thus far we strongly argue that dimensional scores are much more informative and, thus, useful to the "end user". We present further justification of this statement in Sect. 10. This being the case we propose a dimensional method, which produces multiple scores that can be organized into a *dimensional similarity vector*. Let us discuss this now this idea in some detail.

## 7   Combining Similarity Dimensions

As discussed so far, the canonical approach to similarity scoring is to present a single number as a result. Sometimes a range of intermediate results is calculated, in which case a method of combining those results into one, such as a weighted sum, is utilized. In this case, there are many weighting methods including metrics [77, 78], machine learning [47], aggregation operators [79, 80] and others [81].

For instance, in [47] authors used a weighted sum of 5 similarity scores of Word-Net concepts and various machine learning methods. The weights were trained against a (human) survey similarity scores for pairs of concepts. The authors remarked that for the scores from each dimension considered alone (for a test set of 20 pairs), each dimension at least once (i.e. for a specific pair) provided the best score (i.e. closest to training data). This led to the conclusion that the trained weights, even though useful for this specific application, may not be a good fit for a different domain or ontology. Nevertheless, according to [47], results from multiple dimensions are more useful than any individual dimension used separately.

Another work [56], calculated the score as a weighted sum of 4 intermediate (dimensional) scores. The intermediate scores were not deemed to be individually relevant and were only considered as parts of the weighted sum. Unfortunately, although the authors claim that the weights were "determined experimentally", the specific method of choosing weights was not described. The authors also cite problems with choosing good fixed weights [82], some of which are reiterated below.

The advantage of weighted sum is that the final score includes (and combines) a very broad range (possibly all) of available knowledge. Good set of weights offsets the possibility of overlap of dimensions by adjusting overlapping scores. A disadvantage is that there is no indication that weights calculated for one ontology give good results for a different one and recalculation of weights is expensive and requires a good training set (which may not be easy to deliver). Using a predefined metric or operator is less computationally intensive, but suffers from accuracy problems [81]. Overall, to the best of our knowledge, there is no weighing method that would produce a "universal" set of weights. Universality of weights, in the context of similarity scoring, means that one set of weights produces "good" (i.e. as compared to benchmarks) results for any testing set. So far no good method (training or otherwise) applicable to a wide range of ontologies and problems was found. This leads to a conclusion that it is extremely likely that weights are problem-specific, or ontology-specific.

In contrast, the dimensional similarity method proposed here produces a single score for each dimension. Those scores may be presented in a structure of the *dimensional similarity vector*, in which each cell contains a score from one dimension. While such vector can be then weighted and "reduced" to a single number, in the proposed approach the vector itself, as a whole, is to be used as the result.

Even though presenting the dimensional similarity vector as the final score goes against the established methods, it has clear advantages. First, we avoid the aforementioned problems with finding a good set of weights, which is very significant since the existing research suggests there might not be a good universal one. The gain is the amount of information that is contained within each vector cell. As mentioned, in Sect. 5, the interpretation of a similarity dimension is helpful when designing dimensional algorithms, but it also provides useful information about the final dimensional score. Understanding what each score stands for is helpful when deciding what knowledge is relevant to our particular problem. In a sense, it is an avoidance of the universal weights problem, because, assuming that weights represent importance, we don't consider weighting a part of similarity scoring. Instead, the implicit "weighting" is done after scoring in each dimension, when we apply the results to solving a specific problem. From this perspective, we are free to use (or disregard) data from any subset of cells from the full similarity vector. Guided by the interpretation of similarity dimensions we can decide what dimensions of similarity are useful in the context of the problem that is being addressed.

Good understanding of dimensions also helps with correct interpretation of overlap(s) between them. In case of a single final score it is impossible to, for example, "subtract" the impact of a taxonomy, in cases where we are not interested in this dimension of similarity. Moreover, since usually the weights are hidden from the user, it is not possible to know the impact brought about by new data introduced into an ontology, without either experimentation, or analysis of the code (i.e. reverse engineering), or documentation of used algorithm.

For the weighted sum there is also a general question—what is the actual meaning/interpretation of applied weights? If we assume the (intuitive) understanding that weights represent importance of dimensions, we may conclude that in a survey benchmark (like the Miller's one [15]) dimensions had some given importance to the participants. This approach, however, is problematic when it comes to automatic methods, because there usually is more than one set of weights that can produce the same weighted sum, for a single pair of entities. The hypothetical "importance" weights might also change on a pair by pair basis. Since the weights might not be unique, we cannot decidedly say that they represent importance, as viewed by the survey participants. Using a subset selected from cells of a dimensional similarity vector we essentially make a degenerated ad-hoc weighting. What we mean by that, is that choosing that we want only taxonomic score is equivalent to setting the weight of taxonomic score to 1, and rest to 0. For a subset of 3 dimensions, each weight for the chosen ones would be $\frac{1}{3}$, zero for the rest, and so on.

### Example Revisited

To visualize the problem let us propose a few sets of weights for the example from Sect. 6. Table 3 describes an example with three hypothetical sets of weights $w_1, w_2, w_3$. The weights are used to obtain a single similarity score for two cases—comparison of Shark with Dolphin ($Sim_{Total}(S, D)$) and Shark with Ray ($Sim_{Total}(S, R)$). The weighted sum is made from three dimensional scores —taxonomic, descriptive and physical (values calculated in Sect. 6 are recalled in the top part of the table).

The first set of weights $w_1$ assigns approximately equal value to each dimension. The resulting "total" scores are 0.41 and 0.29 for $Sim_{Total}(S, D)$ and $Sim_{Total}(S, R)$ respectively. Second set of weights $w_2$ indicates that the taxonomic dimension is decidedly more important than the other two and results in the scores of 0.34 and 0.42. Lastly, $w_3$ is a set of weights trained so that $Sim_{Total}(S, D)$ and $Sim_{Total}(S, R)$ are close in value (the result is approximately 0.4 for both).

First, notice that for $w_1$, because of the actual dimensional scores, the *physical* dimension has the highest contribution (i.e. highest value) to similarity of Shark and Dolphin, while the *taxonomic* dimension is the strongest in Shark and Ray comparison. This is in no way apparent in any of the $Sim_{Total}$ scores.

More importantly, the final score $Sim_{Total}$ has a very vague interpretation for any set of weights. All we can say about those numbers is that, depending on the weights, similarity of Shark and Dolphin is either greater, smaller or equal to that of Shark and Ray. The $w_3$ case indicates that a Shark is just as similar to the Dolphin as to a Ray, while the other two cases each produce an ordering of the two similarities. Any of those results can be put into question, depending on the perspective. A layman would

**Table 3** Similarity dimension weights example

| Dolphin ray | Shark | | | | $Sim_{Total}(S, D)$ | $Sim_{Total}(S, R)$ |
|---|---|---|---|---|---|---|
| | Taxonomic | Descriptive | Physical | | | |
| | 0.105 | 0.2 | 0.932 | | | |
| | 0.799 | 0.0 | 0.067 | | | |
| $w_1$ | 0.33 | 0.33 | 0.33 | | 0.41 | 0.29 |
| $w_2$ | 0.5 | 0.25 | 0.25 | | 0.34 | 0.42 |
| $w_3$ | 0.47 | 0.20 | 0.33 | | 0.40 | 0.40 |

classify Dolphin as much closer to Shark simply because the Ray looks nothing like the other two creatures. An expert in biology would, however, see much more differences and similarities in all creatures and would give a different score. Finally, an expert working specifically in phylogenetics would say that (phylogenetically) Shark and Ray have more in common than Shark and Dolphin. Traditionally, in a survey, the results would be averaged and produce a number that none of the participants exactly agrees on, but is representative of the average opinion. We conjecture that in similarity scoring, since the "average opinion" does not represent any actual perspective, it has a diminished usability. None of the scores—for $w_1$, $w_2$, $w_3$, or average of those, agrees with any other and yet, since the methodology is formally correct we cannot say that any of them are wrong, unless we adopt a specific perspective, e.g. to solve a specific problem in biology. Moreover, even from a particular perspective it is not possible to learn from the final score what information was most important (i.e. what were the weights for each dimension), if we only look at $Sim_{Total}$. Knowing that none of the total scores is indicative of every of the hypothetical survey participants, led us to believe that modeling different perspectives requires different weights, and none of the weight sets is, in general case, "more correct" than the other.

The "total" weighted sum score is contrasted with the dimensional score. Here the result is the dimensional similarity vector with 3 cells, one for each considered dimension, e.g. $Sim_{Dim}(S, D) = [0.105, 0.2, 0.932]$ for a set of dimensions [*taxonomic, descriptive, physical*]. Separately, each cell contains explicitly a single number, but also (implicitly) an explanation of the score in the form of interpretation (or description) of the dimension. The information contained in the vector lets us discern different kinds of similarity and learn that, for instance, Shark and Dolphin have high *physical* likeness, but their genetic ancestry (*taxonomic* similarity) is low. We can afterward decide whether this similarity dimension is relevant to solving our specific problem, or, in other words, whether it fits our perspective. Notice that this is useful both to an expert, and a layman. The first will learn much more from information about similarity in genetic taxonomy, rather than overall similarity. The latter will find more understanding in information about physical similarity of creatures, rather than some vague "universal" similarity.

To summarize, our recommendation is that the single number score ($Sim_{Total}$) should be used whenever it is required by a methodology—e.g. as an input to another method that accepts single number only, or to compare results with benchmark data (which usually gives only one number for any pair of entities). In other cases we recommend the use of the full similarity vector ($Sim_{Dim}$), or a selected subset of dimensional scores, especially when the similarity score is presented to a user (as opposed to an automated system) that has a specific problem to solve. Dimensional score, simply put, gives the user more information without (possibly overwhelming and gratuitous) technical details of implementation and algorithm structure.

# 8 Properties of Dimensions

None of the relevant works [46–49, 53, 55, 56] that present some form of division of knowledge (collectively labeled as *dimensions*), in the context of similarity scoring, gave any formal reasons to support their specific choice of dimensions. The division of knowledge in each of the works was guided by authors intuitions and was tailored to fit the needs of a particular implementation of a given similarity algorithm. In this section we outline characteristics of dimensions that may support a decision as to, which dimensions to use, as well as guidelines for creation of new dimensions.

In our approach to semantic similarity dimensions, the *meaning* (semantics) of the dimension is absolutely essential. Table 4 summarizes the informal meaning (i.e. interpretation) of dimensions described in Sect. 5.

Ideally, the meaning (semantics) of each dimensions should be easily understandable even to a layman. Note that each dimensions from Table 4 can be summarized in a single sentence. Such concise summary on a high level of abstraction should be accompanied by a more verbose explanation. For instance the meaning of "type" in the *taxonomic* dimension summary is clear to an ontology engineer, but it might be confusing to others. It is crucial that the semantics of every dimension is properly explained. This is because the explanation of the meaning is the main guideline when it comes to actual implementation of the similarity algorithm. As noted in the explanation of the *physical* dimension, we can expect that the structure, roles and even semantics of different ontologies will vary greatly. Despite this, the same simi-

**Table 4** Interpretation of dimensions

| Dimension | Interpretation |
| --- | --- |
| *Lexical* | Entities are *lexically* similar, when the words used to label them (i.e. their names) are similar according to a dictionary |
| *Co-occurence* | Objects are *co-occurrence* similar, when they often appear together |
| *Taxonomic* | Objects are *taxonomicaly* similar, when they are of similar class, kind or type |
| *Descriptive* | Objects are *descriptively* similar, when they have similar properties, attributes or characteristics |
| *Physical* | Objects are *physically* similar, when their physical characteristics and appearance is similar |
| *Compositional* | Objects are *compositionaly* similar, when they have similar set of parts or ingredients |
| *Membership* | Objects are *membership* similar, when they have similar sets of representatives, instances or members |

larity dimensions should be applicable to many ontologies. Even implementations of *taxonomic* similarity might differ considerably, especially when we extend our problem space to systems that place, or relax, specific restrictions on taxonomies (e.g. multiple inheritance, no common root, etc.). The *idea* behind each dimension must be independent of any particular structure and should not make any unnecessary assumptions. It should have meaning outside of computer science and encompass many possible implementations within it.

A common, agreed upon, interpretation of a similarity dimension allows for direct comparisons of similarity scores from different methods. It also allows to distinguish that scores from different dimensions (e.g. *physical* and *compositional*) refer to a different kind of similarity and we can expect that they will not be related to each other, even for the same pair of objects. A "total" similarity score has only a very vague meaning of a "degree of similarity" and even though there is no basis for this, we expect such scores to be close to some idealized target similarity, and, therefore, close to each other. As is apparent from Table 2 the scores (for the same ontology) vary depending on selected algorithm and data fed to it. This shows that there does not exist one universal and ideal similarity, outside of artificially constructed references (sometimes based on averages from a survey).

Even though the general understanding of what a given similarity dimension represents is always the same, its informativeness is improved when we put it in a context of a specific ontology. For instance, in case of example from Sect. 6 the *taxonomic* score has an interpretation of *phylogenetic similarity* (in general terms, evolutionary ancestry) on top of the general one (given in Table 4). In the example ontology the taxonomy contains exclusively classes of living organisms and the position of an entity in this taxonomy is representative of its position in evolutionary tree (phylogeny). Understanding of what phylogeny is and how it is constructed improves the understanding of this dimension even more. Note that there may be many phylogenetic ontologies, each with (slightly) different taxonomy. The general interpretation of the *taxonomic* dimension is the same for any ontology. The *phylogenetic* interpretation of this dimension is the same for any *phylogenetic* ontology. The details of a *very specific* interpretation of the dimensions may differ in different phylogenetic ontologies, but the general interpretation stays the same. A well-defined semantic similarity dimension should be interpretable on many levels. In other words, it should have the *granularity* that is most useful.

### *Granularity*

Let us now consider the fact that the *granularity* of a dimension is directly related to how detailed and specific is the explanation of its interpretation. In other words *granularity* is the amount of information carried in a description of a dimension. The least *granular* (or informative) notion is simply what we referred throughout this text as "universal similarity". The "universal similarity" is mostly understood as an intuitive concept and its meaning may be studied in the field of philosophy, not computer science. "Semantic similarity" is almost as vague of a term, describing the similarity of meaning. There is no formally strict definition of it and, although some define it as a metric. However, mathematical properties of semantic similarity are not set in stone, as mentioned in Sect. 3.

To visualize the *granularity* of dimensions let us use a simple example in the context of the MusicBrainz database [83]. MusicBrainz contains data about worldwide music industry i.e. artists, albums, music companies, music genres etc. It is available in many forms, one of which is LinkedBrainz—a linked data version of the database. Usefulness of LinkedBrainz can be enhanced by exploiting the linked data and connecting it to dbPedia, which contains information that is directly related.

Music albums in MusicBrainz are "releases" (*mb:release*) of type "album" (*mb:album*). A descriptive similarity dimension for MusicBrainz is more informative than just "semantic similarity" and it is applicable to any concept within MusicBrainz in the same way as in any other knowledge base. For a *mb:release*, it denotes similarity of all its properties such as *mb:artist*, *mb:title*, *mb:label*, *mb:format* and others. In particular the *mb:type* is not included here. In simplistic terms the gain of information stems from restricting the fields that we include in the similarity scoring to a smaller set. This is also true if we design a similarity dimension for any specific ontology. Doing that, however, we loose the ability to directly apply our new dimension to any other ontology. An increase of specificity (information) means a decreased range of possible applications. This is particularly apparent when the description of a dimension specifically mentions a property. For instance, in order to group albums by musical era we need to know the similarity of their release date. Such "album-time-of-release" dimension is very specific, because it can only be applied to an ontology that describes music albums and stores release time data. It is also very informative—we know exactly what data is used and, since time data is numerical, we can directly relate it to a syntactic similarity, or closeness of numbers. The possible data and algorithms used in this dimension are very restricted. Separating the data in such dimension does not bring any immediately apparent benefit and is, frankly, not necessary or advised. This is in stark contrast to low granularity dimensions, e.g. *descriptive* similarity. Dimensions of moderate similarity are an attempt to strike a balance. For instance, the *compositional* dimension is only applicable to ontologies with appropriate roles (e.g. *hasPart*), but since many ontologies do in fact have such roles, this requirement is not very restrictive. The granularities of this example are summarized in Table 5.

In summary, a good design of similarity dimensions exhibits a balance between informativeness and applicability. From the point of view of *granularity*, similarity dimensions can be put on a spectrum between very specific syntactic similarity and very vague (semantic) similarity. Low informativeness gives a wide range of possibilities when it comes to implementation. High granularity leaves no doubt when it comes to the meaning of such highly granular, dimensional similarity score. The choice of granularity should be made to best help solve a given problem, but very high granularities are not advised.

*Implementation*

Implementation of dimensions may vary greatly. For instance the *lexical* dimension may be implemented as a string edit distance like in the ASMOV [56] (that uses the

**Table 5** Information in similarity

| Relative informativeness | Similarity description | Similarity interpretation |
| --- | --- | --- |
| 0 | Similarity | Likeness, closeness |
| 1 | Semantic similarity | Similarity of meaning of entities |
| 2 | *Descriptive* semantic similarity | Similarity of meaning of descriptions of entities (attributes and characteristics) |
| 3 | *Descriptive* semantic similarity of music albums | Similarity of meaning of descriptions of music albums |
| 4 | *Descriptive* semantic similarity of music albums in a music ontology (MusicBrainz) | Similarity of meaning of description of *MusicBrainz:album*(s) i.e. similarity of artist, title, label etc. |
| 5 | Semantic similarity of release year of albums from a music ontology | Similarity of meaning of numbers representing years (e.g. numerical similarity) |
| 6 | Semantic similarity of release year of albums from MusicBrainz ontology | Similarity of meaning of *MusicBrainz:Release_event:date* |

Levenshtein distance), or as an *external* thesaurus lookup, like in ASCO [84] (which actually uses both the edit distance and WordNet similarity). As explained before, low granularity leaves a lot of room for different implementations.

In case of ontology matching *taxonomic* and *descriptive* dimensions are often combined into one, called *structural*. There are many different approaches to *structural* similarity. For instance, CIDER [85] uses a feature vector model that combines taxonomy and roles into one set of features. In Anchor-Flood [86], on the other hand, the *structural* similarity is constructed purely from taxonomy. ASMOV [56] has an even more disparate definition of *structural* dimension that involves a weighted sum of the domain and range similarities of roles. This difference of approaches demonstrates the importance of a good description of semantic similarity dimensions. Since *structural* similarity (dimension) lacks a good description, it allows for very different implementations. One possible definition, i.e. a dimension that combines *taxonomic* and *descriptive* similarities would endow it with a very low *granularity* that places it very close to a vague "universal" semantic similarity. In other words the meaning of *structural* similarity is too vague (it is very different in each of the presented examples) and, therefore, it does not provide much information.

Let us reiterate that similarity dimensions are defined primarily by their interpretation and not by implementation, or even type of method used.

# 9 Applications of Dimensional Semantic Similarity

In this paper, we have focused on presenting the idea of similarity dimensions on the examples concerning pairwise comparison of ontological entities. The idea itself can be applied to comparison of other objects, such as full ontologies, entities in semantic graphs, documents, etc. Throughout the text we have already suggested potential applications outside of ontological entities. Let us now reiterate and summarize these considerations.

Analysis of multiple articles and surveys on ontology matching [50, 52, 87] reveals that modern methods usually use multiple kinds of semantic similarity akin to similarity dimensions. In particular there is a strong distinction between *lexical* methods (also called *linguistic*) and others. A popular approach, exemplified in FalconAO [88] is to use *lexical* similarity first, as an input for further parts of the matching algorithm that use some kind of *structural* data (a graph matching algorithm in case of FalconAO). Some methods, such as AgreementMaker [89] and COMA [90] use multiple so-called matchers, some of which use taxonomy, relationship graph or lexical data. Matchers that work in the same dimension use different algorithms (e.g. some lexical matchers use edit distance, some thesaurus lookup or others). It seems that researchers in the field of ontology matching realized that construction of a good matching requires one to look at similarity of ontologies from many different perspectives. We have formalized this idea in the form of semantic similarity dimensions.

In the field of document analysis, semantic similarity means the similarity of meaning (in natural language) of the content of the documents i.e. text similarity [91]. Within this field, similarity of other features of documents, such as author, type of document (e.g. scientific article, a poem, news article, short story, etc.), publishing events and others is usually not considered. Those features are a good candidate for implementation of similarity dimensions (e.g. type of document describes the *taxonomic* dimension), but require *external* ontology (e.g. a taxonomy of document types), so, in some way, similarity of documents is understood as *lexical* similarity of content of documents.

The *lexical* dimension, in the context of document similarity, has many features that may be used to construct subdimensions. Features considered in practice [92] include statistical analysis (e.g. bag of words approach), sentence length, punctuation count, specific names count, synonyms, hypernyms, hyponyms and others. Those may be divided into corpus-level (e.g. TF-IDF), document-level (e.g. bag of words), sentence-level (e.g. extraction of subjects and objects, number of capitalized words) and word-level (e.g. synonyms, edit distance). Phrase-level features are also sometimes considered, although they are used in machine translation [93] rather than in similarity scoring. Even though many researchers have proposed multiple features [92], so far there was no attempt to group those features into classes that would resemble the low granularity dimensional approach described in this paper. Although those are not applied to similarity scoring, there are many dimension-like properties relevant to text and speech analysis. Those include affect [94] (also applied to WordNet [95]), salience, writing style (formal or informal) and others. Theoretically, we might score text with respect to, for instance, affect similarity (dimension), but as said before, such high level properties are not included in current document similarity scoring methods.

Presence of similarity dimensions in semantic graphs is most pronounced in methods that use WordNet. This semantic graph offers many different kinds of edges and sets of features. Different methods use different subsets of available information (e.g. some methods use synsets, others bots synsets and homonyms). This dimensionality is, however, not made explicit and, so far, those methods have not been categorized with respect to dimensions.

## 10 Concluding Remarks

The notion of (semantic) similarity is, by its nature, vague and ambiguous. Many semantic similarity measuring methods have been proposed and work well for ontology-specific or domain-specific applications. Their approaches, however, do not easily generalize across domains (or ontologies). The proposal of *similarity dimensions* address this problem and attempts at rectify the ambiguity of similarity scores.

A single, universal, score suggests how similar two entities are, but does not answer the question: in what way are the entities actually similar? A similarity vector provides such answers by treating each similarity dimension separately. Thus, it is possible to capture the fact that being *descriptively* similar is different from *taxonomicaly* similar, or *lexically* similar, etc. In short, similarity dimensions add extra meaning to similarity. Dimensional scores specify not only how similar entities are, but also why.

Generally speaking, there are two ways of dealing with semantic similarity. First, the overall approach, based on application of similarity dimensions, with separate scores in each, to understand how similar entities are, and in what way. Second, development of domain/ontology specific methods that focus on the nature of the problem at hand. The latter approaches (e.g. [96]) work well when solving a specific problems, but do not transfer well to other application areas. Canonically, similarity calculating methods produce a single score that combines all aspects of semantic similarity. It is a useful simplification that enables direct comparison of results from different methods. However, different methods approach similarity from a "different perspective," use different data and capture different aspect of semantic similarity. Moreover, since any well-defined method is formally correct, no individual score can be said to be formally wrong.

Note also that, comparison of single number results form different methods is, by nature, flawed. Even methods that utilize multiple intermediate similarity scores, in the end provide a single weighted sum, which "flattens" the meaning of similarity. Furthermore, making explicit the considered aspect of semantic similarity can be also useful. For instance, Resnik's method is purely taxonomic. Hence, by explicitly labeling it as such, one gains valuable information. For instance, someone not familiar with details of Resnik's method would not know why similarity does not change, even if one adds a number of roles into the KB. Labeling the method as taxonomic informs that it is insensitive to roles.

On the other hand, the proposed dimensional similarity vector presents a more detailed (expanded) view of similarity, and allows for a more meaningful comparison of results between methods. Another advantage of similarity dimensions is that each one of them has a universal interpretation, that may be refined depending on context and is independent of the data format. As long as this interpretation is preserved, multiple different algorithms may be used to represent each dimension. Furthermore, proper usage and interpretation of a specific dimension is reliant on intuitive understanding of general description of that dimension. In this way, the similarity vector reflects the subjective nature of similarity.

Let us recall that there are many different and correct ways to model any given domain or problem. The multiplicity of modeling paradigms is a well-known and studied subject [97, 98]. It suggests that, for any domain, there is no single, exclusively correct, modeling solution. We believe that the same is true for semantic similarity, i.e. the correct "absolute" / "ultimate" similarity measure does not exist. Instead, the similarity changes with the perspective, from which we calculate it. Here, it should be stressed that the proposed approach recognizes this fact by its inherent flexibility. Specifically, it allows: (i) existence of domain/ontology specific methods to combine separate scores into a single one (as in [46]), (ii) restricting similarity dimensions that are actually considered in a given domain (e.g. only taxonomic and compositional dimensions are to be used), based on the "nature of the application". Moreover, if one is interested in similarity in a taxonomy, one needs to use only a taxonomic method. Alternatively, if one already obtained a dimensional vector, (s)he can utilize any part of it that is of interest in a given context. Here, again, available dimensionality provides information useful both before and after similarity scoring.

In this way, the dimensional similarity vector provides, in a sense, a disentanglement of similarity. A dimensional answer to a question of similarity is more informative not just because one receives more values, but also because each value (i.e. each dimension) has an interpretation. This interpretation adds knowledge about the way, in which entities are similar, on top of a numerical value representing similarity. A single score is much more concise, but it lacks this additional information, i.e. this information is obfuscated, when only one score is available, without any explanation as to how it was arrived at.

In summary, similarity dimensions are a way of introducing *semantics* the into semantic similarity itself. The low-granularity dimensions (presented in Sect. 5) provide a basic understanding of similarity even to a layman. For instance *physical* similarity is immediately understood by everyone. High-granularity dimensions may be created to serve very particular needs of experts in a given field. It is thus our opinion that, in calculating semantic similarity, the most important part is the reason why we calculate it. In conclusion, recognizing similarity dimensions adds meaning to semantic similarity. Dimensional score tells us not only how entities are similar, but also indicates why.

# References

1. https://www.w3.org/TR/owl-guide/
2. Szmeja, P., Ganzha, M., Paprzycki, M., Pawlowski, W.: Dimensions of ontological similarity. In: 2016 IEEE Tenth International Conference on Semantic Computing (ICSC), pp. 246–249. IEEE, February 2016
3. Baader, F., Calvanese, D., McGuinness, D., Nardi, D., Patel-Schneider, P.: The Description Logic Handbook. Cambridge University Press (2003)
4. Cohen, W., Borgida, A., Hirsh, H.: Computing least common subsumers in description logics. In: Proceedings of the 10th National Conference on Artificial Intelligence, pp. 754–760. MIT Press (1992)
5. Baader, F.: Least Common Subsumers and Most Specific Concepts in a Description Logic with Existential Restrictions and Terminological Cycles (2003)
6. Chang, C., Lee, R.: Symbolic Logic and Mechanical Theorem Proving. Academic Press, San Diego (1973)
7. d'Amato, C., Fanizzi, N., Esposito, F.: A semantic similarity measure for expressive description logics. In: Proceedings of convegno italiano di logica computazionale (2005)
8. Shepard, Roger N.: The analysis of proximities: Multidimensional scaling with an unknown distance function. I. Psychometrika. **27**(2), 125–140 (1962). doi:10.1007/BF02289630
9. Hahn, Ulrike, Chater, Nick, Richardson, Lucy B.: Similarity as transformation. Cognition **87**(1), 1–32 (2003). doi:10.1016/S0010-0277(02)00184-1
10. Asl, M.E., et al.: Similitude analysis of composite I-beams with application to subcomponent testing of wind turbine blades. In: Experimental and Applied Mechanics, vo. 4, pp. 115–126. Springer International Publishing (2016)
11. Tversky, A.: Features of similarity. Psycholog. Rev. **84**, 327–352 (1977)
12. Nothdurft, Hans-Christoph: Feature analysis and the role of similarity in preattentive vision. Atten. Percept. Psychophys. **52**(4), 355–375 (1992)
13. Santini, Simone: Jain, Ramesh: The graphical specification of similarity queries. J. Vis. Lang. Comput. **7**(4), 403–421 (1996)
14. Rubenstein, Herbert, Goodenough, John: Contextual cor-relates of synonymy. CACM **8**(10), 627–633 (1965)
15. Miller, G.A., Charles, W.G.: Contextual correlates of semantic similarity. Lang. Cognit. Processes **6**, 1–28 (1991)
16. Resnik, P.: Using information content to evaluate semantic similarity in a taxonomy. In: Proceedings of the 14th international joint conference on Artificial intelligence, pp. 448–453 (1995)
17. Milne, D., Witten, I.: An effective, low-cost measure of semantic relatedness obtained from wikipedia links. In: Proceedings of the AAAI Workshop on Wikipedia and Artificial Intelligence: an Evolving Synergy, pp. 25–30 (2008)
18. Hliaoutakis, A., Varelas, G., Voutsakis, E., Petrakis, E.G., Milios, E.: Information retrieval by semantic similarity. IJSWIS **2**(3), 55–73 (2006)
19. Sanchez, D., Batet, M., Isern, D., Valls, A.: Ontology-based semantic similarity: a new feature-based approach. Expert Syst. Appl. **39**(9), 7718–7728 (2012)
20. Ceccarelli, D., Lucchese, C., Orlando, S., Perego, R., Trani, S.: Learning relatedness measures for entity linking. In: Proceedings of the 22nd ACM international Conference on Information and Knowledge Management, pp. 139–148 (2013)
21. De Nies, T., et al.: A distance-based approach for semantic dissimilarity in knowledge graphs. In: 2016 IEEE Tenth International Conference on Semantic Computing (ICSC). IEEE (2016)
22. Ontology Alignment Evaluation Initiative. http://oaei.ontologymatching.org/
23. http://davis.wpi.edu/xmdv/datasets/ohsumed.html
24. http://qwone.com/~jason/20Newsgroups/
25. Bohm, C., Berchtold, S., Keim, D.A.: Searching in high-dimensional spaces: index structures for improving the performance of multi-media databases. ACM Comput. Surv. **33**(3), 322–373 (2001)

26. Harispe, S., Ranwez, S., Janaqi, S., Montmain, J.: Semantic Measures for the Comparison of Units of Language. Concepts or Instances from Text and Knowledge Representation Analysis, CoRR (2013)
27. Semantic Measures Library. http://www.semantic-measures-library.org/sml/
28. Bollegala, D., Matsuo, Y., Ishizuka, M.: A relational model of semantic similarity between words using automatically extracted lexical pattern clusters from the web. In: Conference on Empirical Methods in Natural Language Processing, EMNLP 2009, pp. 803–812. ACL and AFNLP (2009)
29. Wan, S., Angryk, R.A.: Measuring semantic similarity using wordnet-based context vectors. In: El-Hawary, M. (ed.) IEEE International Conference on Systems, Man and Cybernetics, SMC 2007, pp. 908–913. IEEE Computer Society, Montreal, Quebec, Canada (2007)
30. Rada, R., Mili, H., Bicknell, E., Blettner, M.: Development and application of a metric on semantic nets. IEEE Trans. Syst. Man Cybern. **19**(1), 17–19 (1989)
31. Wu, Z., Palmer, M.: Verb semantics and lexical selection. In: Proceedings of the 32Nd Annual Meeting on Association for Computational Linguistics, pp. 133–138 (1994)
32. Rhee, S.K., Lee, J., Park, M.-W., Szymczak, M.: Frąckowiak, G., Ganzha, M., Paprzycki, M.: Measuring semantic closeness of ontologically demarcated resources. Fundam. Inform. **96**(4), 395–418 (2009)
33. Zhang, L., et al.: FSIM: a feature similarity index for image quality assessment. IEEE Trans. Image Process. **20**(8), 2378–2386 (2011)
34. Aha, D., Kibler, D., Albert, M.: Instance-based learning algorithms. Mach. Learn. **6**(1), 37–66 (1991)
35. Fellbaum, C.: WordNet: An Electronic Lexical Database. MIT Press, Cambridge, UK (1998)
36. Jaccard, P.: Distribution de la flore alpine dans le bassin des Dranses et dans quelques régions voisines. Bull. de la Société Vaudoise des Sci. Nat. **37**, 241–272 (1901)
37. Rodriguez, M.A., Egenhofer, M.J.: Determining semantic similarity among entity classes from different ontologies. IEEE Trans. Knowl. Data Eng. **15**, 442–456 (2003)
38. Petrakis, E.G.M., Varelas, G., Hliaoutakis, A., Raftopoulou, P.: X-similarity: computing semantic similarity between concepts from different ontologies. J. Digit. Inf. Manag. **4**, 233–237 (2006)
39. Shannon, Claude Elwood: A mathematical theory of communication. ACM SIGMOBILE Mob. Comput. Commun. Rev. **5**(1), 3–5 (2001)
40. Pirró, G., Seco, N.: Design, implementation and evaluation of a new semantic similarity metric combining features and intrinsic information content. In: Meersman, R., Tari, Z. (eds.) OTM 2008 Confederated International Conferences CoopIS, DOA, GADA, IS, and ODBASE 2008, Monterrey, Mexico, vol. 5332, pp. 1271–1288. Springer, Heidelberg (2008)
41. Zhou, Z., Wang, Y., Gu, J.: A new model of information content for semantic similarity in WordNet. In: 2008 Second International Conference on Future Generation Communication and Networking Symposia, FGCNS'08, vol. 3. IEEE (2008)
42. Seco, N., Veale, T., Hayes, J.: An intrinsic information content metric for semantic similarity in WordNet. In: Proceedings of the 16th European conference on artificial intelligence. IOS Press (2004)
43. Sánchez, D., Batet, M., Isern, D.: Ontology-based information content computation. Knowl.-Based Syst. **24**(2), 297–303 (2011)
44. Pirró, G.: A semantic similarity metric combining features and intrinsic information content. Data Knowl. Eng. **68**, 1289–1308 (2009)
45. Hamming, Richard W.: Error detecting and error correcting codes. Bell Syst. Tech. J. **29**(2), 147–160 (1950). doi:10.1002/j.1538-7305.1950.tb00463.x,MR0035935
46. Calle, F.J., Castro,E., Cuadra, D.: Ontological Dimensions Applied to Natural Interaction. In: ONTORACT '08 Proceedings of the 2008 First International Workshop on Ontologies in Interactive Systems, p. 91–96
47. Albacete, E., Calle, J., Castro, E., Cuadra, D.: Semantic similarity measures applied to an ontology for human-like interaction. J. Artif. Intell. Res. **44**, 397–421 (2012)

48. Euzenat, J., Valtchev, P.: Similarity-based ontology alignment in OWL-lite. In: ECAI. vol. 16 (2004)
49. Rahm, Erhard: Bernstein, Philip: A survey of approaches to auto-matic schema matching. VLDB J. **10**(4), 334–350 (2001)
50. Euzenat, J., Shvaiko, P.: Ontology Matching. Springer, Berlin (2007)
51. Lin, F.: State of the art: automatic ontology matching. Tekniska Högskolan (2007)
52. Shvaiko, Pavel: Euzenat, Jérôme: Ontology matching: state of the art and future challenges. IEEE Trans. Knowl. Data Eng. **25**(1), 158–176 (2013)
53. Lin, D.: An information-theoretic definition of similarity. In Proceedings of the Fifteenth International Conference on Machine Learning, pp. 296–304 (1998)
54. Ashburner, M., Ball, C.A., Blake, J.A., Botstein, D., Butler, H., Cherry, J.M.: et. al.: Gene ontology: tool for the unification of biology. The Gene Ontology Consortium. Nature genetics. 25(1), pp. 25–29. Stanford University School of Medicine, California, USA, Department of Genetics (2000)
55. Pesquita, C., Faria, D., Falca, A.O., Lord, P., Couto, F.M.: Semantic Similarity in Biomedical Ontologies (2009)
56. Jean-Mary, Y.R., Shironoshita, E.P., Kabuka, M.R.: Ontology Matching with Semantic Ver-ification. INFOTECH Soft, Inc., 9200 S Dadeland Blvd. Suite 620, Miami, FL 33156, USA 1 University of Miami, Coral Gables, FL 33124, USA
57. Vargas-Vera, M., Nagy, M., Motta, E.: DSSim—managing uncertainty on the semantic web, pp. 1–11 (2011). http://oro.open.ac.uk/23598/1/10.1.1.104.99635B15D.pdf
58. Ichise, R.: Machine learning approach for ontology mapping using multiple concept similarity measures. In: Seventh IEEE/ACIS International Conference on Computer and Information Science (icis 2008), Portland/Oregon. IEEE
59. http://linkeddata.org/
60. Navigli, R.: Word sense disambiguation: A survey. ACM Comput. Surv. (CSUR) **41**(2), 10 (2009)
61. Bunescu, R.C., Pasca, M.: Using encyclopedic knowledge for named entity disambiguation. In: EACL, vol. 6 (2006)
62. Lund, Kevin: Burgess, Curt: Producing high-dimensional semantic spaces from lexical co-occurrence. Behav. Res. Methods Instrum. Comput. **28**(2), 203–208 (1996)
63. Pekar, V., Staab, S.: Taxonomy learning: factoring the structure of a taxonomy into a semantic classification decision. In: Proceedings of 19th International Conference on Computational Linguistics, pp. 1–7 (2012)
64. Jiang, J.J., Conrath, D.W.: Semantic similarity based on corpus statistics and lexical taxon-omy. In: International Conference on Research on Computational Linguistics (1997)
65. Maguitman, A.G., Menczer, F., Roinestad, H., Vespignani, A.: Algorithmic detection of semantic similarity. In: Proceedings of the 14th International Conference on World Wide Web, pp. 107–116 (2005)
66. Harispe, S., Sánchez, D., Ranweza, S., Janaqia, S., Montmaina, J.: A framework for unifying ontology-based semantic similarity measures: a study in the biomedical domain. J. Biomed. Inform. **48**, 38–53 (2014)
67. Horrocks, I., Sattler, U.: A description logic with transitive and inverse roles and role hierar-chies. J. Log. Comput. **9**(3), 385–410 (1999)
68. Blank, A.: Words and concepts in time: towards diachronic cognitive onomasiology. In: Eckardt, R., von Heusinger, K., Schwarze, C. (eds.) Words in Time, pp. 37–66. Mouton de Gruyter, Berlin, Germany (2013)
69. http://purl.oclc.org/NET/ssnx/ssn
70. Lehmann, K.: A Framework for Semantic Invariant Similarity Measures for ELH Concept Descriptions. Diplomarbeit, Technishe Universitat Dresden (2012)
71. https://en.wikipedia.org/wiki/Short-beaked_common_dolphin
72. https://en.wikipedia.org/wiki/Silvertip_shark
73. https://en.wikipedia.org/wiki/Lesser_electric_ray
74. http://www.flmnh.ufl.edu/fish/gallery/descript/silvertipshark/silvertipshark.html

75. http://www.arkive.org/lesser-electric-ray/narcine-brasiliensis/
76. http://www.marinebiodiversity.ca/skatesandrays/external
77. Cunningham, P.: A taxonomy of similarity mechanisms for case-based reasoning. IEEE Trans. Knowl. Data Eng. **21**(11), 1532–1543 (2009)
78. Zerzucha, P., Walczak, B.: Concept of (dis)similarity in data analysis. Trends Anal. Chem. **38**, 116–128 (2012)
79. Detyniecki, M.: Mathematical aggregation operators and their application to video querying. Research Report, LIP6, Paris (2001)
80. Dubois, D., Prade, H.: On the use of aggregation operations in information fusion processes. Fuzzy Sets Syst. **142**, 143–161 (2004)
81. Younes, A.A., Blanchard, F., Herbin, M.: New similarity index based on the aggregation of membership functions through OWA operator. In: 2015 Federated Conference on Computer Science and Information Systems (FedCSIS). IEEE (2015)
82. Bach, T., Dieng-Kuntz, R.: Measuring similarity of elements in owl DL ontologies. In: Theory, Practice and Applications, Workshop on Contexts and Ontologies (2005)
83. MusicBrainz—The Open Music Encyclopedia. https://musicbrainz.org/
84. Le B.T., Dieng-Kuntz R., Gandon F.: Ontology matching: A machine learning approach for building a corporate semantic web in a multi-communities organization, 14–17 April 2004
85. Gracia, J., Asooja, K.: Monolingual and cross-lingual ontology matching with CIDER-CL: evaluation report for OAEI 2013. In: Proceedings of the 8th Ontology Matching Workshop (OM'13), at 12th International Semantic Web Conference (ISWC'13), Syndey (Australia), CEUR-WS, vol. 1111 October 2013. ISSN-1613-0073
86. Seddiqui, M.H., Aono. M.: Anchor-flood: results for OAEI 2009. In: Proceedings of the 4th International Conference on Ontology Matching-Volume 551. CEUR-WS. org (2009)
87. Otero-Cerdeira, Lorena, Rodríguez-Martínez, Francisco J., Gómez-Rodríguez, Alma: Ontology matching: a literature review. Expert Syst. Appl. **42**(2), 949–971 (2015)
88. Hu, Wei: Yuzhong, Qu: Falcon-AO: a practical ontology matching system. Web Semant. Sci. Serv. Agents. World Wide Web **6**(3), 237–239 (2008)
89. Cruz, I.F., Antonelli, F.P.: Stroe. C.: AgreementMaker: efficient matching for large real-world schemas and ontologies. Proc. VLDB Endow. **2**(2), 1586–1589 (2009)
90. Massmann, S., et al.: Evolution of the COMA match system. In: Proceedings of the 6th International Conference on Ontology Matching-Volume 814. CEUR-WS. org (2011)
91. Mihalcea, R., Corley, C., Strapparava, C.: Corpus-based and knowledge-based measures of text semantic similarity. In: AAAI, vol. 6 (2006)
92. Murphy, M. L.: Semantic relations and the lexicon: antonymy, synonymy and other paradigms. Cambridge University Press (2003)
93. Li, J., Resnik, P., Daumé III.H.: Modeling syntactic and semantic structures in hierarchical phrase-based translation. In: HLT-NAACL (2013)
94. Besnier, Niko: Language and affect. Annu. Rev. Anthropol. **19**, 419–451 (1990)
95. Strapparava, C., Valitutti. A.: WordNet Affect: an Affective Extension of WordNet. In: LREC, vol. 4 (2004)
96. Benabderrahmane, S., Smail-Tabbone, M., Poch, O., Napoli, A., Devignes, M-D.: IntelliGO a new vector-based semantic similarity measure including annotation origin. BMC Bioinform. **11**(1) (2010)
97. Goldkuhl, G.: Design theories in information systems-a need for multi-grounding. JITTA J. Inf. Technol. Theor. Appl. **6**(2), 59 (2004)
98. Dietz, J.L.G.: What is Enterprise Ontology?. Springer, Heidelberg (2006)
99. Google Knowledge Graph. https://developers.google.com/structured-data/customize/overview
100. Open Directory Project. https://www.dmoz.org/

# Some Interesting Phenomenon Occurring During Self-learning Process with Its Psychological Interpretation

Ryszard Tadeusiewicz

## 1 Introduction

This book is dedicated for the eminent scientist, former president of IEEE and candidate for 2018 IEEE President-Elect, wonderful man and—last but not least—my friend, professor Jacek Zurada. Professor Zurada is one of the best experts in (among other) computational intelligence [1], neural networks [2] and machine learning areas [3]. Therefore selecting the material for this chapter I must prefer scientific results related to quoted areas.

Neural networks are useful tools for solving many practical problem (e.g. [4–7]). But every of such solution is interesting for limited number of readers, working with similar problems and similar applications. Therefore we select more interesting observations, which are related to the phenomena observed during the neural network self-learning process. Because of same similarity to psychological processes [8], observed during natural activity in our own mind, we call such phenomena "artificial dreams" [9]. This name is similar to the title of Hamid Ekbia's book [10], but the meaning of this term in our works is slightly different. In Ekbia's book "artificial dreams" are presented as unrealized and unrealizable projects related to Artificial Intelligence. In our research we do observe "artificial dreams" as **spontaneous and unexpected processes, emerging automatically from the natural self-learning procedures**.

R. Tadeusiewicz (✉)
AGH University of Science and Technology, Krakow, Poland
e-mail: rtad@agh.edu.pl
URL: http://www.tadeusiewicz.pl

© Springer International Publishing AG 2018
A.E. Gawęda et al. (eds.), *Advances in Data Analysis with Computational Intelligence Methods*, Studies in Computational Intelligence 738,
https://doi.org/10.1007/978-3-319-67946-4_4

The phenomena under consideration are very interesting and exciting, therefore can be mysterious, why so rare are reported by Artificial Intelligence or Computational Intelligence researchers? Yet so many people perform self-learning processes for many purposes—so why such phenomena are still not discovered and described?

The answer is simple. Most papers describing methods and results of the self-learning (even in neural networks, which are main tool considered in this work) are mainly **goal-oriented**. The researcher or practitioner are concentrate on the applications, not on the tool and its behavior. Authors of almost all papers first try to obtain the best result in terms of solving of specified problem (e.g. building of neural network based model of some process or finding the neural solution of the pattern recognition problem). Therefore the discussion of the self-learning results taking into account only the final result (e.g. quality of the model or correctness of classification), while the phenomena discussed in this paper occur when the self-learning system is not learned enough. In all works known to the author at this time nobody see on the details of network (or other self-learning system) behavior during the learning process. Meanwhile some phenomena observed during the self-learning process are really interesting, because totally unexpected.

## 2   Self-learning and Learning

In this paper we take into consideration **self-learning** process, instead of more known and more useful (from technical point of view) machine learning process. Let us describe the main difference between such two processes, because it will be important from the main thesis of this paper.

During the regular learning process we have the "teacher", who teach "pupil" (in fact it is machine) on the base of examples of properly solved tasks. In machine learning teacher is an algorithm, powered with examples database, but the main idea of teaching is based on simple scheme: get the knowledge from teacher and put it the pupil. After learning process "artificial pupil" can take an exam, where quality of learned knowledge can be evaluated and assessed. On Fig. 1 you can see how it works on the base of gender recognition problem.

In contrast to this scheme self-learning process is based on the knowledge discovery methods. The pupil (in fact it is still machine) can accept input data, but there are no teacher, who can explain, what the data means. Therefore self-learning must not only accumulate knowledge, but it must **discover** this knowledge without any external help. It is in general difficult task, but many successful applications prove this way effective. On Fig. 2 you can see how it works also on the base of gender recognition problem. Self-learning system after connecting with many input data can differentiate man from women, but off course cannot give the proper names to the genders. During the exam self-learning system can give classification for new person (sometimes proper, and sometimes not—as every artificial classification system) using symbols of classes instead names.

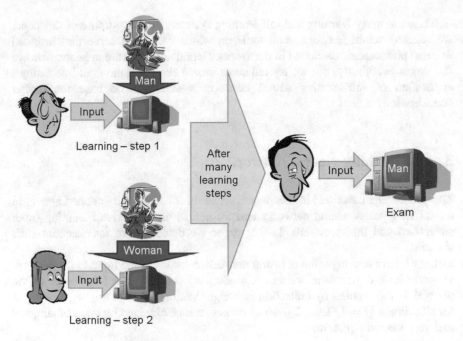

**Fig. 1** Learning and exam in supervised learning

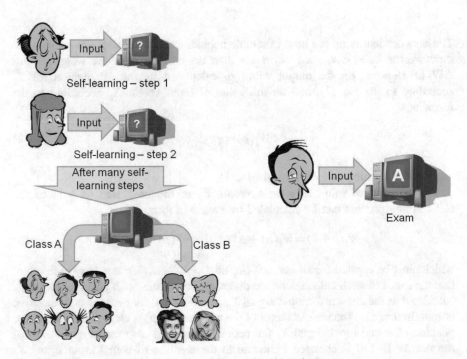

**Fig. 2** Learning and exam in unsupervised learning (self-learning)

There are many learning and self-learning systems, but for purpose of this paper we selected neural networks as a tools, in which we can observe the "artificial dreams" phenomena, discussed in our works. Neural networks are in general known for almost everybody, but we try tell some words about simple (and interesting!) application of self-learning neural network, which will be base for further consideration.

## 3  Self-learning Neural Network

The phenomena described in this paper can be discovered, as mentioned above, in almost all types of neural networks and for almost all methods of learning (both supervised and unsupervised). In this paper we decided take into account such situation:

Let we have one-layer linear neural network. It means as the input to the network we consider $n$-dimensional vectors $X = \,<\, x_1, x_2, ..., x_n >$, the knowledge of the network is represented by collection of weight vectors $W_j = \,<\, w_{1j}, w_{2j}, ..., w_{nj} >$ for all neurons ($j = 1, 2, ..., L$), which outputs can be obtained by mans of simplest and very known equation:

$$y_j = \sum_{i=1}^{n} w_{ij} x_i \tag{1}$$

The network learns on the base of simple hebbian rule: If on step $p$ we obtain the input vector $X_p = \,<\, x_{1p}, x_{2p}, ..., x_{np} >$ than the correction of the weight vector $\Delta W_j (p)$ depends on the output value $y_{jp}$ calculated by the $j$-th neuron for $X_p$ according to the Eq. (1), and on the value of input vector $X_p$ according to the formula:

$$\Delta W_j(p) = \eta\, y_{jp} X_p \tag{2}$$

where $\eta$ is the learning rate coefficient ($\eta < 1$).

Of course new value of weight vector $W_j$ at the next step ($p + 1$) of the self-learning process can be calculated by means of formula:

$$W_j(p+1) = W_j(p) + \Delta W_j(p) = W_j(p) + \eta\, y_{jp} X_p \tag{3}$$

which must be applied for all neurons (for all $j = 1, 2, ..., L$). It is easy to find out, that the result of such calculations are different for neurons with positive output $y_{jp}$ calculated as the answer for input signal $X_p$, and different for neurons with negative output. In first case the weight vector of the neuron $W_j (p)$ is changed toward to the position of actual input signal $X_p$ (attraction), in second case the weight vector of the neuron $W_j (p)$ is changed backward to the position of actual input signal $X_p$

**Fig. 3** Migration of the
weight vectors during one
step at the self-learning
process

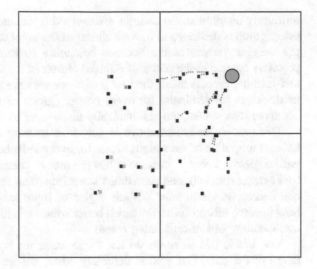

(repulsion). This process is presented on Fig. 3, where big ring denotes position of
input signal $X_p$, and the small squares denotes positions of weight vectors of the
neurons. The "migration" of the weight vectors can be observed on this plot—one
are attracted toward the input signal, where the other are pushed in opposite
direction.

The same process performed by big populations of self-learned neurons is
presented on Fig. 4.

Everybody know, what results after many steps of such self-learning process,
performed by the network connected with a real data stream. If the data are not

**Fig. 4** Migration of the
weight vectors in the biggest
self-learned network

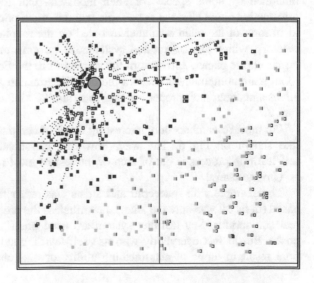

uniformly distributed, the neurons are divided (spontaneously!) onto groups, when every group is dedicated to the one cluster of the input data. Moreover the values of the weights vectors of the neurons belonging to each group are more or less precisely located in the center of selected cluster of the data. It means, that after the self-learning process inside the neural network we have neurons, which can be used as detectors (or sentinels) for every cluster (group of similar signals), present in observed data stream and automatically discovered by the network.

This process described above is not ideal, because as everybody know, spontaneous migration of the weight vector for every independent neuron leads to many pathologies: every attractor have many neurons as the detectors (over-representation), and sometimes some important attractors can be omitted (no one neuron decide to point out this region of input space). Everybody know also, how to solve this problem: the much better solution is to use Kohonen network and methodology of self-organizing maps.

Yes, but in this work we do not try to made the best self-organized representation of the data. Our goal is definitely other: we are searching for very simple model of the learning of neural network, because on the base of this model we try to show, how (and why) the learned network sometimes presents behavior, which can be interpreted as "artificial dreams".

## 4 How and Where Artificial Dreams Phenomena Can Be Discovered?

Let assume we must design spacecraft for discovery mysterious world of distant star and planetary robot, which will be send on the ground of totally unknown planet, inhabited by some species of alien monsters. Our robot must collect as many information about aliens as is possible without any a'priori knowledge (Fig. 5). The ideal form of the main computer installed on the robot desk is self-learning neural network, which can collect and systematize information about all creatures found on the exotic planet. After return the spacecraft to the Earth we can obtain from the robot main computer self-learned memory information about number of species of aliens and about their properties, thanks to similar kind classification like shown on Fig. 2.

For most researchers only interesting result of computer memory investigation is like shown on Fig. 6. The way, how this classification was obtained by the self-learning process is out of area of interest of most researchers.

Unfortunately!

The example with spacecraft and aliens was rather fantastic and science-fiction based (in fact it was only the joke!), nevertheless the problem under consideration is real and serious. Self-learning system are used often, eagerly and for many purposes. But in fact everybody who use self-learning systems is interested only on final result in terms of classification ability or data clustering, when the way of

**Fig. 5** Hypothetical spacecraft robot powered by self-learning neuro-computer

**Fig. 6** Content of the
memory of self-learning
spacecraft after discovery
alien planet

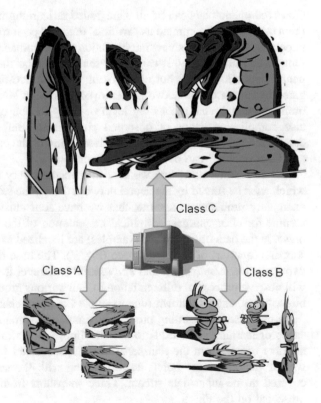

learning process is disregarded. Meanwhile we try show in this paper, that the unstable and transitory phenomena, observed in neural networks **during** the self-learning process, are also very interesting, impressive and inspiring.

Such phenomena can be discovered long time after the start of learning, when the network knows nothing because of random values assigned to all it weights. Self-learning process goes then automatically, so typical researcher starts performing another job or goes home. At the same time moment, when such unusual phenomena can be observed, occur long time before final point of learning process, when the network knows (almost) everything and can be exploited according to the plan. Such phenomena can be classified as errors of not matured enough self-learning neural network and therefore can be disregarded. But try give them some psychological interpretations.

## 5   How Manifest Artificial Dreams?

Observed phenomena can be all disregarded as learning imperfections, but some of them can also be interpreted as "artificial dreams" performed by the artificial neural networks. It can give us new interpretation of the human ability to the imagination, fantasy and also poetry. It can be presented even on the base of the very simple neural network models, but of course the most interesting results can be investigated by means of the networks deployed with high level of similarity to the real brain structures what means big level of complication of the neural structure and also complicated forms of observed phenomena. Before we show and discuss considered phenomena we must shot description of the example problem, in which "artificial dreams" can be very easy encountered.

Let us assume now, that we take into account very simple example problem, which must be solved by the neural network during the self-learning process. In this exemplary problem we assume, that we have four clusters in the input data. Let assume for clear and easy graphical presentation of the results, that the attractors preset in the data (most typical examples) are localized exactly at the centers of four subparts (quarter) of the input space (Fig. 7). The base of this space is defined by two parameters: *body form* and *body shape* (whatever it means). In such space we will observe process of differentiation of four various groups living beings (women, birds, fishes, snakes) shown (one example for every class) on Fig. 7.

In this case self-learning process in simulated neural network after some thousands of learning steps leads to the situation, when almost every neuron become member of one from the four separate groups, located (in sense of localization of weight vectors) at the points corresponding with the centers of the clusters discovered in the input data stream. Three snapshots from the learning process are presented on the Fig. 8.

Typical user of the neural network takes into account mainly last snapshot, presenting, how many neurons are located in proper positions after the learning process and how precisely the real values of attractors coordinates are reproduced

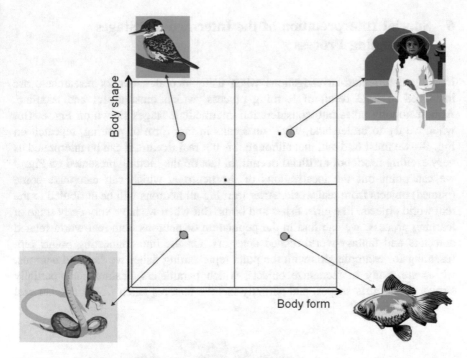

**Fig. 7** The example problem. Detail description in the text

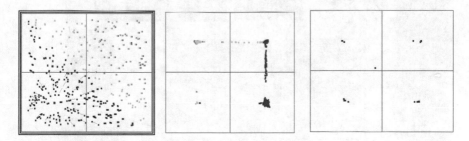

**Fig. 8** Three stages of the self-learning process

by the neurons parameters. For our consideration the medium snapshot will be most interesting, because in presents something strange: situation, when knowledge of the network is definitely not complete, but also the initial chaos was partially removed. This stage of learning process is usually skipped by neural network researchers, because apparently man cannot find anything interesting in this plots: the learning process is not ready yet, it's all.

Apparently.

In fact what we see on the central plot on Fig. 8 is registration of "artificial dream". We must only think in terms of special interpretation...

## 6   Special Interpretation of the Intermediate Stages of Learning Process

In all goal oriented investigations when using neural networks researchers are interested in final result of learning process, which must useful and accurate. Almost nobody takes into consideration intermediate stages shown on Fig. 8. But when we try to understand, what can means in fact form of plotting, repeated on Fig. 9—we must find out, that although it is not real **dream**, it can be interpreted as very exciting **model of artificial dream**. In fact on the plotting presented on Fig. 9 we can point out the localizations of the neurons, which can recognize some (named) objects from real world. After learning all neurons will be attributed to the real world objects, like girls, fishes and birds. But when we have very early stage of learning process, we can find in the population of neurons both real-world related detectors and fantasy-world related detectors. On the line connecting points representing for example girls with the point representing fishes we can find neurons, which are ready to recognize objects, which parameters (features) are partially similar to the girls shapes, and partially include features taken from the other real

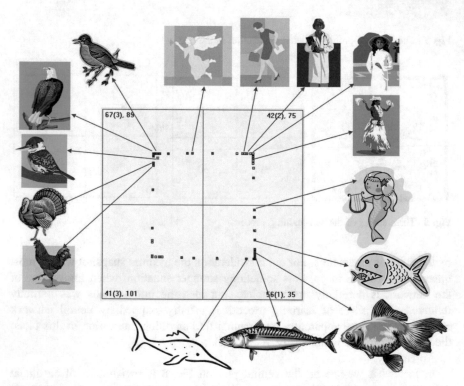

**Fig. 9** Parameters of self-learning neural network shows after encoding, that some of neurons spontaneously produce imaginations of non-existing beings. There are the "artificial dreams"!

objects, for example fishes (e.g. tiles). Another hybrid imagination is creature having features taken from girls and from birds. Perhaps it can be angel?

Isn't it something known in the plots shown on Fig. 9? Obviously in real world object like some of plotted here cannot exist. The objects of such properties cannot also be elements of learning data stream, because input information for the network is every time taken from the real world examples. Nevertheless in neural network structure learning process forms neurons, which want to observe and recognize such not real objects.

Isn't it some kind of "artificial dreams"?

Very interesting is fact, that the fantasy-oriented objects, like presented on Fig. 9, encountered during the learning process, never are unrestricted or simply random. We can find only such neurons, which are able to recognize some hybrids, fantastic, but build from the real elements. Isn't it analogy to the tell-stories or myths?

Limited volume of this presentation not allows us to present many other examples of the "artificial dreams" encountered during the learning processes in neural networks. But one more example can be also interesting, because it shows another kind of fantasy identified in neural network behavior. This form of fantasy can be called "making giants". Example of such behavior of the learned network is presented on the Fig. 10. When the network is learned by means of examples of real world object—in the neural structures the prototypes of these objects are formed and enhanced. This process goes over the big population of neurons and leads to the forming of internal representation (in neural structures) of particular real objects. Neurons belonging to these representation can recognize every real object of the type under consideration. It is very known and regular process.

But sometimes in contrast to this regular pattern we can observe single neurons, which parameters are formed in such way, which leads to the surprise after interpretation. Let us assume, that real objects on the base of which the network was learned during the experiment illustrated on Fig. 10, was lion. The network can "see" many lions (of course as a collections of parameters, representing selected data about lions—e.g. how toll is lion, how long and sharp is lion's tooth and so on). After some learning period inside the network we have some imagination of real lion. This imagination, given as collection of parameters (neurons weights), enable us to recognize every real lion. But some neurons have parameters, which enable to recognize surreal lion, much bigger than real one, with biggest tooth and with much more dangerous claws. The relations and proportions between parameters are the same, as for real lions (see on Fig. 10 relations between parameters of real objects and relations between parameters of the imprinted in weights of refugee neuron imagination of the "giant"—both belonging to the same line, coming from the root of coordination system), but so big lion cannot exist. Nevertheless we can find neuron ready for recognition of this giant, although it not exists!

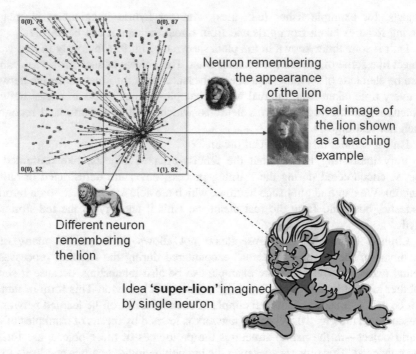

Neuron remembering the appearance of the lion

Real image of the lion shown as a teaching example

Different neuron remembering the lion

Idea **'super-lion'** imagined by single neuron

**Fig. 10** Another form of "artificial dream". Description in the text

## 7 Concluding Remarks

Facts and comments presented in this paper definitely aren't very important from the scientific point of view and also are not applicable to the practical problem solving using neural networks. But as long as we use neural networks as the artificial systems very similar to the structures discovered in human brain—we still thinking about analogies between processes in our psychic and in neurocomputers. Results of simulations presented in this paper gives us new point to such considerations and we hope can be interesting for many neural network researchers bored with new learning paradigms, new network structures and new neurocomputing applications and searching for something absolutely different from the serious and boring standards. This paper is something for him!

## References

1. Zurada, J.M., Marks, R.J., Robinson, C.J. (eds.): Computational Intelligence: Imitating Life. IEEE Press, New York (1994)
2. Zurada, J.M.: Introduction to Artificial Neural Systems. West Publishing Company, St. Paul, Minnesota (1992)

3. Cloete, I., Zurada, J.M. (eds.): Knowledge-Based Neurocomputing. MIT Press, Cambridge, Massachusetts (2000)
4. Sasiada, M., Fraczek-Szczypta, A., Tadeusiewicz, R.: Efficiency testing of artificial neural networks in predicting the properties of carbon nanomaterials as potential systems for nervous tissue stimulation and regeneration. Bio-Algorithms and Med-Systems (2017). doi:10.1515/bams-2016-0025
5. Mazurkiewicz, E., Tomecka-Suchoń, S., Tadeusiewicz, R.: Application of neural network enhanced ground penetrating radar to localization of burial sites. Appl. Artif. Intell. **30**(9), 844–860 (2016). doi:10.1080/08839514.2016.1274250
6. Smyczyńska, J., Hilczer, M., Smyczyńska, U., Stawerska, R., Tadeusiewicz, R., Lewiński, A.: Artificial neural models—a novel tool for predictying the efficacy of growth hormone (GH) therapy in children with short stature. Neuroendocrinol. Lett. **36**(4), 348–353 (2015). ISSN 0172-780X; ISSN-L 0172-780X
7. Tadeusiewicz, R.: Neural networks in mining sciences—general overview and some representative examples. Arch. Min. Sci. **60**(4), 971–984 (2015). doi:10.1515/amsc-2015-0064
8. Tadeusiewicz, R.: Using neural networks for simplified discovery of some psychological phenomena. In: Rutkowski, L. et al. ( eds.) Artificial Intelligence and Soft Computing, LNAI 6114, pp. 104–123. Springer, Berlin, Heidelberg, New York (2010)
9. Tadeusiewicz, R., Izworski, A.: Learning in neural network—unusual effects of "Artificial Dreams". In: King et al. (eds.) Neural Information Processing, Lecture Notes in Computer Science, Part I, vol. 4232, pp. 211–218, Springer, Berlin, Heidelberg, New York (2006)
10. Ekbia, H.: Artificial Dreams: The Quest for Non-Biological Intelligence, Cambridge University Press (2008)

# Part II
# Neural Networks and Connectionist Systems

# On the Interpretation and Characterization of Echo State Networks Dynamics: A Complex Systems Perspective

Filippo Maria Bianchi, Lorenzo Livi and Cesare Alippi

**Abstract** In this chapter, we discuss recently developed methods for characterizing the dynamics of recurrent neural networks. Such methods rely on theory and concepts coming from the field of complex systems. We focus on a class of recurrent networks called echo state networks. First, we present a method to analyze and characterize the evolution of its internal state. This allows to provide a qualitative interpretation of the network dynamics. In addition, it allows to assess the stability of the system, a necessary requirement in many practical applications. Successively, we focus on the identification of the onset of criticality in such networks. We discuss an unsupervised method based on Fisher information, which can be used to tune the network hyperparameters. With respect to standard supervised techniques, we show that the proposed approach offers several advantages and is effective on a number of tasks.

**Keywords** Echo state networks · Criticality · Recurrence quantification analysis · Fisher information matrix · Unsupervised learning

F.M. Bianchi
Machine Learning Group, Department of Physics and Technology,
UiT the Arctic University of Norway, Tromsø, Norway
e-mail: filippo.m.bianchi@uit.no

L. Livi · C. Alippi (✉)
Department of Electronics, Information, and Bioengineering,
Politecnico di Milano, Milan, Italy
e-mail: cesare.alippi@polimi.it

L. Livi
e-mail: lorenz.livi@gmail.com

L. Livi · C. Alippi
Faculty of Informatics, Università della Svizzera Italiana, Lugano, Switzerland

© Springer International Publishing AG 2018
A.E. Gawęda et al. (eds.), *Advances in Data Analysis with Computational Intelligence Methods*, Studies in Computational Intelligence 738,
https://doi.org/10.1007/978-3-319-67946-4_5

# 1   Introduction

Since the very first recurrent neural network (RNN) architectures, several attempts have been made to describe and understand their internal dynamics [64]. Nowadays, such efforts found renewed interest by those researchers trying to "open the black-box" [26, 45, 46, 49]. This is mostly motivated by recent advances in various fields, such as neuroscience [10]. In fact, understanding the inner mechanisms that drive the inductive inference is of utmost importance for deriving novel scientific results [48].

Research on complex dynamical systems is focusing more and more on networks characterized by time-varying properties [2], which can be related to the topology and/or features associated with vertices and edges (e.g., states of networked dynamic systems). Of particular interest are those systems that also perform a computation when driven by an external stimulus. RNNs, initially proposed in the 80s [12, 42, 60], offer an example of those systems. RNNs are universal approximators of Lebesgue measurable dynamical systems [15], with the capability of storing the history of input signals and utilize such information for prediction [8, 23, 40, 50]. While in principle RNNs are characterized by a simple, yet powerful and flexible model, in practice they are hard to train. In fact, in order to learn the internal connection weights, the network designer has to face a series of technical issues [36]. The most important obstacles are due to the vanishing and exploding gradient [3].

In this chapter, we focus on a particular class of RNN, called Echo State Network (ESN). The main peculiarity of ESNs is that the recurrent part, called reservoir, is randomly generated and the connection weights are kept fixed. The only part that is trained is the so-called readout, a memory-less component that combines the neuron activations of the reservoir in order to reproduce a suitable output, according to the specified task at hand. ESNs not only benefit from the presence of feedbacks like any other RNN (the feature which gives to the system the capability to model any complex dynamic behavior) but their sparsely interconnected reservoir of neurons leads to a very fast and simple training procedure. In fact, unlike the complicated and time consuming training process required by standard RNNs, a simple linear readout can be used to solve efficiently a great variety of tasks. On the downside, ESN is characterized by a short-term memory, making it unsuitable for application when long-term correlations must be modeled [37].

Even if ESNs offer an important simplification for what concerns training, they depend on hyperparameters affecting their behavior; additionally, their modus operandi is still not fully understood and it represents an actual object of study [6, 49]. An ESN can generate complex dynamics characterized by sharp transitions between ordered and chaotic regimes. Several experimental results suggest that ESNs achieve the highest information processing capabilities exactly on the edge of this transition, called edge of criticality, resulting in high memory capacity (storage of past events) and good performance on the modeling/prediction task at hand (low prediction errors) [1, 5, 21, 39, 54, 58]. To determine such "critical" network configurations, an ESN requires fine tuning of its controlling hyperparameters. This general behavior is in agreement with the widely-discussed "criticality hypothesis" observed in many biological (complex) systems [14, 16, 41, 43, 51], including the brain [9, 32, 35, 52, 53]. In fact, it was noted [34] that such complex systems tend to self-organize and operate in a critical regime. Investigating weather a given complex system operates more efficiently in the critical regime or not, requires theoretically sound methods for detecting the onset of criticality [44].

Best-performing network configurations are typically identified through supervised methods, such as cross-validation and alike. In this chapter, we present recent research results [6, 22] that focus on unsupervised approaches to characterize ESN dynamics and to identify the edge of criticality. These approaches do not require a validation set, an important limitation in several applications, with scarce amount of data. Another issue of validation procedures is the need to repeat training for each hyperparameter configuration taken into account. Through the proposed unsupervised approaches, hyperparameters are tuned in advance and training is performed just once, at the end. Finally, cross validation considers only the performance obtained on the given task, treating the network as black box. Instead, the presented methods offer insights on the functioning, by modeling dynamics with more easily interpretable tools.

Different unsupervised approaches to identify configurations that maximize ESN computation capability have been proposed in the literature. These, are quickly reviewed in Sect. 2 after an overview on the ESN architecture. In Sect. 3, we address the issue of interpretability of ESN dynamics by relying on recurrence plots and recurrence quantification analysis [6] to characterize the evolution of the internal states. When the network is driven by a specific input signal, these instruments can be used to monitor its degree of stability, for a given configuration of its hyperparameters. In Sect. 4, we define an unsupervised methodology for tuning ESN hyperparameters by means of sensitivity analyses [22]. In particular, we present a theoretical framework based on Fisher information matrix [55, 62] and its related connection with criticality. Conclusions and future research directions are provided in Sect. 5.

## 2 Echo State Networks

A schematic representation of an ESN is shown in Fig. 1. An ESN consists of a reservoir of $N_r$ nodes characterized by a non-linear transfer function $f(\cdot)$. At time $t$, the network is driven by the input $\mathbf{x}[t] \in \mathbb{R}^{N_i}$ and produces the output $\mathbf{y}[t] \in \mathbb{R}^{N_o}$, being $N_i$ and $N_o$ the dimensionalities of input and output, respectively. The weight matrices $\mathbf{W}_r^r \in \mathbb{R}^{N_r \times N_r}$ (reservoir internal connections), $\mathbf{W}_i^r \in \mathbb{R}^{N_i \times N_r}$ (input-to-reservoir connections), and $\mathbf{W}_o^r \in \mathbb{R}^{N_o \times N_r}$ (output-to-reservoir feedback connections) contain values in the $[-1, 1]$ interval drawn from a uniform distribution.

ESN is a discrete-time nonlinear system with feedback, whose model reads:

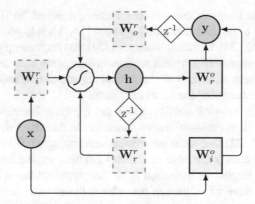

**Fig. 1** Schematic depiction of the ESN architecture. The circles represent input **x**, state, **h**, and output, **y**, respectively. Solid squares $\mathbf{W}_r^o$ and $\mathbf{W}_i^o$, are the trainable matrices, respectively, of the readout, while dashed squares, $\mathbf{W}_r^r$, $\mathbf{W}_o^r$, and $\mathbf{W}_i^r$, are randomly initialized matrices. The polygon represents the non-linear transformation performed by neurons and $z^{-1}$ is the unit delay operator

$$\mathbf{h}[t] = f\left(\mathbf{W}_r^r\mathbf{h}[t-1] + \mathbf{W}_i^r\mathbf{x}[k] + \mathbf{W}_o^r\mathbf{y}[t-1]\right);\tag{1}$$

$$\mathbf{y}[t] = g\left(\mathbf{W}_r^o\mathbf{h}[t] + \mathbf{W}_i^o\mathbf{x}[k]\right).\tag{2}$$

Activation functions $f(\cdot)$ and $g(\cdot)$, both applied component-wise, are typically implemented as a sigmoidal (*tanh*) and identity function, respectively. The output weight matrices $\mathbf{W}_r^o \in \mathbb{R}^{N_r \times N_o}$ and $\mathbf{W}_i^o \in \mathbb{R}^{N_i \times N_o}$, which connect, respectively, reservoir and input to the output, represent the readout of the network. The standard training procedure for such matrices requires solving a straightforward regularized least-square problem [18].

Even though the three matrices $\mathbf{W}_r^r$, $\mathbf{W}_o^r$, and $\mathbf{W}_i^r$ are generated randomly, they can be modified in order to obtain desired properties. For instance, $\mathbf{W}_r^o$ is controlled by a multiplicative constant, which in this work is set to 0 to remove the output feedback connection. $\mathbf{W}_i^r$ is controlled by scalar parameter $\theta_{IS}$, which determines the amount of non-linearity introduced by the sigmoid processing units that is largest around the origin. In particular, inputs far from zero tend to drive the activation of the neurons towards saturation where they show more non-linearity. Finally, the parameter $\theta_{RC}$ defines the percentage of non-zero connections in $\mathbf{W}_r^r$, while its spectral radius $\theta_{SR}$ controls important properties, as discussed in the sequel.

## 2.1 ESN Dynamics and Stability Measures

An ESN is typically designed so that the influence of past inputs gradually fades away and the initial state of the reservoir is eventually washed out. This is formalized by the Echo State Property (ESP), which ensures that, given any input sequence taken from a compact set, trajectories of any two different initial states become eventually indistinguishable. ESP was originally investigated in [18] and successively in [61]; we refer the interested reader to [25] for a more recent definition, where also the influence of input is explicitly accounted for. In ESNs with no output feedback, as in our case, the state update of Eq. (1) reduces to:

$$\mathbf{h}[t] = f(\mathbf{W}_r^r\mathbf{h}[t-1] + \mathbf{W}_i^r\mathbf{x}[k]). \tag{3}$$

In order to study the stability of the network, we compute the maximal local Lyapunov exponent ($\lambda$) from the Jacobian of the state update (3) of the reservoir. This quantity is used to approximate (for an autonomous system) the separation rate in phase space of trajectories having very similar initial conditions. $\lambda$ is derived from the Jacobian at time $t$, which can be conveniently expressed if neurons are implemented with a *tanh* activation function as

$$\mathbf{J}(\mathbf{h}[t]) = \mathbb{1}_{N_r} \cdot \left[ 1 - (h_1[t])^2, 1 - (h_2[t])^2, \dots, 1 - (h_{N_r}[t])^2 \right]^T. \tag{4}$$

where $h_l[t]$ is the activation of the $l$-th neuron, with $l = 1, 2, \dots, N_r$. $\lambda$ is then computed as

$$\lambda = \max_{n=1,\dots,N_r} \frac{1}{t_{\max}} \sum_{t=1}^{t_{\max}} \log\left(r_n[t]\right), \tag{5}$$

being $r_n[t]$ the module of $n$-th eigenvalue of $\mathbf{J}(h[t])$ and $t_{\max}$ the total number of time-steps in the considered trajectory.

Local, first-order approximations provided by Eq. 4 are useful to study the stability of a (simplified) reservoir operating around the zero state, $\mathbf{0}$. In fact, implementing $f(\cdot)$ as a *tanh* assures $f(\mathbf{0}) = \mathbf{0}$, i.e., $\mathbf{0}$ is a fixed point of the ESN dynamics. Therefore, by linearizing (3) around $\mathbf{0}$ and assuming a zero-input, we obtain from (4)

$$\mathbf{h}[t] = \mathbf{J}(\mathbf{0})\mathbf{h}[t-1] = \mathbf{W}_r^r\mathbf{h}[t-1]. \tag{6}$$

Linear stability analysis of (6) suggests that, if $\theta_{SR} < 1$, the dynamic around $\mathbf{0}$ is stable. In the more general case, the non-linearity of the sigmoid functions in (3) forces the norm of the state vector of the reservoir to remain bounded. Therefore, the condition $\theta_{SR} < 1$ looses its significance and does not guarantee stability when the system deviates from a small region around $\mathbf{0}$ [57]. Notably, it is possible to find reservoirs (3) having $\theta_{SR} > 1$, which still possess the ESP. In fact, the effective local gain decreases when the operating point of the neurons shifts toward the positive/negative branch of the sigmoid, where stabilizing saturation effects start to influence the excitability of reservoir dynamics [61]. In the more realistic and useful scenario where the input driving the network is a generic (non-zero) signal, a sufficient condition for the ESP is met if $\mathbf{W}_r^r$ is diagonally Schur-stable, i.e., if there exists a positive definite diagonal matrix, $\mathbf{P}$, such that $(\mathbf{W}_r^r)^T\mathbf{P}\mathbf{W}_r^r - \mathbf{P}$ is negative definite [61]. However, this recipe is fairly restrictive in practice as this condition might generate reservoirs that are not rich enough in terms of provided dynamics, since the use of a conservative scaling factor might compromise the amount of memory in the network and thus the ability to accurately model a given problem. Therefore, for most practical purposes, the necessary condition $\theta_{SR} < 1$ is considered "sufficient in practice", since the state update map is contractive with high probability, regardless of the input and given a sufficiently large reservoir [63].

## 2.2 Edge of Criticality

The number of reservoir neurons and the bounds on $\theta_{SR}$ can be used for a naïve quantification of the computational capability of a reservoir [61]. However, those are static measures that only consider the algebraic properties of $\mathbf{W}_r^r$, without taking into account other factors, such as the input scaling $\theta_{IS}$ and the particular properties of the given input signals. Moreover, it is still not clear how, in a mathematical sense, these stability bounds relate to the actual ESN dynamics when processing non-trivial input signals [25]. In this context, the idea of pushing the system toward the edge of criticality has been explored. In [5, 20, 21] it is shown that several dynamical systems, among which randomly connected RNNs, achieved the highest computational capabilities when moving toward the unstable (sometime even chaotic) regime, where the ESP is lost and the system enters into an oscillatory behavior. This justifies the use of spectral radii above the unity in some practical applications.

The stable–unstable transition can be detected numerically by considering the sign of $\lambda$ (5). In fact, in autonomous systems, $\lambda > 0$ indicates that the dynamics is chaotic. Relative to ESNs, $\lambda$ was proposed to characterize reservoir dynamics and it demonstrated its efficacy in designing a suitable network configuration in several applications [56, 57]. Further descriptors used for characterizing the dynamics of a reservoir are based on information-theoretic quantities, such as (average) transfer entropy and active information storage [7]. The authors have shown that such quantities peak right when $\lambda > 0$. In addition, the minimal singular value of the Jacobian (4), denoted as $\eta$, was demonstrated to be an accurate predictor of ESN performance, providing more accurate information regarding the ESN dynamics than both $\lambda$ and $\theta_{SR}$ [56]. Hyperparameters that maximize $\eta$ generate a dynamical system that is far from singularity, it has many degrees of freedom, a good excitability, and it separates well the input signals in phase space [56].

## 3  Interpreting and Tuning ESN Through Recurrence Quantification Analysis

Poincaré recurrence provides fundamental information for the analysis of dynamical systems [29]. This follows from Poincaré's theorem, which guarantees that the states of a dynamic system must recur during its evolution. Recurrences contain all relevant information regarding a system behavior in phase space and can be linked also with dynamical invariants (e.g., metric entropy) and features related to stability. However, especially for high-dimensional complex systems, the recurrence time elapsed between recurring states is difficult to calculate, even when assuming full analytical knowledge of the system.

Recurrence Plots (RPs) [11, 27, 29, 30], together with the computation of dynamical invariants and heuristic complexity measures called Recurrence Quantification Analysis (RQA), offer a simple yet effective tool to analyze such recurrences starting from a time-series derived from the system under analysis. RP provides a visual representation of recurrence time and its line patterns contain information about the duration of the recurrence [28]. RPs are constructed by considering a suitable distance in the phase space and a threshold $\tau_{RP}$ is used to determine the recurrence/similarity of states during the evolution of the system.

In the following, we address the interpretability issue of ESNs by analyzing the dynamics of the reservoir neuron activations with RPs and RQA complexity measures. Techniques based on RPs and RQA allow the designer to visualize and characterize (high-dimensional) dynamical systems starting from a matrix encoding the recurrences of the system states over time.

## 3.1 Representing ESN Dynamics with RP

The sequence of ESN states can be seen as a multivariate time-series $\mathbf{h}$, relative to the $N_r$ neuron activations. An RP is constructed by calculating a $t_{max} \times t_{max}$ binary matrix $\mathbf{R}$. The generic element $R_{ij}$ is defined as

$$R_{ij} = \Theta(\tau_{RP} - d(\mathbf{h}[i], \mathbf{h}[j])), \quad 1 \leq i, j \leq t_{max}, \tag{7}$$

where $d(\cdot, \cdot)$ is a dissimilarity measure operating in phase space (e.g., Euclidean, Manhattan, or max-norm distance), $\Theta(\cdot)$ is the Heaviside function and $\tau_{RP} > 0$ is a user-defined threshold used to identify recurrences. $\tau_{RP}$ can be defined in different ways, but typically chosen to be proportional to a percentage of the average or the maximum phase space distance between the states. Figure 2 depicts the algorithmic steps required to generate an RP on ESN states.

Depending on the properties of the analyzed time-series, different line patterns emerge in a RP [28]. Besides providing an immediate visualization of the system properties, from $\mathbf{R}$ it is possible to derive several complexity measures, those associated with an RQA. Such measures are defined by the distribution of both vertical/horizontal and diagonal line structures present in the RP and provide a numerical characterization of the underlying dynamics. Several RQA measures are based on the histograms $P(l)$ and $P(v)$, counting, respectively, the diagonal and vertical lines having lengths $l$ and $v$,

$$P(l) = \sum_{i,j=1}^{t_{\max}-l} (1 - R_{i-1,j-1})(1 - R_{i+l,j+l}) \prod_{k=0}^{l-1} R_{i+k,j+k};$$

$$P(v) = \sum_{i,j=1}^{t_{\max}-v} (1 - R_{i,j})(1 - R_{i,j+v}) \prod_{k=0}^{v-1} R_{i,j+k}.$$

The RQA measures considered here are summarized in Table 1; abbreviations and notation are kept consistent with [29].

## 3.2 Visualize and Classify Reservoir Dynamics

In the following, we show how RPs permit to visualize, and hence classify, reservoir dynamics when ESN is fed with inputs possessing well-known characteristics. We consider a stable ESN described by (3); RPs are constructed following the procedure depicted in Fig. 2. Although many classes of signals/systems exist (with related sub-classes) [29], here we focus on the ability to discriminate between important classes for the input signals: (i) with/without time-dependence, (ii) periodic/non-periodic

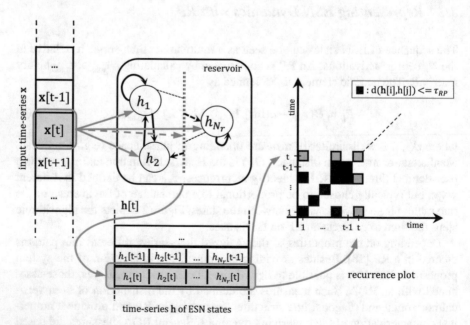

**Fig. 2** When $\mathbf{x}[t]$ is fed as input to the $N_r$ neurons of the ESN reservoir, the internal state is updated to $\mathbf{h}[t] = [h_1[t], h_2[t], \dots, h_{N_r}[t]]^T$, where $h_n[t]$ is the output of the $n$-th neuron. Once the time-series $\mathbf{h}$ is generated, the RP is constructed by using a threshold $\tau_{RP}$ and a dissimilarity measure $d(\cdot, \cdot)$. If $d(\mathbf{h}[t], \mathbf{h}[i]) \leq \tau_{RP}$, the cell of the RP in position $(t, i)$ is colored in black, otherwise it is left white. The elements in gray highlight the operations performed at time-step $t$. Taken from [6]

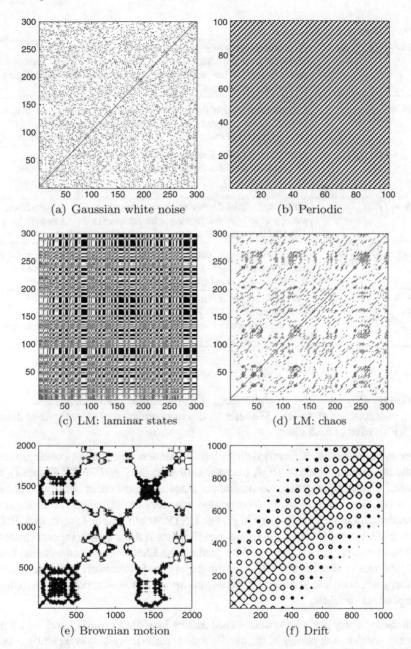

**Fig. 3** RPs generated by state sequences **h** of ESNs fed with input signals taken into account. Both axes represent time. Taken from [6]

**Table 1** Definition of RQA measures

| | |
|---|---|
| $RR = \frac{1}{t_{max}^2} \sum_{i,j=1}^{t_{max}} R_{ij}$ | Recurrence rate, a measure of density of recurrences in **R**. It corresponds to the correlation sum, an important concept used in chaos theory. RR can help to select $\tau_{RP}$ when performing multiple tests on different conditions, e.g., by preserving the rate |
| $DET = \frac{\sum_{l=l_{min}}^{l_{max}} lP(l)}{\sum_{l=1}^{l_{max}} lP(l)}$ | Determinism level of the system, based on the percentage of diagonal lines of minimum length $l_{min}$. A periodic system would have DET close to unity and close to zero for a signal with no time-dependency |
| $L_{max} = \max\{l_i\}_{i=1}^{N_l}$ | Maximum diagonal line length, with $1 \leq L_{max} \leq \sqrt{2}t_{max}$. $l_i$ is $i$th diagonal line length and $N_l$ is the total number of diagonal lines, defined as $N_l = \sum_{l \geq l_{min}} P(l)$ |
| $DIV = 1/L_{max}$ | Mean exponential divergence in phase space, related to correlation entropy of the system. Notably, chaotic systems do not present long diagonal lines, as trajectories diverge exponentially fast |
| $LAM = \frac{\sum_{v=v_{min}}^{l_{max}} vP(v)}{\sum_{v=1}^{l_{max}} vP(v)}$ | Presence of laminar phases, which denote states of the system that do not change or change very slowly for a number of consecutive time-steps. $v_{min}$ is the minimal vertical line length considered |
| $ENTR = -\sum_{l=1}^{t_{max}} p(l)\ln(p(l))$ | Diagonal lines distribution, with $p(l) = P(l)/N_l$. In absence of time-dependence, $ENTR \simeq 0$, i.e., the diagonal lines distribution is fully concentrated on very short lines. Conversely, ENTR increases when the diagonal lines distribution become heterogeneous |

motions, (iii) laminar behaviours, (iv) chaotic dynamics, and finally (v) non-stationary processes. We refer to the examples depicted in Fig. 3 to discuss the RP relative to each class.

*Time-dependency*: a uniformly distributed RP denotes absence of time-dependence in the time-series. Specific RQA measures, such as DET and ENTR (Table 1), can be used to numerically investigate the presence of time-dependency, as their values is very low if the signal is uncorrelated. For periodic signal with a strong time-dependency, DET would be very high, but ENTR would still be low. In fact, ENTR measures the complexity of the signal, which is low if there is no temporal structure. Figure 3a depicts the RP generated by feeding the ESN with Gaussian white noise, a typical example of signal with no time-dependency. Reservoir states generates a uniform RP, which is peculiar of signals composed by realizations of statistically independent variables.

*Periodicity*: every periodic system would induce long diagonal lines and the vertical spacing provides the period of the oscillation. A periodic system is typically accompanied by high values for DET and $L_{max}$, while its low complexity is expressed by ENTR. In Fig. 3b, we show an example of periodic motion generated by reservoir neurons, when ESN is fed with a sinusoid having a single dominating frequency. The regularity of the diagonal lines can be immediately recognized from the figure.

*Laminarity*: a system presents laminar phases if its state does not change or change very slowly over a number of successive time-steps. Laminar phases can be visually recognized in an RP by the presence of black rectangles. Every system possessing laminar phases is characterized by high values for LAM. To provide an example, we consider the logistic map (LM), defined by the differential equation $\mathbf{x}[t+1] = \tau_{LM}\mathbf{x}[t](1 - \mathbf{x}[t])$, where usually $\tau_{LM} \in (0, 4]$; here we set the initial condition $\mathbf{x}[1] = 0.5$. Figure 3c depicts RP obtained for $\tau_{LM} = 3.679$, where the system exhibits chaos-chaos transitions. In fact, such a RP is compatible with the one of a (mildly) chaotic system, showing the presence of laminar phases (large black rectangles).

*Chaoticity*: RPs offer a particularly useful visual tool in the case of chaotic dynamics, which are characterized by the presence of erratic and very short diagonal lines. As a consequence, RR would be very low. ENTR is also useful to determine the degree of chaoticity: the higher its value, the more chaotic/complex the system. Chaos is characterized by trajectories diverging exponentially fast. This can be quantified with $L_{max}$ and DIV, whose values would be respectively very low and close to one for systems with a high degree of chaoticity. As an example, we consider a chaotic system obtained through LM set with $\tau_{LM} = 4$. The reservoir dynamics, as shown in the RP in Fig. 3d, denotes fully developed chaos, as indicated by the presence of short and erratic diagonal lines.

*Non-stationarity*: Peculiar line patterns observed for all nonstationary signals include large white areas with irregular patterns denoting abrupt changes in the dynamics. Drift is a typical form of nonstationarity, which is visually recognized in an RP by the fading of recurrences in the upper-left and lower-right corners. In Fig. 3e, we show an example by feeding the ESN with a well-known nonstationary signal: Brownian motion, a random walk resulting in a nonstationary stochastic process; whose increments correspond to Gaussian white noise, a stationary process. In Fig. 3f we show an example of drift, obtained by adding a linear trend to a sinusoid. Nonstationarity can be numerically detected by considering an RQA measure called TREND (not used in our study) and by analyzing the variation of RQA measures when time-delay is applied to the signal (see [29] for technical details).

## 3.3 Recurrence Analysis to Determine ESN Stability

In this section, we show how recurrence analysis can be used to assess stability for a given configuration. We perform two experiments: in the first one, we use RPs to visualize reservoir dynamics when driven by a given input signal. When the reservoir operates in a stable regime, RPs of reservoir and input show similar line patterns. In a second experiment, We show that $L_{max}$ is anticorrelated with $\lambda$ and hence it can be considered as a reliable indicator for the (input-dependent) degree of network stability.

To test our methodology, we consider two time-series generated respectively by an oscillatory and by the Mackey-Glass (MG) dynamical system [47]. We chose these two signals since both of them are often considered as benchmarks for prediction in the ESN literature [18, 57] and they exemplify a very regular and a mildly chaotic system, respectively. In both experiments, we consider an ESN with no output feedback, configured with a standard setting: uniformly distributed weights in $[-1, 1]$ for $W_i^r$ and $W_r^r$, percentage of non-zero reservoir connections $\theta_{RC} = 25\%$. The readout is trained by setting the regularization parameter in the ridge regression to 0.1. According to the standard drop-out procedure, we discarded the first 100 elements of $h$ in order to get rid of the ESN transient states. The number of reservoir neurons is set to $N_r = 75$. We used the Manhattan distance for evaluating the dissimilarity in the phase space. The threshold $\tau_{RP}$ has been calculated by using a percentage of the average dissimilarity value between the states in $h$. Our results are easily reproducible by using the ESN[1] and RP[2] toolboxes available online.

The first experiment consists in generating the RP relative to the input sequence $\{x[t]\}_{t=1}^{t_{max}}$ (sinusoid or MG time-series) and the ones relative to neuron activations $\{h[t]\}_{t=1}^{t_{max}}$, when the reservoir is configured with a spectral radius $\theta_{SR}$ that determines a ordered or a chaotic dynamics.

In Fig. 4, we report the RPs relative to the input signal and the reservoir states, generated for two different values of $\theta_{SR}$. The left column is relative to the ESN fed with a sinusoid and the right column to the ESN fed with the MG time-series. As we can see, when $\theta_{SR} = 0.9$ the ESN is stable and the dynamics of the input, represented by the RPs in Fig. 4a and b produce very similar line patterns in the RPs of the reservoirs, reported in Fig. 4c and d. Instead, when the spectral radius is pushed far beyond unity, the ESN dynamics become unstable and the similarity in the reservoir RPs is lost, as we can see from Fig. 4e and f.

In the second experiment, we evaluate the effectiveness of $L_{max}$ and DIV in determining the degree of stability in the ESN. Specifically, the higher the value of $L_{max}$, the more stable the system. The opposite holds for DIV, which is computed as the reciprocal of $L_{max}$ (see Table 1). Our evaluation consists in comparing, $\lambda$, a global indicator of stability (see Eq. 5), with $L_{max}$, the value of the longest diagonal line in an RP, and with DIV. As before, we consider two ESN fed with the sinusoid and the MG time-series. The correlations of these measures are reported in Table 2.

To visually assess the agreement of $\lambda$ with $L_{max}$ and DIV, in Fig. 5 we show a 2D depiction obtained by selecting a specific input scaling $\theta_{IS} = 0.8$ and by varying $\theta_{SR}$ in the interval $[0.1, 2]$. For the sinusoidal input, $\lambda$ and $L_{max}$ are anticorrelated with (Pearson) correlation equal to $-0.74$: the value of $L_{max}$ decreases as $\theta_{SR}$ increases, while $\lambda$, as expected, increases with $\theta_{SR}$. Additionally, it is possible to observe that there exists a positive correlation (0.53) between $\lambda$ and DIV. Also for the MG time-series, $\lambda$ and $L_{max}$ show a good anticorrelation, with a value of $-0.65$. Analogously, $\lambda$ and DIV are correlated with a slightly lower value of 0.57.

---

[1]http://www.reservoir-computing.org/node/129.
[2]http://www.recurrence-plot.tk/.

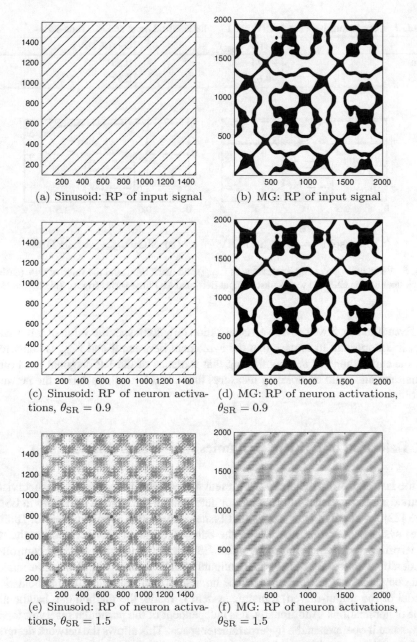

(a) Sinusoid: RP of input signal    (b) MG: RP of input signal

(c) Sinusoid: RP of neuron activa-    (d) MG: RP of neuron activations,
tions, $\theta_{SR} = 0.9$    $\theta_{SR} = 0.9$

(e) Sinusoid: RP of neuron activa-    (f) MG: RP of neuron activations,
tions, $\theta_{SR} = 1.5$    $\theta_{SR} = 1.5$

**Fig. 4** RPs of input signal and sequence of states of the reservoir. When $\theta_{SR} = 0.9$, the ESN is stable and the activations are compatible with the input dynamics. When $\theta_{SR}$ exceeds one, the activations denote instability. Taken from [6]

**Table 2** Correlations between $\lambda$, DIV, and $L_{max}$ for sinusoid input and MG time-series

|      | corr($\lambda, L_{max}$) | corr($\lambda, DIV$) |
|------|-------------------------|----------------------|
| Sin  | -0.74                   | 0.53                 |
| MG   | -0.65                   | 0.57                 |

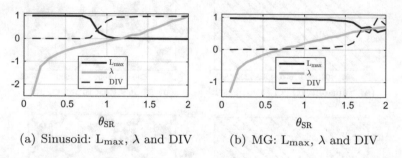

(a) Sinusoid: $L_{max}$, $\lambda$ and DIV         (b) MG: $L_{max}$, $\lambda$ and DIV

**Fig. 5** Value of $\lambda$ (gray solid line), value of $L_{max}$ (solid black line), and the value of DIV (dashed black line) for the ESN fed with sinusoid input (left) and MG time-series (right). Taken from [6]

Even if in Fig. 5 we provide a visualization only for a specific value of input scaling, it is important to remark that the agreement between $\lambda$ and $L_{max}$ is consistent for the entire range of $\theta_{IS}$, confirming that statistics of the RP diagonal lines offer consistent and solid complexity measures that are able to characterize the network stability.

## 4 Detection of Critical Dynamics with Fisher Information

In the last part of this chapter, we present a theoretically motivated, unsupervised method based on Fisher information for determining the edge of criticality in ESNs (see [22] for details). It is proven that Fisher information is maximized for (finite-size) systems operating near or on the edge of criticality [38]. Accordingly, the hyperparameters, which indirectly affect ESN performance, are suitably controlled to identify a collection of network configurations that maximize Fisher information and computational performance. Since no assumption regarding the mathematical model of the (input-driven) dynamic system is made, the method can handle any type of applications. Additionally, it is independent of the particular reservoir topology, since it operates in the hyperparameter space. This allows the network designer to instantiate a specific architecture based on problem-dependent design choices. However, Fisher information is notoriously difficult to compute and either requires the probability density function or the conditional dependence of the system states with respect to the model parameters. In the proposed framework, we take advantage of a recently-developed non-parametric estimator of the Fisher information matrix [4].

## 4.1 Fisher Information Matrix and the Non-parametric Estimator

Fisher information matrix (FIM) [62] is a symmetric positive semi-definite (PD) matrix, whose elements are defined as follows:

$$F_{ij}(p_\theta(\cdot)) = \int_D p_\theta(\mathbf{u}) \left( \frac{\partial \ln p_\theta(\mathbf{u})}{\partial \theta_i} \right) \left( \frac{\partial \ln p_\theta(\mathbf{u})}{\partial \theta_j} \right) d\mathbf{u}, \tag{8}$$

where $p_\theta(\cdot)$ is a parametric probability density function (PDF), which depends on $d$ parameters $\theta = [\theta_1, \theta_2, \dots, \theta_d]^T \in \Theta \subset \mathbb{R}^d$; $\Theta$ is the parameter space. In the ESN framework, $\theta$ contains the hyperparameters under consideration. In (8), $\ln p_\theta(\cdot)$ is the log-likelihood function and $D \subseteq \mathbb{R}^D$ denotes the domain of the PDF. To simplify notation, we denote $\mathbf{F}(p_\theta(\cdot))$ as $\mathbf{F}(\theta)$. The FIM contains $d(d + 1)/2$ distinct entries encoding the sensitivity of the PDF with respect to the parameters $\theta$.

Fisher information is tightly linked with statistical mechanics and, in particular, with the field of (continuous) phase transitions. In fact, it is possible to provide a thermodynamic interpretation of Fisher information in terms of rate of change of the order parameter [38], quantities used to discriminate the different phases of a system. This fact provides an important link between the concept of criticality and statistical modeling of complex systems. It emerges that the critical phase of a thermodynamic system can be mathematically described as that region of the phase space where the order parameters vanish and their derivatives diverge. This implies that, on the critical region, FIM diverges as well, hence providing a quantitative, well-justified tool for detecting the onset of criticality in both theoretical models and computational simulations [59]. In the ESN framework considered here, we identify the edge of criticality as the region in parameter space where the Fisher information is maximized. Figure 6 provides an intuitive illustration, linking criticality and ESNs.

Calculation of the FIM (8) requires full analytical knowledge of the PDF. However, in many experimental settings either the PDF underlying the observed data is unknown or the relation linking the variation of the control parameters $\theta$ and the resulting $p_\theta(\cdot)$ depends on an unknown function. Recently, a non-parametric estimator of the FIM based on divergence measure

$$D_\alpha(p, q) = \frac{1}{4\alpha(1 - \alpha)} \int_D \frac{(\alpha p(\mathbf{u})(1 - \alpha)q(\mathbf{u}))^2}{\alpha p(\mathbf{u})(1 - \alpha)q(\mathbf{u})} d\mathbf{u} - (2\alpha - 1)^2, \tag{9}$$

was proposed [4], with $\alpha \in (0, 1)$; $p(\cdot)$ and $q(\cdot)$ are PDFs both supported on $D$. $D_\alpha$ belongs to the family of $f$-divergences and it can be computed directly by means of an extension of the Friedman-Rafsky multi-variate two-sample test statistic [13].

FIM can be approximated by using a proper $f$-divergence measure computed between the parametric PDF of interest and a perturbed version of it [17]. Notably, by expanding Eq. 9 up to the second order we obtain:

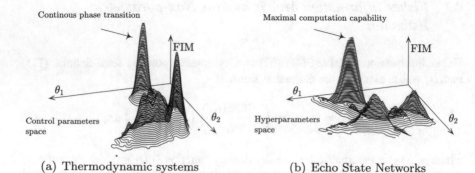

(a) Thermodynamic systems        (b) Echo State Networks

**Fig. 6** The approach based on FIM maximization used to identify a continuous phase transition can be adopted also to characterize dynamics in ESNs. In this context, ESN hyperparameters (e.g., spectral radius, input scaling) play the same role of the control parameters in a thermodynamic system. FIM can be used to identify the critical region in the ESN hyperparameter space, where the computational capability is maximized. Taken from [22]

$$D_\alpha(p_\theta, p_{\hat\theta}) \simeq \frac{1}{2}\mathbf{r}^T\mathbf{F}(\theta)\mathbf{r}, \tag{10}$$

where $\hat\theta = \theta + \mathbf{r}$, being $\mathbf{r} \sim \mathcal{N}(\mathbf{0}, \sigma^2\mathbf{I}_{d\times d})$ a small normally distributed perturbation vector with standard deviation $\sigma$.

In the following, we omit $\theta$ and we refer to the estimated FIM as $\hat{\mathbf{F}}$. According to [4], FIM can be estimated through least-square optimization:

$$\hat{\mathbf{F}}_{\mathrm{hvec}} = (\mathbf{R}^T\mathbf{R})^{-1}\mathbf{R}^T\mathbf{v}_\theta, \tag{11}$$

where $\mathbf{v}_\theta = [v_\theta(\mathbf{r}_1), \dots, v_\theta(\mathbf{r}_M)]^T$, with $v_\theta(\mathbf{r}_i) = 2D_\alpha(p_\theta, p_{\hat\theta_i})$, $i = 1, \dots, M$, and $D_\alpha(\cdot, \cdot)$ is computed by means of the Friedman-Rafsky test. $\mathbf{R}$ is a matrix containing all $M$ perturbation vectors $\mathbf{r}_i$ arranged as column vectors, and $\hat{\mathbf{F}}_{\mathrm{hvec}}$ is the half-vector representation of $\hat{\mathbf{F}}$. Note that a vector representation $\hat{\mathbf{F}}_{\mathrm{vec}}$ of $\hat{\mathbf{F}}$ reads as $[f_{11}, \dots, f_{m1}, f_{12}, \dots, f_{mn}]^T$. Since $\hat{\mathbf{F}}$ is symmetric, it can be represented through the half-vector representation, $\hat{\mathbf{F}}_{\mathrm{hvec}}$, which is obtained by eliminating all superdiagonal elements of $\hat{\mathbf{F}}$ from $\hat{\mathbf{F}}_{\mathrm{vec}}$ [24]. $\hat{\mathbf{F}}_{\mathrm{hvec}}$ in Eq. 11 is hence defined as $[\hat{f}_{11}, \dots, \hat{f}_{dd}, \hat{f}_{12}, \dots, \hat{f}_{d(d-1)}]^T$, where the diagonal elements are located in the first components of the vector.

## 4.2 Tuning ESN by Exploiting FIM Properties

In the following, we define the procedure to identify the edge of criticality, here defined as parameter configurations $\mathcal{K} \subset \Theta$ that maximize the ESN computational

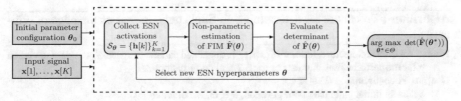

**Fig. 7**   Schematic, high-level description of the proposed procedure. Taken from [22]

capability. Figure 7 shows a schematic description of the main phases involved in the proposed method.

In order to determine $\mathcal{K}$, we introduce an algorithm that take advantage of the FIM properties on a system undergoing a continuous phase transition. FIM defines a metric tensor for the smooth manifold of parametric PDFs embedded in $\Theta$ [38], providing thus a geometric characterization of the system under analysis. It is possible to prove [33] that $\mathcal{K}$ corresponds to a region in $\Theta$ characterized by the largest volume (high concentration of parametric PDFs). This geometric result is reflected in the determinant $\det(\mathbf{F}(\theta))$, which is monotonically related to the aforementioned volume in parameter space. Therefore, considering that the FIM is a PD matrix, and hence its determinant is always non-negative, we identify $\mathcal{K}$ with all those hyperparameters $\theta^*$ for which:

$$\theta^* = \arg \max_{\theta \in \Theta} \det(\mathbf{F}(\theta)). \tag{12}$$

Algorithm 1 delivers the pseudo-code of the proposed procedure. The impact provided by the variation of the control parameters $\theta$ on the resulting ESN state cannot be described analytically without making further assumptions [31]: the (unknown) input signal driving the network plays an important role in the resulting ESN dynamics. Therefore, in order to calculate $\mathbf{F}(\theta)$, in Algorithm 1 we rely on the nonparametric FIM estimator described in Sect. 4.1. The estimation of the FIM for a given $\theta$ is performed by analyzing the sequence $S_\theta = \{\mathbf{h}[t]\}_{t=1}^{t_{\max}}$ of reservoir neuron activations. Since $\mathbf{h}[t] \in [-1, 1]^{N_r}$, the domain of the PDF in (8) is defined as $D = [-1, 1]^{N_r}$. Additional sequences of activations, $S_{\hat{\theta}_j}$, are considered (see line 7), which are obtained by perturbing $M$ times the current network configuration $\theta$ under analysis, and processing the same input $\mathbf{x}$. Perturbations are introduced by means of a small zero-mean noise with spherical covariance matrix, thus characterized by a single scalar parameter $\sigma$ controlling the magnitude of the perturbation. FIM is estimated according to Eq. 11. In order to make the estimation more robust, we follow an ensemble approach and perform a number of trials (see line 3). The determinant is computed only once on the resulting average FIM, which is obtained by using $T$ different (and independent) random realizations of the ESN architecture (see line 16).

**Algorithm 1** Procedure for determining an ESN configuration on the edge of criticality.

---

**Input:** An ESN architecture, input $\mathbf{x} = \{x[i]\}_{i=1}^{t_{max}}$, quantized parameter space $\Theta$, standard deviation $\sigma$ for the perturbations, number of trials $T$ and perturbations $M$.

**Output:** A configuration $\theta^* \in \mathcal{K}$

1: Select an initial parameter configuration, $\theta \in \Theta$; maximum $\eta = 0$
2: **loop**
3:    **for** $t = 1$ to $T$ **do**
4:       Randomly initialize the ESN weight matrices
5:       Configure ESN with $\theta$ and process input $\mathbf{x}$
6:       Collect the related activations $S_\theta = \{\mathbf{h}[i]\}_{i=1}^{t_{max}}$
7:       **for** $j = 1$ to $M$ **do**
8:          Generate a perturbation vector $\mathbf{r}_j \sim \mathcal{N}(\mathbf{0}, \sigma^2 \mathbf{I}_{d \times d})$
9:          Randomly initialize the ESN weight matrices
10:         Configure ESN with perturbed version $\hat{\theta}_j = \theta + \mathbf{r}_j$ and process input $\mathbf{x}$
11:         Collect the related activations $S_{\hat{\theta}_j} = \{\mathbf{h}[i]\}_{i=1}^{t_{max}}$
12:       **end for**
13:       Define $S_{\hat{\theta}} = \cup_{j=1}^{M} S_{\hat{\theta}_j}$
14:       Estimate the FIM $\mathbf{F}^{(t)}(\theta)$ of trial $t$ using $S_\theta$ and $S_{\hat{\theta}}$ with the non-parametric estimator introduced in Sect. 4.1
15:    **end for**
16:    Compute the average FIM, $\mathbf{F}(\theta)$, using all $\mathbf{F}^{(t)}(\theta), t = 1, \ldots, T$
17:    **if** $\det(\mathbf{F}(\theta)) > \eta$ **then**
18:       Update $\eta = \det(\mathbf{F}(\theta))$ and $\theta^* = \theta$
19:    **end if**
20:    **if** Stop criterion is met **then**
21:       **return** $\theta^*$
22:    **else**
23:       Select a new $\theta \in \Theta$ based on a suitable search scheme
24:    **end if**
25: **end loop**

---

## 4.3 Results

In the following, we compare the agreement between the hyperparameter configurations identified by the unsupervised FIM-based approach as the edge of criticality, with the configurations where supervised performance measures are maximized. Specifically, we consider the prediction accuracy, defined as $\gamma = \max\{1 - \text{NRMSE}, 0\}$, where NRMSE is the Normalized Root Mean Squared Error of the ESN. Then, we account the memory capacity (MC), which quantifies the capability of ESN to remember previous inputs, relative to an i.i.d. signal. MC is measured as the squared correlation coefficient between the desired output, which is the input signal delayed by different delays $\delta > 0$, and the observed network output $\mathbf{y}[t]$:

$$\text{MC} = \sum_{\delta=1}^{\delta_{max}} \frac{\text{cov}^2(\mathbf{x}[t-\delta], \mathbf{y}[t])}{\text{var}(\mathbf{x}[t-\delta]) \, \text{var}(\mathbf{y}[t])}. \tag{13}$$

MC is computed by training several readout layers, one for each delay $\delta \in \{1, 10, \ldots, 100\}$, while keeping fixed input and reservoir layers.

To test the effectiveness of the identified edge of criticality in terms of forecast accuracy, we consider the prediction of the sinusoid and the MG time-series. We also take into account the NARMA task,

$$\mathbf{y}[t+1] = 0.3\mathbf{y}[t] + 0.05\mathbf{y}[t] \left( \sum_{i=0}^{r-1} \mathbf{y}[t-i] \right) + 1.5\mathbf{x}[t-r]\mathbf{x}[t] + 0.1, \qquad (14)$$

being $\mathbf{x}[t]$ an i.i.d. uniform noise in $[0, 1]$.

In addition to the spectral radius $\theta_{SR}$ and the input scaling $\theta_{IS}$, we consider also the effect of the density of the reservoir connections $\theta_{RC}$ as a core hyperparameter. The hyperparameters are searched in a discretized space through a grid search, which considers 10 different configurations for each parameter. Specifically, we search for the spectral radius $\theta_{SR}$ in $[0.4, 1.6]$, input scaling $\theta_{IS}$ in $[0.3, 0.8]$, and reservoir connectivity $\theta_{RC}$ in $[0.1, 0.7]$, evaluating a total of 1000 hyperparameter configurations. Since we considered a parameter space with three dimensions, the related edge of criticality $\mathcal{K}$ is a two-dimensional manifold embedded in such a three-dimensional space. For each hyperparameter configuration, in Algorithm 1 we perform $T = 10$ independent trials and $M = 80$ perturbations to compute the ensemble average of the FIM; the variance for the perturbations is set to $\sigma^2 = 0.25$. In each trial, we sample new (and independent) input and reservoir connection weights ($\mathbf{W}_i^r$ and $\mathbf{W}_r^r$).

In Fig. 8, we report the critical regions of the parameter space identified in each test by: maximization of FIM determinant, denoted by $\phi$, zero-crossing of MLLE ($\lambda$), and maximization of minimum singular value of the Jacobian ($\eta$). The light gray manifold corresponds to the regions in parameter space where the performance of the network is maximized and the dark gray manifolds represent $\phi$, $\lambda$, and $\eta$. In Table 3, we report the numerical values of the correlations between the light gray manifold and the dark gray ones.

The numerical values of the correlations are reported in Table 3. As it is possible to notice in Fig. 8a, the critical regions identified by each one of the three methods follow with good accuracy the region of the hyperparameter space where MC is maximized. The degrees of correlation for the MC task are described in Table 3. It is interesting to note that $\lambda$ shows a very high correlation (81%) preforming better than $\eta$ for this task. The correlation between $\phi$ and the region with maximum MC is also very high (75%), showing that both $\phi$ and $\lambda$ can be used as reliable indicators to identify the optimal configurations that enhance the short-term memory capacity of ESNs. The $p$-values for each correlation measure are lower than 0.05, indicating statistical significance of the results.

Relative to the prediction of the sinusoid, as it is possible to observe in Fig. 8b, both $\phi$ and $\eta$ are consistent with $\gamma$, while $\lambda$ shows a lower agreement. From Table 3, we see that $\phi$ achieves the best results, all the measures have positive degrees of correlation with $\gamma$ and small $p$-values (hence statistical significance).

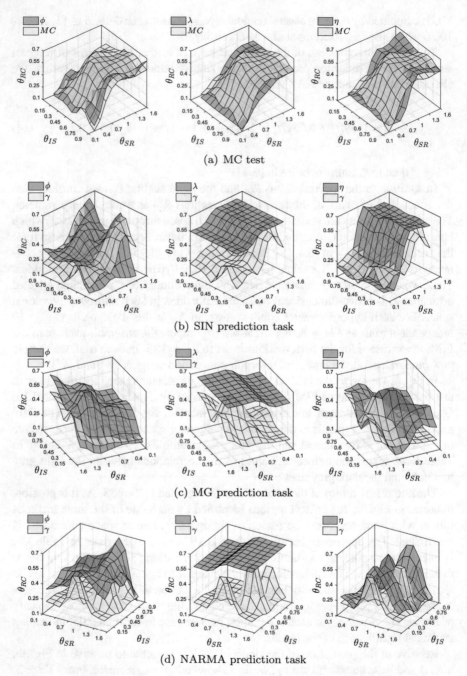

**Fig. 8** In each figure, the light gray manifold represents configurations of spectral radius ($\theta_{SR}$), input scaling ($\theta_{IS}$), and reservoir connectivity ($\theta_{RC}$) that maximize Memory Capacity (MC) or prediction accuracy ($\gamma$). The dark gray manifolds represent (from left to right): configurations where the FIM determinant is maximized ($\phi$); configurations where MLLE crosses zero ($\lambda$); configurations where mSVJ is maximized ($\eta$). Taken from [22]

**Table 3** Correlations between the regions where FIM determinant is maximized ($\phi$), MLLE crosses zero ($\lambda$), minimum singular value of the Jacobian is maximized ($\eta$) and performances are maximized ($\gamma$/MC). Best results are shown in bold, $p$-values are reported in brackets

| Test | Corr ($\phi, \gamma$/MC) | ($\lambda, \gamma$/MC) | Corr ($\eta, \gamma$/MC) |
|------|------|------|------|
| MC | 0.75 (1e-5) | **0.81** (1e-8) | 0.65 (1e-4) |
| Predict—SIN | **0.58** (0.02) | 0.52 (1e-3) | 0.56 (1e-3) |
| Predict—MG | **0.71** (1e-5) | 0.66 (1e-4) | 0.38 (0.06) |
| Predict—NARMA | **0.52** (0.01) | 0.25 (0.22) | 0.48 (0.02) |

In MG test, both $\phi$ and $\lambda$ provide better results than $\eta$ to identify the optimal configuration, as we can see from Fig. 8c and the results in the table. Notably, the correlation between $\gamma$ and $\eta$ has a $p$-value beyond the confidence level 0.05, suggesting that correlations are not different from zero.

According to the results shown in Fig. 8d and Table 3, in the NARMA task $\phi$ and $\eta$ perform significantly better than $\lambda$ for identifying the critical region. If fact, the correlation between $\gamma$ and $\lambda$ is low and not statistically significant. Even in this case, the best results in terms of correlation are achieved by $\phi$.

## 5 Concluding Remarks and Future Research Perspectives

In this chapter, we presented recent research developments for the characterization and tuning of echo state networks. We have shown how recurrence plots can be generated from reservoir neurons activations and exploited by the designer as visual tools to analyze the response of the network to a specific input. Recurrence plots provide an immediate visual interpretation of network stability: short and erratic diagonal lines denote instability/chaoticity, while long diagonal lines denote regularity (e.g., a periodic motion). Through the recurrence quantification analysis, the designer can deduce important and consistent conclusions about the behavior of the network, depending on the actual input driving the system and the current configurations of the hyperparameters.

Successively, we discussed a method that establishes a connection between the notion of continuous phase transition, echo state networks, and Fisher information. Based on this interplay, we have developed a principled approach to configure ESNs on the edge of criticality, where computational capability (defined in terms of prediction performance and short-term memory capacity) is maximized. The proposed methodology is completely unsupervised and it opens new perspectives for analyzing the dynamics of driven recurrent neural networks. Fisher information requires analytic knowledge of the distribution ruling the system. To address this issue, we have followed an ensemble estimation approach based on a recently proposed nonparametric FIM estimator, which, thanks to a graph-based representation of the data, is also applicable to high-dimensional densities. This last aspect plays a fundamental role in our domain of application, since we analyze the network through a multivariate sequence of reservoir neuron activations; hence the number of dimensions is determined by the number of reservoir neurons. We evaluated the proposed method on benchmarks of short-term memory capacity and prediction accuracy, to identify the ESN hyperparameters maximizing the computational capability. We compared our method with established criteria based on the sign of the maximum local Lyapunov exponent and the minimum singular value of the Jacobian. Our experiments demonstrated that the FIM-based approach achieves comparable or even better accuracy than the two other indicators in identifying the onset of criticality.

The methodologies discussed here are independent of the particular task at hand and offer an insight on the dynamics and actual functioning of the network. In this sense, the proposed framework of analysis represents a step forward to the understanding of these systems that, even if are capable of solving efficiently a variety of tasks, are often treated as black boxes. We believe that, the linkage of methods from the complex systems field with recurrent neural networks offers the potential to disclose a whole new set of opportunities for further studies and applications. Our future directions point toward graph-based approaches, which demonstrated to be powerful tools to represent complex systems and to model their dynamics when observed through time-series [19].

# References

1. Aljadeff, J., Stern, M., Sharpee, T.: Transition to chaos in random networks with cell-type-specific connectivity. Phys. Rev. Lett. **114**, 088101 (2015). doi:10.1103/PhysRevLett.114. 088101
2. Barzel, B., Barabási, A.-L.: Universality in network dynamics. Nat. Phys. **9**(10), 673–681 (2013). doi:10.1038/nphys2741
3. Bengio, Y., Simard, P., Frasconi, P.: Learning long-term dependencies with gradient descent is difficult. IEEE Trans. Neural Netw. **5**(2), 157–166 (1994). ISSN 1045-9227. doi:10.1109/72. 279181
4. Berisha, V., Hero, A.Q. III.: Empirical non-parametric estimation of the Fisher information. IEEE Signal Process. Lett. **22**(7), 988–992 (2015). ISSN 1070-9908. doi:10.1109/LSP.2014. 2378514
5. Bertschinger, N., Natschläger, T.: Real-time computation at the edge of chaos in recurrent neural networks. Neural Comput. **16**(7), 1413–1436 (2004). doi:10.1162/ 089976604323057443
6. Bianchi, F.M., Livi, L., Alippi, C.: Investigating echo state networks dynamics by means of recurrence analysis. IEEE Trans. Neural Netw. Learn. Syst. 1–13 (2016). doi:10.1109/TNNLS. 2016.2630802
7. Boedecker, J., Obst, O., Lizier, J.T., Mayer, N.M., Asada, M.: Information processing in echo state networks at the edge of chaos. Theory Biosci. **131**(3), 205–213 (2012). doi:10.1007/ s12064-011-0146-8
8. Charles, A., Yin, D., Rozell, C.: Distributed sequence memory of multidimensional inputs in recurrent networks. arXiv:1605.08346 (2016)
9. De Arcangelis, L., Lombardi, F., Herrmann, H.J.: Criticality in the brain. J. Stat. Mech. Theory Exp. **2014**(3), P03026 (2014). doi:10.1088/1742-5468/2014/03/P03026

10. Enel, P., Procyk, E., Quilodran, R., Dominey, P.F.: Reservoir computing properties of neural dynamics in prefrontal cortex. PLoS Comput. Biol. **12**(6), e1004967 (2016). doi:10.1371/journal.pcbi.1004967
11. Eroglu, D., Peron, T.K.D.M., Marwan, N., Rodrigues, F.A., da Costa, L.F., Sebek, M., Kiss, I.Z., Kurths, J.: Entropy of weighted recurrence plots. Phys. Rev. E **90**(4), 042919 (2014). doi:10.1103/PhysRevE.90.042919
12. Elman, J.L.: Finding structure in time. Cogn. Sci. **14**(2), 179–211 (1990). ISSN 0364-0213. doi:10.1016/0364-0213(90)90002-E
13. Friedman, J.H., Rafsky, L.C.: Multivariate generalizations of the Wald-Wolfowitz and Smirnov two-sample tests. Ann. Stat. **7**(4), 697–717 (1979)
14. Grigolini, P.: Emergence of biological complexity: criticality, renewal and memory. Chaos, Solitons Fractals (2015). doi:10.1016/j.chaos.2015.07.025
15. Hammer, B., Micheli, A., Sperduti, A., Strickert, M.: Recursive self-organizing network models. Neural Netw. **17**(8–9), 1061–1085 (2004). ISSN 0893-6080. doi:10.1016/j.neunet.2004.06.009
16. Hidalgo, J., Grilli, J., Suweis, S., Muñoz, M.A., Banavar, J.R., Maritan, A.: Information-based fitness and the emergence of criticality in living systems. Proc. Natl. Acad. Sci. **111**(28), 10095–10100 (2014). doi:10.1073/pnas.1319166111
17. Hidalgo, J., Grilli, J., Suweis, S., Maritan, A., Muñoz, M.A.: Cooperation, competition and the emergence of criticality in communities of adaptive systems. J. Stat. Mech. Theory Exp. **2016**(3), 033203 (2016). doi:10.1088/1742-5468/2016/03/033203
18. Jaeger, H.: The "echo state" approach to analysing and training recurrent neural networks-with an erratum note. Bonn, Germany: German National Research Center for Information Technology GMD Technical Report, vol. 148, p. 34 (2001)
19. Lacasa, L., Nicosia, V., Latora, V.: Network structure of multivariate time series. Sci. Rep. **5**, (2015). doi:10.1038/srep15508
20. Langton, C.G.: Computation at the edge of chaos: Phase transitions and emergent computation. Phys. D Nonlinear Phenom. **42**(1), 12–37 (1990). doi:10.1016/0167-2789(90)90064-V
21. Legenstein, R., Maass, W.: Edge of chaos and prediction of computational performance for neural circuit models. Neural Netw. **20**(3), 323–334 (2007). doi:10.1016/j.neunet.2007.04.017
22. Livi, L., Bianchi, F.M., Alippi, C.: Determination of the edge of criticality in echo state networks through Fisher information maximization. IEEE Trans. Neural Netw. Learn. Syst. 1–12 (2017). doi:10.1109/TNNLS.2016.2644268
23. Maass, W., Joshi, P., Sontag, E.D.: Computational aspects of feedback in neural circuits. PLoS Comput. Biol. **3**(1), e165 (2007). doi:10.1371/journal.pcbi.0020165.eor
24. Magnus, J.R., Neudecker, H.: Matrix Differential Calculus with Applications in Statistics and Econometrics. Wiley, New York (1995)
25. Manjunath, G., Jaeger, H.: Echo state property linked to an input: Exploring a fundamental characteristic of recurrent neural networks. Neural Comput. **25**(3), 671–696 (2013). doi:10.1162/NECO_a_00411
26. Marichal, R.L., Piñeiro, J.D.: Analysis of multiple quasi-periodic orbits in recurrent neural networks. Neurocomputing **162**, 85–95 (2015). doi:10.1016/j.neucom.2015.04.001
27. Marwan, N.: How to avoid potential pitfalls in recurrence plot based data analysis. Int. J. Bifurcat. Chaos **21**(04), 1003–1017 (2011). doi:10.1142/S0218127411029008
28. Marwan, N., Kurths, J.: Line structures in recurrence plots. Phys. Lett. A **336**(4), 349–357 (2005). doi:10.1016/j.physleta.2004.12.056
29. Marwan, N., Carmen, M., Thiel, R.M., Kurths, J.: Recurrence plots for the analysis of complex systems. Phys. Rep. **438**(5), 237–329 (2007). doi:10.1016/j.physrep.2006.11.001
30. Marwan, N., Schinkel, S., Kurths, J.: Recurrence plots 25 years later-Gaining confidence in dynamical transitions. EPL (Europhys. Lett.) **101**(2), 20007 (2013). doi:10.1209/0295-5075/101/20007
31. Massar, M., Massar, S.: Mean-field theory of echo state networks. Phys. Rev. E **87**(4), 042809 (2013). doi:10.1103/PhysRevE.87.042809

32. Massobrio, P., de Arcangelis, L., Pasquale, V., Jensen, H.J., Plenz, D.: Criticality as a signature of healthy neural systems. Front. Syst. Neurosci. **9**, 22 (2015). doi:10.3389/fnsys.2015.00022
33. Mastromatteo, I., Marsili, M.: On the criticality of inferred models. J. Stat. Mech. Theory Exp. **2011**(10), P10012 (2011). doi:10.1088/1742-5468/2011/10/P10012
34. Mora, T., Bialek, W.: Are biological systems poised at criticality? J. Stat. Phys. **144**(2), 268–302 (2011). doi:10.1007/s10955-011-0229-4
35. Mora, T., Deny, S., Marre, O.: Dynamical criticality in the collective activity of a population of retinal neurons. Phys. Rev. Lett. **114**(7), 078105 (2015). doi:10.1103/PhysRevLett.114.078105
36. Pascanu, R., Mikolov, T., Bengio, Y.: On the difficulty of training recurrent neural networks. arXiv:1211.5063 (2012)
37. Peng, Y., Lei, M., Li, J.-B., Peng, X.-Y.: A novel hybridization of echo state networks and multiplicative seasonal ARIMA model for mobile communication traffic series forecasting. Neural Comput. Appl. **24**(3–4), 883–890 (2014)
38. Prokopenko, M., Lizier, J.T., Obst, O., Wang, X.R.: Relating Fisher information to order parameters. Phys. Rev. E **84**(4), 041116 (2011). doi:10.1103/PhysRevE.84.041116
39. Rajan, K., Abbott, L.F., Sompolinsky, H.: Stimulus-dependent suppression of chaos in recurrent neural networks. Phys. Rev. E **82**(1), 011903 (2010). doi:10.1103/PhysRevE.82.011903
40. Reinhart, R.F., Steil, J.J.: Regularization and stability in reservoir networks with output feedback. Neurocomputing **90**, 96–105 (2012). doi:10.1016/j.neucom.2012.01.032
41. Roli, A., Villani, M., Filisetti, A., Serra, R.: Dynamical criticality: overview and open questions. arXiv:1512.05259 (2015)
42. Rumelhart, D.E., Smolensky, P., McClelland, J.L., Hinton, G.: Sequential thought processes in pdp models. V **2**, 3–57 (1986)
43. Scheffer, M., Bascompte, J., Brock, W.A., Brovkin, V., Carpenter, S.R., Dakos, V., Held, H., Van Nes, E.H., Rietkerk, M., Sugihara, G.: Early-warning signals for critical transitions. Nature **461**(7260), 53–59 (2009). doi:10.1038/nature08227
44. Scheffer, M., Carpenter, S.R., Lenton, T.M., Bascompte, J., Brock, W., Dakos, V., van De Koppel, J., van De Leemput, I.A., Levin, S.A., van Nes, E.H., Pascual, M., Vandermeer, J.: Anticipating critical transitions. Science **338**(6105), 344–348 (2012). doi:10.1126/science.1225244
45. Schiller, U.D., Steil, J.J.: Analyzing the weight dynamics of recurrent learning algorithms. Neurocomputing **63**, 5–23 (2005). doi:10.1016/j.neucom.2004.04.006
46. Shen, Y., Wang, J.: An improved algebraic criterion for global exponential stability of recurrent neural networks with time-varying delays. IEEE Trans. Neural Netw. **19**(3), 528–531 (2008). ISSN 1045-9227. doi:10.1109/TNN.2007.911751
47. Steil, J.J.: Memory in backpropagation-decorrelation o(n) efficient online recurrent learning. In: Duch, W., Kacprzyk, J., Oja, E., Zadrożny, S. (eds.) Artificial Neural Networks: Formal Models and Their Applications-ICANN 2005, pp. 649–654. Springer, Berlin, Heidelberg (2005)
48. Sussillo, D.: Neural circuits as computational dynamical systems. Curr. Opin. Neurobiol. **25**, 156–163 (2014). doi:10.1016/j.conb.2014.01.008
49. Sussillo, D., Barak, O.: Opening the black box: low-dimensional dynamics in high-dimensional recurrent neural networks. Neural Comput. **25**(3), 626–649 (2013). doi:10.1162/NECO_a_00409
50. Tiňo, P., Rodan, A.: Short term memory in input-driven linear dynamical systems. Neurocomputing **112**, 58–63 (2013). doi:10.1016/j.neucom.2012.12.041
51. Tkačik, G., Bialek, W.: Information processing in living systems. Ann. Rev. Condens. Matter Phys. **7**(1), 89–117 (2016). doi:10.1146/annurev-conmatphys-031214-014803
52. Tkačik, G., Mora, T., Marre, O., Amodei, D., Palmer, S.E., Berry, M.J., Bialek, W.: Thermodynamics and signatures of criticality in a network of neurons. Proc. Natl. Acad. Sci. **112**(37), 11508–11513 (2015). doi:10.1073/pnas.1514188112
53. Torres, J.J., Marro, J.: Brain performance versus phase transitions. Sci. Rep. **5** (2015). doi:10.1038/srep12216
54. Toyoizumi, T., Abbott, L.F.: Beyond the edge of chaos: amplification and temporal integration by recurrent networks in the chaotic regime. Phys. Rev. E **84**(5), 051908 (2011). doi:10.1103/PhysRevE.84.051908

55. Toyoizumi, T., Aihara, K., Amari, S.-I.: Fisher information for spike-based population decoding. Phys. Rev. Lett. **97**(9), 098102 (2006). doi:10.1103/PhysRevLett.97.098102
56. Verstraeten, D., Schrauwen, B.: On the quantification of dynamics in reservoir computing. In: Artificial Neural Networks–ICANN 2009, pp. 985–994. Springer, Berlin (2009). doi:10.1007/978-3-642-04274-4_101
57. Verstraeten, D., Schrauwen, B., D'Haene, M., Stroobandt, D.: An experimental unification of reservoir computing methods. Neural Netw. **20**(3), 391–403 (2007). ISSN 0893-6080. doi:10.1016/j.neunet.2007.04.003. Echo State Networks and Liquid State Machines
58. Wainrib, G., Touboul, J.: Topological and dynamical complexity of random neural networks. Phys. Rev. Lett. **110**, 118101 (2013). doi:10.1103/PhysRevLett.110.118101
59. Wang, X., Lizier, J., Prokopenko, M.: Fisher information at the edge of chaos in random boolean networks. Artif. Life **17**(4), 315–329 (2011). ISSN 1064-5462. doi:10.1162/artl_a_00041
60. Werbos, P.J.: Backpropagation: past and future. Proc. IEEE Int. Conf. Neural Netw. **1**, 343–353 (1988). doi:10.1109/ICNN.1988.23866
61. Yildiz, I.B., Jaeger, H., Kiebel, S.J.: Re-visiting the echo state property. Neural Netw. **35**, 1–9 (2012). doi:10.1016/j.neunet.2012.07.005
62. Zegers, P.: Fisher information properties. Entropy **17**(7), 4918–4939 (2015). doi:10.3390/e17074918
63. Zhang, B., Miller, D.J., Wang, Y.: Nonlinear system modeling with random matrices: echo state networks revisited. IEEE Trans. Neural Netw. Learn. Syst. **23**(1), 175–182 (2012). ISSN 2162-237X. doi:10.1109/TNNLS.2011.2178562
64. Zhang, Y., Wang, J.: Global exponential stability of recurrent neural networks for synthesizing linear feedback control systems via pole assignment. IEEE Trans. Neural Netw. **13**(3), 633–644 (2002). ISSN 1045-9227. doi:10.1109/TNN.2002.1000129

# Optimization of Ensemble Neural Networks with Type-1 and Interval Type-2 Fuzzy Integration for Forecasting the Taiwan Stock Exchange

Martha Pulido, Patricia Melin and Olivia Mendoza

**Abstract** This paper describes an optimization method based on particle swarm optimization for ensemble neural networks with type-1 and type-2 fuzzy aggregation for forecasting complex time series. The time series that was considered in this paper to compare the hybrid approach with traditional methods is the Taiwan Stock Exchange (TAIEX), and the results shown are for the optimization of the structure of the ensemble neural network with type-1 and type-2 fuzzy integration. Simulation results show that ensemble approach produces good prediction of the Taiwan Stock Exchange.

**Keywords** Ensemble neural networks · Time series · Particle swarm · Fuzzy system

## 1 Introduction

Time series are usually analyzed to understand the past and to predict the future, enabling managers or policy makers to make properly informed decisions. Time series analysis quantifies the main features in data, like the random variation. These facts, combined with improved computing power, have made time series methods widely applicable in government, industry, and commerce. In most branches of science, engineering, and commerce, there are variables measured sequentially in time. Reserve banks record interest rates and exchange rates each day. The government statistics department will compute the country's gross domestic product on a yearly basis. Newspapers publish yesterday's noon temperatures for capital cities from around the world. Meteorological offices record rainfall at many different sites with differing resolutions. When a variable is measured sequentially in time over or at a fixed interval, known as the sampling interval, the resulting data form a time series [1].

M. Pulido · P. Melin (✉) · O. Mendoza
Tijuana Institute of Technology, Tijuana, México, USA
e-mail: pmelin@tectijuana.edu.mx; pmelin@tectijuana.mx

© Springer International Publishing AG 2018
A.E. Gawęda et al. (eds.), *Advances in Data Analysis with Computational Intelligence Methods*, Studies in Computational Intelligence 738,
https://doi.org/10.1007/978-3-319-67946-4_6

Time series predictions are very important because based on them we can analyze past events to know the possible behavior of futures events and thus can take preventive or corrective decisions to help avoid unwanted circumstances.

The choice and implementation of an appropriate method for prediction has always been a major issue for enterprises that seek to ensure the profitability and survival of business. The predictions give the company the ability to make decisions in the medium and long term, and due to the accuracy or inaccuracy of data this could mean predicted growth or profits and financial losses. It is very important for companies to know the behavior that will be the future development of their business, and thus be able to make decisions that improve the company's activities, and avoid unwanted situations, which in some cases can lead to the company's failure. In this paper we propose a hybrid approach for time series prediction by using an ensemble neural network and its with optimization with particle swarm optimization. In the literature there have been recent produced work of time series [2–10].

## 2 Preliminaries

In this section we present basic concepts that are used in this proposed method:

### 2.1 Time Series and Prediction

The word "prediction" comes from the Latin prognosticum, which means I know in advance. Prediction is to issue a statement about what is likely to happen in the future, based on analysis and considerations of experiments. Making a forecast is to obtain knowledge about uncertain events that are important in decision-making [6]. Time series prediction tries to predict the future based on past data, it take a series of real data $x_t - n, \ldots, x_t - 2, 0 x_t - 1, x_t$ and then obtains the prediction of the data $x_t + 1, x_t + 2, \ldots, x_n + n$. The goal of time series prediction or a model is to observe the series of real data, so that future data may be accurately predicted [1, 11].

### 2.2 Neural Networks

Neural networks Neural networks (NNs) are composed of many elements (Artificial Neurons), grouped into layers and are highly interconnected (with the synapses), this structure has several inputs and outputs, which are trained to react (or give values) in a way you want to input stimuli. These systems emulate in some way, the human brain. Neural networks are required to learn to behave (Learning) and someone should be responsible for the teaching or training (Training), based on prior knowledge of the environment problem [12, 13].

## 2.3 Ensemble Neural Networks

An Ensemble Neural Network is a learning paradigm where many neural networks are jointly used to solve a problem [14]. A Neural network ensemble is a learning paradigm where a collection of a finite number of neural networks is trained for the same task [15]. It originates from Hansen and Salamon's work [16], which shows that the generalization ability of a neural network system can be significantly improved through ensembling a number of neural networks, i.e. training many neural networks and then combining their predictions. Since this technology behaves remarkably well, recently it has become a very hot topic in both neural networks and machine learning communities [17], and has already been successfully applied to diverse areas such as face recognition [18, 19], optical character recognition [20–22], scientific image analysis [23], medical diagnosis [24, 25], seismic signals classification [26], etc.

In general, a neural network ensemble is constructed in two steps, i.e. training a number of component neural networks and then combining the component predictions.

There are also many other approaches for training the component neural networks. Examples are as follows. Hampshire and Waibel [22] utilize different object functions to train distinct component neural networks.

## 2.4 Fuzzy Systems as Methods of Integration

There exists a diversity of methods of integration or aggregation of information, and we mention some of these methods below.

Fuzzy logic was proposed for the first time in the mid-sixties at the University of California Berkeley by the brilliant engineer Lofty A. Zadeh., who proposed what it's called the principle of incompatibility: "As the complexity of system increases, our ability to be precise instructions and build on their behavior decreases to the threshold beyond which the accuracy and meaning are mutually exclusive characteristics." Then introduced the concept of a fuzzy set, under which lies the idea that the elements on which to build human thinking are not numbers but linguistic labels. Fuzzy logic can represent the common knowledge as a form of language that is mostly qualitative and not necessarily a quantity in a mathematical language that means of fuzzy set theory and function characteristics associated with them [12].

## 2.5   Optimization

The process of optimization is the process of obtaining the 'best', if it is possible to measure and change what is 'good' or 'bad'. In practice, one wishes the 'most' or 'maximum' (e.g., salary) or the 'least' or 'minimum' (e.g., expenses). Therefore, the word 'optimum' is takes the meaning of 'maximum' or 'minimum' de pending on the circumstances; 'optimum' is a technical term which implies quantitative measurement and is a stronger word than 'best' which is more appropriate for everyday use. Likewise, the word 'optimize', which means to achieve an optimum, is a stronger word than 'improve'. Optimization theory is the branch of mathematics encompassing the quantitative study of optima and methods for finding them. Optimization practice, on the other hand, is the collection of techniques, methods, procedures, and algorithms that can be used to find the optima [27].

## 2.6   Particle Swarm Optimization

Particle Swarm Optimization (PSO) is a bio-inspired optimization method proposed by R. Eberhart and J. Kennedy [28] in 1995. PSO is a search algorithm based on the behavior of biological communities that exhibits individual and social behavior [29], and examples of these communities are groups of birds, schools of fish and swarms of bees [29].

A PSO algorithm maintains a swarm of particles, where each particle represents a potential solution. In analogy with the paradigms of evolutionary computation, a swarm is similar to a population, while a particle is similar to an individual. In simple terms, the particles are "flown" through a multidimensional search space, where the position of each particle is adjusted according to its own experience and that of its neighbors. Let $x_i$ denote the position $i$ in the search space at time step $t$, unless otherwise stated, $t$ denotes discrete time steps. The position of the particle is changed by adding a velocity, $v_i(t)$, to the current position, i.e.

$$x_i(t+1) = x_i(t) + v_i(t+1)$$
$$\text{with } x_i(0) \sim U(X_{min}, X_{max}).$$

(1)

## 3   Problem Statement and Proposed Method

The goal of this work was to implement Particle Swarm Optimization to optimize the ensemble neural network architectures. In this cases the optimization is for each of the modules, and thus to find a neural network architecture that yields optimum results in each of the Time Series to be considered. In Fig. 1 we have the historical data of each time series prediction, then the data is provided to the modules that will

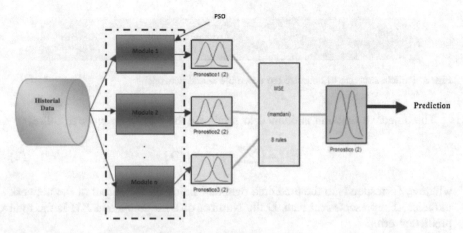

**Fig. 1** General architecture of the proposed ensemble model

be optimized with the particle swarm optimization for the ensemble network, and then these modules are integrated with integration based on type-1 and type-2 Fuzzy Integration.

Historical data of the Taiwan Stock Exchange time series was used for the ensemble neural network trainings, where each module was fed with the same information, unlike the modular networks, where each module is fed with different data, which leads to architectures that are not uniform.

The Taiwan Stock Exchange (Taiwan Stock Exchange Corporation) is a financial institution that was founded in 1961 in Taipei and began to operate as stock exchange on 9 February 1962. The Financial Supervisory Commission regulates it. The index of the Taiwan Stock Exchange is the TWSE [30].

Data of the Taiwan Stock Exchange time series: We are using 800 points that correspond to a period from 03/04/2011 to 05/07/2014 (as shown in Fig. 2). We used 70% of the data for the ensemble neural network trainings and 30% to test the network [30].

**Fig. 2** Taiwan Stock Exchange

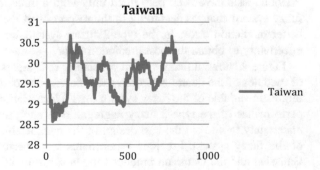

| Number of Modules | Number of Layers | Neurons 1 | ... | Neurons n |
|---|---|---|---|---|

**Fig. 3** Particle structure to optimize the ensemble neural network

The objective function is defined to minimize the prediction error as follows:

$$EM = \left( \sum_{i=1}^{D} |a_i - x_i| / D \right) \tag{2}$$

where *a,* corresponds to the predicted data depending on the output of the network modules, $X$ represents real data, $D$ the Number of Data points and *EM* is the total prediction error.

The corresponding particle structure is shown in Fig. 3.

Figure 3 represents the Particle Structure to optimize the ensemble neural network, where the parameters that are optimized are the number of modules, number of layers, and number of neurons of the ensemble neural network. PSO determines the number of modules, number of layers and number of neurons per layer that the neural network ensemble should have, to meet the objective of achieving the better Prediction error.

The parameters for the particle swarm optimization algorithm are: 100 Particles, 100 iterations, Cognitive Component (C1) = 2, Social Component (C2) = 2, Constriction coefficient of linear increase (C) = (0–0.9) and Inertia weight with linear decrease (W) = (0.9–0). We consider a number of 1–5 modules, number of layers of 1–3 and neurons number from 1 to 30.

The aggregation of the responses of the optimized ensemble neural network is performed with type-1 and type-2 fuzzy systems. In this work the fuzzy system consists of 5 inputs depending on the number of modules of the neural network ensemble and one output is used. Each input and output linguistic variable of the fuzzy system uses 2 Gaussian membership functions. The performance of the type-2 fuzzy aggregators is analyzed under different levels of uncertainty to find out the best design of the membership functions for the 32 rules of the fuzzy system. Previous tests have been performed only with a three input fuzzy system and the fuzzy system changes according to the responses of the neural network to give us better prediction error. In the type-2 fuzzy system we also change the levels of uncertainty to obtain the best prediction error.

Figure 4 shows a fuzzy system consisting of 5 inputs depending on the number of modules of the neural network ensemble and one output. Each input and output linguistic variable of the fuzzy system uses 2 Gaussian membership functions. The performance of the type-2 fuzzy aggregators is analyzed under different levels of uncertainty to find out the best design of the membership functions for the 32 rules of the fuzzy system. Previous experiments were performed with triangular, and Gaussian and the Gaussian produced the best results of the prediction.

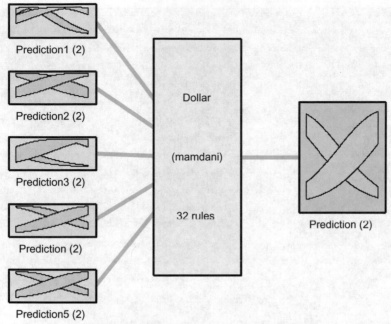

Prediction1 (2)

Prediction2 (2)

Prediction3 (2)

Prediction (2)

Prediction5 (2)

Dollar

(mamdani)

32 rules

Prediction (2)

System Dollar: 5 inputs, 1 outputs, 32 rules

**Fig. 4** Fuzzy inference system for integration of the ensemble neural network

Figure 5 represents the 32 possible rules of the fuzzy system; we have 5 inputs in the fuzzy system with 2 membership functions, and the outputs with 2 membership functions. These fuzzy rules are used for both the type-1 and type-2 fuzzy systems. In previous work several tests were performed with 3 inputs, and the prediction error obtained was significant and the number of rules was greater, and this is why we changed to 2 inputs.

# 4  Simulation Results

In this section we present the simulation results obtained with the genetic algorithm and particle swarm optimization for the Taiwan Stock Exchange.

We consider working with a genetic algorithm to optimize the structure of an ensemble neural network and the best architecture obtained was the following (shown in Fig. 6).

In this architecture we have two layers in each module. In module 1, in the first layer we have 23 neurons and the second layer we have 9 neurons, and In module 2 we used 9 neurons in the first layer and the second layer we have 15 neurons the

**Fig. 5** Rules of the type-2 fuzzy system

Levenberg-Marquardt (LM) training method was used; 3 delays for the network were considered.

Table 1 shows the particle swarm optimization results (as shown in Fig. 6) where the prediction error is of 0.0013066.

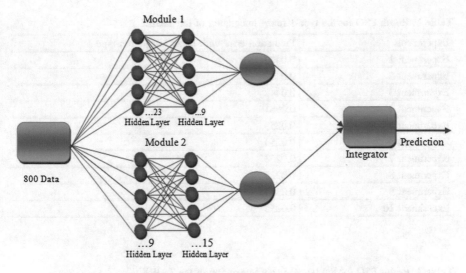

**Fig. 6** Prediction with the optimized ensemble neural network with GA of the TAIEX

**Table 1** Particle swarm optimization result for the ensemble neural network

| No. | Iterations | Particles | Number of modules | Number of layers | Number of neurons | Duration | Prediction error |
|---|---|---|---|---|---|---|---|
| 1 | 100 | 100 | 2 | 3 | 13, 16, 2 18, 20, 18 | 01:48:30 | 0.002147 |
| 2 | 100 | 100 | 2 | 2 | 3, 9 14, 19 | 01:03:09 | 0.0021653 |
| 3 | 100 | 100 | 2 | 2 | 20, 4 10, 7 | 01:21:02 | 0.0024006 |
| 4 | 100 | 100 | 2 | 2 | 16, 19 3, 12 | 01:29:02 | 0.0019454 |
| 5 | 100 | 100 | 2 | 2 | 19, 19 24, 17 | 02:20:22 | 0.0024575 |
| 6 | 100 | 100 | 2 | 3 | 21, 14, 23 14, 24, 20 | 01:21:07 | 0.0018404 |
| 7 | 100 | 100 | 2 | 2 | 23, 9 9, 15 | 01:19:08 | 0.0013065 |
| 8 | 100 | 100 | 2 | 2 | 15, 17 9, 22 | 01:13:20 | 0.0018956 |
| 9 | 100 | 100 | 2 | 2 | 20, 16 | 01:13:35 | 0.0023377 |
| 10 | 100 | 100 | 2 | 2 | 23, 8 10, 17 | 01:04:23 | 0.0023204 |

Fuzzy integration is performed initially by implementing a type-1 fuzzy system in which the best result is in experiment of row number 8 of Table 2 with an error of: 0.0235.

**Table 2** Result PSO for the type-1 fuzzy integration of the TAIEX

| Experiments | Prediction error with fuzzy integration type-1 |
| --- | --- |
| Experiment 1 | 0.0473 |
| Experiment 2 | 0.0422 |
| Experiment 3 | 0.0442 |
| Experiment 4 | 0.0981 |
| Experiment 5 | 0.0253 |
| Experiment 6 | 0.0253 |
| Experiment 7 | 0.0253 |
| Experiment 8 | 0.0235 |
| Experiment 9 | 0.0253 |
| Experiment 10 | 0.0253 |

**Table 3** Result PSO for the type-2 fuzzy integration of the TAIEX

| Experiment | Prediction error 0.3 uncertainty | Prediction error 0.4 uncertainty | Prediction error 0.5 uncertainty |
| --- | --- | --- | --- |
| Experiment 1 | 0.0335 | 0.033 | 0.0372 |
| Experiment 2 | 0.0299 | 0.5494 | 0.01968 |
| Experiment 3 | 0.0382 | 0.0382 | 0.0387 |
| Experiment 4 | 0.0197 | 0.0222 | 0.0243 |
| Experiment 5 | 0.0433 | 0.0435 | 0.0488 |
| Experiment 6 | 0.0121 | 0.0119 | 0.0131 |
| Experiment 7 | 0.01098 | 0.01122 | 0.01244 |
| Experiment 8 | 0.0387 | 0.0277 | 0.0368 |
| Experiment 9 | 0.0435 | 0.0499 | 0.0485 |
| Experiment 10 | 0.0227 | 0.0229 | 0.0239 |

As a second phase, to integrate the results of the optimized ensemble neural network a type-2 fuzzy system is implemented, where the best results that are obtained are as follows: with a degree uncertainty of 0.3 a forecast error of 0.01098 is obtained, with a degree of uncertainty of 0.4 the error is of 0.01122 and with a degree of uncertainty of 0.5 the error is of 0.001244, as shown in Table 3.

Figure 7 shows the plot of real data against the predicted data generated by the ensemble neural network optimized with the particle swarm optimization.

**Fig. 7** Prediction with the optimized ensemble neural network with PSO of the TAIEX

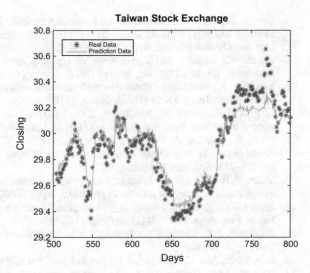

## 5 Conclusions

The best result when applying the particle swarm to optimize the ensemble neural network was: 0.0013066 (as shown in Fig. 6 and Table 1). Implemented a type 2 fuzzy system for ensemble neural network, in which the results where for the best evolution as obtained a degree of uncertainty of 0.3 yielded a forecast error of 0.01098, with an 0.4 uncertainty error: 0.01122, and 0.5 uncertainty error of 0.01244, as shown in Table 3. After achieving these results, we have verified efficiency of the algorithms applied to optimize the neural network ensemble architecture. In this case, the method was efficient but it also has certain disadvantages, sometimes the results are not as good, but genetic algorithms can be considered as good technique a for solving search and optimization problems.

**Acknowledgements** We would like to express our gratitude to the CONACYT, Tijuana Institute of Technology for the facilities and resources granted for the development of this research.

## References

1. Cowpertwait, P., Metcalfe, A.: Time Series. In: Introductory Time Series with R, pp. 2–5. Springer, Heidelberg (2009)
2. Castillo, O., Melin, P.: Hybrid intelligent systems for time series prediction using neural networks, fuzzy logic, and fractal theory. IEEE Trans. Neural Netw. **13**(6), 1395–1408 (2002)
3. Castillo, O., Melin, P.: Simulation and forecasting complex economic time series using neural networks and fuzzy logic. In: Proceeding of the International Neural Networks Conference, vol. 3, pp. 1805–1810 (2001)

4. Castillo, O., Melin, P.: Simulation and forecasting complex financial time series using neural networks and fuzzy logic. In: Proceedings the IEEE the International Conference on Systems, Man and Cybernetics, vol. 4, pp. 2664–2669 (2001)
5. Karnik, N., Mendel, M.: Applications of type-2 fuzzy logic systems to forecasting of time-series. Inf. Sci. 120(1–4), 89–111 (1999)
6. Kehagias, A., Petridis, V.: Predictive modular neural networks for time series classification. Neural Netw. 10(1), 31–49 (2000), 245–250 (1997)
7. Maguire, L.P., Roche, B., McGinnity, T.M., McDaid, L.J.: Predicting a chaotic time series using a fuzzy neural network. Inf. Sci. 112(1–4), 125–136 (1998)
8. Melin, P., Castillo, O., Gonzalez, S., Cota, J., Trujillo, W., Osuna, P.: Design of Modular Neural Networks with Fuzzy Integration Applied to Time Series Prediction, vol. 41/2007, pp. 265–273. Springer, Heidelberg (2007)
9. Yadav, R.N., Kalra, P.K., John, J.: Time series prediction with single multiplicative neuron model, soft computing for time series prediction. Appl. Soft Comput. 7(4), 1157–1163 (2007)
10. Zhao, L., Yang, Y.: PSO-based single multiplicative neuron model for time series prediction. Expert Syst. Appl. Part 2 36(2), 2805–2812 (2009)
11. Brockwell, P.T., Davis, R.A.: Introduction to Time Series and Forecasting. Springer. New York, pp 1–219 (2002)
12. Jang, J.S.R., Sun, C.T., Mizutani, E.: Neuro-Fuzzy and Soft Computing. Prentice Hall (1996)
13. Multaba, I.M., Hussain, M.A.: Application of neural networks and other learning. In: Technologies in Process Engineering. Imperial Collage Press (2001)
14. Sharkey, A.: Combining Artificial Neural Nets: Ensemble And Modular Multi-net Systems. Springer, London (1999)
15. Sollich, P., Krogh, A.: Learning with ensembles: how over-fitting can be useful. In: Touretzky, D.S., Mozer, M.C., Hasselmo, M.E. (eds.) Advances in Neural Information Processing Systems, Denver, CO, vol. 8, pp. 190–196. MIT Press, Cambridge, MA (1996)
16. Hansen, L.K., Salomon, P.: Neural network ensembles. IEEE Trans. Pattern Anal. Mach. Intell. 12(10), 993–1001 (1990)
17. Sharkey, A.: One combining Artificial of Neural Nets. Department of Computer Science University of Sheffield, U.K. (1996)
18. Gutta, S., Wechsler, H.: Face recognition using hybrid classifier systems. In: Proceedings of the ICNN-96, Washington, DC, pp. 1017–1022. IEEE Computer Society Press, Los Alamitos, CA (1996)
19. Huang, F.J., Huang, Z., Zhang, H.-J., Chen, T.H.: Pose invariant face recognition. In: Proceedings of the 4th IEEE International Conference on Automatic Face and Gesture Recognition, Grenoble, France. IEEE Computer Society Press, Los Alamitos, CA (2000)
20. Drucker, H., Schapire, R., Simard, P.: Improving performance in neural networks using a boosting algorithm. In: Hanson, S.J., Cowan Giles, J.D. (eds.) Advances in Neural Information Processing Systems, Denver, CO, vol. 5, pp. 42–49. Morgan Kaufmann, San Mateo, CA (1993)
21. Hampshire, J., Waibel, A.: A novel objective function for improved phoneme recognition using time- delay neural networks. IEEE Trans. Neural Netw. 1(2), 216–228 (1990)
22. Mao, J.: A case study on bagging, boosting and basic ensembles of neural networks for OCR. In: Proceedings of the IJCNN-98, Anchorage, AK, vol. 3, pp. 1828–1833. IEEE Computer Society Press, Los Alamitos, CA (1998)
23. Cherkauer, K.J.: Human expert level performance on a scientific image analysis task by a system using combined artificial neural networks. In: Chan, P., Stolfo, S., Wolpert, D. (eds.) Proceedings of the AAAI-96 Workshop on Integrating Multiple Learned Models for Improving and Scaling Machine Learning Algorithms, Portland, OR, AAAI, pp. 15–21. Press, Menlo Park, CA (1996)
24. Cunningham, P., Carney, J., Jacob, S.: Stability problems with artificial neural networks and the ensemble solution. Artif. Intell. Med. 20(3), 217–225 (2000)
25. Zhou, Z.-H., Jiang, Y., Yang, Y.-B., Chen, S.-F.: Lung cancer cell identification based on artificial neural network ensembles. Artif. Intell. Med. 24(1), 25–36 (2002)

26. Shimshon, Y.N.: Intrator classification of seimic signal by integrating ensemble of neural networks. IEEE Trans. Signal Process. **461**(5), 1194–1201 (1998)
27. Antoniou, A., Sheng, W. (eds.): Practical optimization algorithms and engineering applications. In: Introduction Optimization. Springer, pp. 1–4 (2007)
28. Eberhart, R., Kennedy, J.: A new optimizer using swarm theory. In: Proceedings of the 6th International Symposium Micro Machine and Human Science (MHS), pp. 39–43, October 1995
29. Kennedy, J., Eberhart, R.: Particle swarm optimization. In: Proceedings of the IEEE International Conference Neural Network (ICNN), Nov. 1995, vol. 4, pp. 1942–1948
30. Taiwan Bank Database: www.twse.com.tw/en (April 03, 2011)

# Deep Neural Networks—A Brief History

Krzysztof J. Cios

**Abstract** In this chapter we describe Deep Neural Networks (DNN), their history, and some related work.

## 1 Introduction

DNN are one of the most efficient tools that belong to a broader area called deep learning. DNN process input information in a hierarchical way, where each subsequent level of processing extracts more abstract/global/invariant features. In other words, DNN (semi) automatically learn key features from data and then aggregate them for some purpose, such as recognizing objects in the images.

We shall illustrate how DNN work by the use of an example from the area of face recognition. There, the inputs are images from which at the first level (first hidden layer) of processing simple image characteristics such as edges are extracted. At the second and subsequent levels, more complex parts of an image are formed to finally, at the output layer, recognize human faces. This is in contrast to using a traditional approach where in the first step, known as preprocessing, an expert guides the process of extracting key features, and then they are used for recognizing faces. The common part of these two, very different, approaches is that at the output layer the labeled data are needed to perform supervised learning, i.e., assign names/labels to faces.

Although DNN can in general work in all three basic learning modes, namely, supervised, unsupervised, and semi-supervised, so far the majority of successful DNN applications used the semi-supervised mode where (almost) unsupervised

K.J. Cios (✉)
Department of Computer Science, Virginia Commonwealth University,
Richmond, VA 23284, USA
e-mail: kcios@vcu.edu

K.J. Cios
Institute of Theoretical and Applied Informatics, Polish Academy of Sciences,
Bałtycka 5, 44-100 Gliwice, Poland

© Springer International Publishing AG 2018
A.E. Gawęda et al. (eds.), *Advances in Data Analysis with Computational Intelligence Methods*, Studies in Computational Intelligence 738,
https://doi.org/10.1007/978-3-319-67946-4_7

183

extracting of key features by the hidden layers was followed by a supervised learning at the output layer. In the fully supervised DNN mode the most frequently used algorithm is backpropagation with a ramp/rectifier activation function, which is very efficient in networks with many layers. The supervised approach, however, contradicts the very idea of deep learning as it is just a classical backpropagation learning with a sigmoid replaced by the ramp function, $(f(x) = \max f(0, x))$. At the other end of the spectrum, fully unsupervised DNN, little progress has been reported so far.

DNN, as well as other types of neural networks, were inspired by the need to solve difficult for computers problems, such as image recognition but that are easily solvable by humans. Specifically, they were inspired by our, although still very vague, understanding of how human brain processes information. Depending on a goal of brain modeling we distinguish two approaches. If the goal is to model brain's neural circuits, the area called neuroinformatics or computational neuro-science, a key question validating the generated model is: How well does it fit the experimental biological data? In this approach, the neuron model frequently used is the spiking one with the appropriate learning rule. On the other hand, if the goal is to solve a practical problem, such as face recognition, then the validation question changes to: Is the model efficient? As in the latter case it is not important whether a simple or complex neuron model or any specific learning rule is used. This type of modeling is known as neuromorphic computing.

A digression about capacity of a human brain. It has about $10^{-11}$ neurons and trillions of synaptic connections, which endows it with enormous storage capacity. If we define storage capacity as the ratio of the number of patterns that can be stored and retrieved, to the size of the network, then a network consisting of N neurons can retrieve correctly P stored patterns, according to this formula: $P < N/(4 * \ln(N))$. Thus, a network with $10^4$ neurons can store only 271 patterns and with $10^{11}$ neurons it grows to $10^9$ patterns. The latter number of patterns is more than enough for a human to store and remember every single image, word, situation etc. encountered during a lifetime. In fact, the human brain has even bigger storage capacity because a group of neurons can store not just one but many different patterns, the phenomenon known as polysynchrony [10]. Fortunately, most people do not remember everything from the time they are born, although there are well documented cases of individuals who remembered everything from their past, day by day. By comparison, current artificial neural networks are incomparably smaller, with the largest using up to tens of thousands of neurons. One of the reasons for the size is that neural networks are designed to solve domain- specific problems, versus solving problems for many domains at the same time. For example, one network is designed to solve an image recognition problem while another a natural language processing problem, but there are no attempts to design a single network for solving problems from both domains.

Our focus here is on DNN, including those that use spiking neuron models and the corresponding learning rules. We start by defining key building blocks of all DNN. They are: (a) a neuron model, which performs basic computations, (b) a learning rule, which updates the weights/synapses between the neurons, and (c) a network architecture, which specifies how the neurons are topologically arranged and interconnected.

## 2 Neuron Models

A wide spectrum of neuron models from very simple to spiking ones is described next. Notice that increasing biological detail of an artificial neuron model also increases its computational complexity.

The first simple model of a neuron, called the *threshold neuron*, was developed by McCulloch and Pitts [19]. It calculates a dot product between the input vector and the weight vector of a neuron, and if it is higher than its transfer function (like a step function) it fires/generates an output of 1 (otherwise 0).

The first *spiking neuron* model was developed by Hodgkin and Huxley [7], for which they later received a Nobel Prize. They modeled squid's giant neuron and they treated each component of the neuron, including its membrane, as electrical component. The model is described by:

$$C\frac{dV}{dt} = I_e - \overbrace{\bar{g}_K n^4 (V - E_K)}^{I_K} - \overbrace{\bar{g}_{Na} m^3 h (V - E_{Na})}^{I_{Na}} - \overbrace{g_L (V - E_L)}^{I_L}$$

where:

| | |
|---|---|
| $I_e$ | stimulus/injected current |
| $V$ | voltage/membrane potential |
| $L$ | leakage current |
| $K$ | potassium and $Na$ = sodium channels |
| $g$ | conductances, e.g., $g_{Na}$ = 120 mS/cm$^2$; $g_K$ = 36 mS/cm$^2$; $g_L$ = 0.3 mS/cm$^2$ |
| $E$ | reversal potentials, e.g., $E_{Na}$ = 115 mV, $E_K$ = −12 mV, $E_L$ = 10.6 mV |
| $n, m, h$ | channel gating/activation variables: $n = n(t)$, $m = m(t)$, $h = h(t)$. |

To better understand it let us look at its equivalent electric circuit, shown in Fig. 1. The membrane is modeled as capacitor, $C_m$, while potassium and sodium ion channels as conductances ($g = 1/R$; R being resistance). V is the neuron's membrane potential, i.e., difference between its intracellular (inside of the neuron) and extracellular potentials. According to Kirchhoff's law the sum of the currents is zero so the current through the membrane, C dV/dt, can be written in a shorter form as:

$$C\,dV/dt = I_e - I_K - I_{Na} - I_L$$

Figure 2 illustrates generation of an action potential/spike by the flows of sodium and potassium ions, represented as conductances.

**Fig. 1** Hodgkin and
Huxley's model circuit
representation

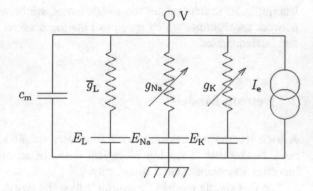

**Fig. 2** Generation of an
action potential by sodium
and potassium ions flows

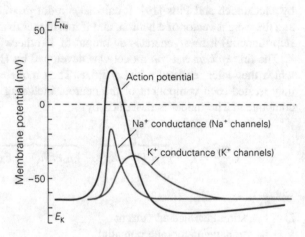

McGregor [20] defined a simpler than Hodgkin-Huxley spiking neuron model,
one that belongs to a group of *integrate-and-fire* models.

It is described by these equations:

$$S = \begin{cases} 1 & E \geq T_h \\ 0 & E < T_h \end{cases}$$

$$\frac{dE}{dt} = \frac{-E + G_K \cdot (E_K - E) + G_e \cdot (E_e - E) + G_i \cdot (E_i - E) + SCN}{T_{mem}}$$

$$\frac{dG_K}{dt} = \frac{-G_K + B \cdot S}{T_{GK}}$$

$$\frac{dT_h}{dt} = \frac{-(T_h - T_{h0}) + c \cdot E}{T_{Th}}$$

where:

V       membrane potential
$V_r$     membrane resting potential

$V_K$     potassium resting potential
$V_i$     inhibitory resting potential
$V_e$     excitatory resting potential

Transmembrane potentials: $E = V - V_r$; $E_K = V_K - V_r$; $E_i = V_i - V_r$; $E_e = V_e - V_r$
Transmembrane conductances: $G_K = g_K/G$; $G_i = g_{si}/G$; $G_e = g_{se}/G$

G       membrane resting conductance
$g_K$     potassium resting conductance
$g_{si}$     inhibitory resting conductance
$g_{se}$     excitatory resting conductance
$T_{GK}$     decay of GK time constant
$T_h$     threshold value
$T_{h0}$     resting value of threshold
$T_{th}$     decay of threshold constant
$T_{mem}$     membrane time constant
    $T_{mem} = C/G$
    Current through membrane: $SCN = SC/G$
SC      current injected to cell (corresponds to $I_e$ in the HH model)
c       rise of threshold $c \in [0, 1]$
C       membrane capacitance
B       postfiring potassium increment.

Its corresponding electric circuit, shown in Fig. 3, is similar to Hodgkin and Huxley's. It models the potassium channel, refractory properties, adaptation to stimuli, and mimics excitatory and inhibitory post synaptic potentials (EPSP and IPSP, respectively) of a neuron. The PSPs are illustrated in Fig. 4. We will refer back to these potentials when we later describe learning rules.

The working of the McGregor's model is illustrated in Fig. 5, using a "network" of only three neurons: two pre-synaptic (one excitatory and one inhibitory) that feed

**Fig. 3** McGregor's model
circuit representation

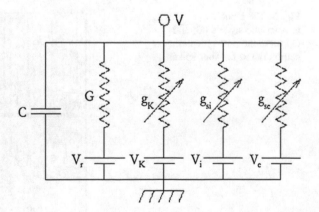

**Fig. 4** Example excitatory
and inhibitory post synaptic
potentials

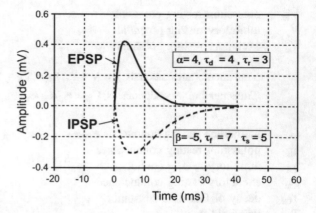

into one post-synaptic neuron [26]. We can see the spikes that are generated by both types of pre-synaptic neurons, shown in the bottom panel of Fig. 5. The positive excitatory ($G_e$) and negative inhibitory ($G_i$) inputs feed into a post-synaptic neuron that integrates them and when the sum rises above its threshold (Th) the post-synaptic neuron fires a spike; four such spikes are generated by the post-synaptic neuron, which is shown in the top panel (E)) of Fig. 5. Notice that the threshold of the neuron (Th) changes over time.

The simplest spiking neuron model was developed by Izhikevich [9]. It does not model any of the biological neuron functions except that it accurately mimics several types/shapes of the postsynaptic potentials (spikes) generated by human brain neurons. It is described by:

$$\begin{cases} v' = 0.04v^2 + 5v + 140 - u + I \\ u' = a(bv - u), \end{cases}$$

where
    if $v > 30$, then $v = c$, $u = u + d$,

**Fig. 5** The input
pre-synaptic signals ($G_e$ and
$G_i$) make the post-synaptic
neuron (E) to fire four spikes

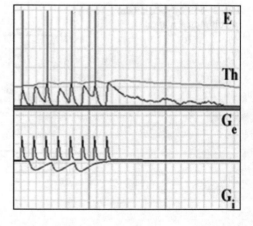

- v is membrane potential
- u is membrane recovery variable
- I is input current

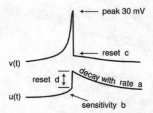

It models over a dozen different post-synaptic firing patterns, four of which are shown in the bottom part of Fig. 6; they correspond to the parameter settings shown in the top part of Fig. 6.

Izhikevich model became very popular because its simplicity allows for building networks consisting of thousands of such neurons. While using it, however, we found that increasing the strength of the stimulus caused it to fire with higher and higher frequency (no upper bound). This is not biologically plausible as neurons cannot fire during the absolute refractory period, needed for restoration of their membrane potentials, no matter the strength of the input. We thus corrected the condition for the neuron firing [29] by changing it

from

**if** $v > 30$, **then** $v = c, u = u + d$

to

**if** $v > 30$:
**if**

| Parameter | Excitatory (RS, IB) | Inhibitory (FS,LTS) |
|-----------|---------------------|---------------------|
| a | 0.02 | $0.02 + 0.08\gamma$ |
| b | 0.2 | $0.25 - 0.05\gamma$ |
| c | $-65 + 15\gamma$ | -65 |
| d | $8 - 6\gamma$ | 2 |
| $\gamma$ is a uniform random variable between 0 and 1 |||

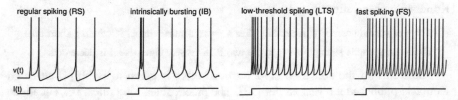

**Fig. 6** Two types of excitatory (the first two) and two types of inhibitory neurons firing patterns (taken from http://www.izhikevich.org/publications/spikes.htm)

**Fig. 7**  **a** Unbounded firing of
the original Izhikevich model
neurons; **b** Firing of the
neurons after Strack et al. [30]
modification accounting for
absolute refractory periods

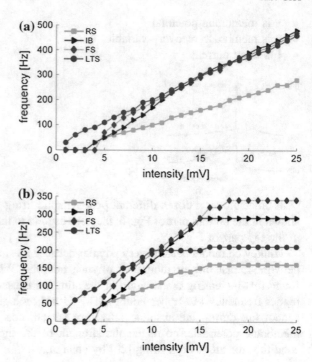

*dt > dtmin:* $v = c, u = u + d$, **spike**
**else**
$v = 30$, **no spike**

That is, we added additional check (ifdt > dtmin) to account for refractory
property of neurons. Figure 7a illustrates firings of the four types of original
Izhikevich neurons, while Fig. 7b shows firings of neurons after our modification.
The modified model was used for modeling multi-column multi-layer model of
neocortex, which was not possible to do using the original Izhikevich model [30].

# 3   Learning Rules

Let us start by noticing that almost all learning rules are based to some degree on
Konorski's observation:

IF a presynaptic neuron "j" repeatedly fires a postsynaptic neuron "i" within a short time

THEN the synaptic strength between the two is increased, otherwise it is decreased.

The credit for the above observation most often is given to Hebb [4] although
Konorski published it a year earlier [13]. The practical learning rules, i.e., equations
corresponding to the above observation were specified much later by computational
scientists [28, 31].

Similar case, of not giving credit to the original inventor, involves a popular *backpropagation learning rule* that was first specified by statisticians Robbins and Monroe [23]: they called it a stochastic approximation method. However, the credit for the rule in neural networks literature was given to Rumelhart et al. [24] before it was found that Werbos [33] specified the rule, a dozen years before them.

The simplest learning rule, called *Perceptron*, for one-layer feed-forward neural networks, was defined by Rosenblatt [25]. Backpropagation rule is in fact the Perceptron's rule extension to many-layer networks. Extending it to such networks, however, became possible only after the step threshold function used in the Perceptron was replaced with a differentiable sigmoid function. This seemingly small change led to an explosion in neural networks research that stagnated for almost 20 years after Minsky and Papert [21] stated that neural networks were useless for solving complex problems.

Kohonen [12] specified *winner-takes-all learning* rule. This rule more closely than Perceptron or backpropagation mimics the learning processes taking place in biological neural circuits. It states that only the neuron whose weight vector (synapse) is the closest to the input's vector is the winner and as such increases its weight to get it even closer to the input pattern vector. Often, a number of neurons in close neighborhood of the winning neuron also adjust their weights.

The first rule for networks of spiking neurons, called *Spike Time-Dependent Plasticity* (STDP), was specified by Song et al. [28]. Swiercz et al. [31] specified another rule for spiking neurons called *Synaptic Activity Plasticity Rule* (SAPR). The two rules are compared in Fig. 8. Konorski's observation is translated, in both rules, into the following recipe:

> The adjustment of the strength of synaptic connections between pre-synaptic neuron "j" and post-synaptic neuron "i" takes place every time the postsynaptic neuron "i" fires, according to the function specified either by STDP or SAPR. If $\Delta t$ is positive that means the pre-synaptic neuron fired **before** the post-synaptic neuron and the strength between the two is increased. If $\Delta t$ is negative it means that the pre-synaptic neuron fired **after** the post-synaptic neuron fired and the strength between the two is decreased.

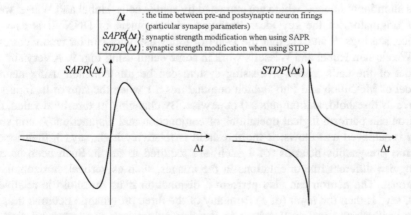

**Fig. 8** Comparison of the SAPR and STDP: the latter is fixed while the former depends on the shape of excitatory and inhibitory post synaptic functions of neurons

The difference between the two rules is that SAPR uses a function that is continuous and differentiable (important in several applications); it is also dynamic because it uses actual post-synaptic potential functions to modify the connection strengths between the neurons. In other words, the adjustments depend on the shape of SAPR, which in turn depends on the shape of the chosen postsynaptic functions in a given neural circuit. The left part of the SAPR function in Fig. 8 (to the left of the y axis) is the chosen inhibitory PSP while the right part is the chosen excitatory PSP; see again the two function shapes in Fig. 4. In contrast, the STDP rule uses a static function meaning that the adjustments are always the same; they do not depend on the shape of inhibitory/excitatory PSPs for a given $\Delta t$.

## 4  Network Architecture

As stated above, DNN use a hierarchical architecture, vaguely mimicking the brain's hierarchical way of performing cognitive tasks. This architecture is one of the key distinguishing factors between several types of neural networks and DNN. It follows that neural networks with just one hidden layer, such as SVM, RBF, or Kohonen's self-organizing feature map, are not DNN. As a digression, a popular decision tree algorithm does not perform deep learning either, in spite of its hierarchical architecture, since it uses original features and not a hierarchy of transformed features.

Hierarchical processing of information in the brain was first discovered by neurophysiologists Hubel and Wiesel [8] who studied the cat's visual system; for this work they were awarded a Nobel Prize. Not only they observed the brain's hierarchical way of processing information but also that at each level of processing the brain extracts more general features performed by *complex* cells, that aggregate the features extracted at the previous level to, at the end of this process, recognize some objects in the input image. At the first level, the brain focuses on recognizing specific simple patterns in the input images, such as vertical or horizontal elements present in input images, which are extracted by *simple* cells. Hubel and Wiesel were thus originators of the key ideas leading to development of DNN. It is easy to notice, see Figs. 9 and 10, that the DNN of today use very similar architectures.

We explain Hubel and Wiesel's work in some detail using Fig. 9. A very simple model of the cat's visual processing system can be implemented using neuron model of McCulloh and Pitts, which outputs/fires a 1 when the sum of its inputs is above its threshold, and outputs a 0 otherwise. By changing its threshold value, the neuron can perform logical operations of conjunction and disjunction. A conjunction is achieved as follows: if the threshold is relatively high, say, 3, then inputs from 3 presynaptic neurons (of 1 each) are required to fire it. Such neurons can recognize different line orientations in the images, such as vertical, horizontal, or diagonal. The neuron can also perform a disjunction if its threshold is relatively low, say, 1; then the input (of 1) from any of the three presynaptic neurons fires it. This is illustrated in Fig. 9, where in the first column we see image of digit 2.

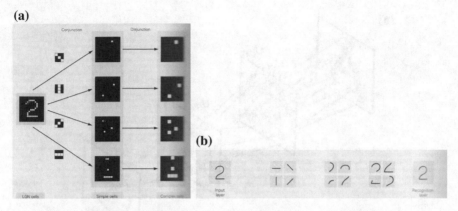

**Fig. 9 a** Illustration of how the simple and complex cells extract specific features from input images; **b** Implementation of how the features are extracted and aggregated (using three hidden layers) in Neocognitron to recognize digit 2 (both pictures are taken from Kandel et al. Principles of Neural Science, 5th edition, [11]

The four neurons, the simple cells, in the first (hidden) layer perform conjunctions to recognize three-element line patterns, while the four neurons, the complex cells, in the second layer perform disjunctions that aggregate the simple patterns into more complex ones until, at the output layer, digit 2 is recognized. In the parlance of DNN the conjunction is called convolution, the disjunction a spatial pooling, the simple cell a feature extractor/detector, and the complex cell a feature aggregator/analyzer. The difference between the just described very simple scheme and DNN is that feature extraction in DNN happens (almost) without human intervention (we describe later how DNN do it).

The first researcher to design a direct precursor of DNN, using Hubel and Wiesel's discoveries, was Fukushima [3] who called his network Neocognitron. Figure 10a illustrates how key features of an image of letter A are first picked up by simple cells (S) and then aggregated by complex cells (C), in order to recognize letter A at the output. S-layer of simple cells extracts features from the previous stage in the hierarchy, while the C-layer of complex cells ensures tolerance for shifts of features extracted by the S-layer.

DNN became popular and the term "deep learning" was coined and widely accepted around 2010 due to the development of efficient learning algorithms and hardware speed-ups such as the use of GPUs. In particular, LeCun [16–18], Hinton [5, 6] and Krizhevski [14] made significant impact on the field. Comparison of architecture of Neocognitron shown in Fig. 10a with the DNN architecture of LeCun's convolutional network shown in Fig. 10b shows their great similarity.

As aforementioned, the first few layers of DNN perform feature extraction using unsupervised learning, and only the top layer weights (i.e., those between the last hidden and output layer) are trained in a supervised mode. In DNN the most often used approach to perform feature extraction between the input and hidden layer(s)

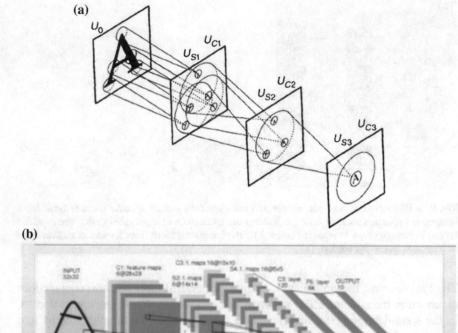

**Fig. 10 a** Fukushima's Neocognitron architecture, and **b** LeCun's convolutional neural network architecture

is to use an idea of autoencoder. Other method often used to perform unsupervised learning (always a form of clustering) is a Boltzman machine.

We now explain how an autoencoder works using a feed-forward neural network with backpropagation learning in a vertical composition, meaning that the same operation that is performed by the first hidden layer on the original input, is also performed by the second layer on the output of the first hidden layer, etc. Let us also assume that our input is an image of size n × n and that the number of neurons in the first hidden layer is p, with p smaller than $n^2$ (this condition is not required but using it makes it easier to understand and explain). The task of the autoencoder is to learn outputs of the first hidden layer in such a way that after learning we can reconstruct (using outputs of the hidden layer) the inputs with a very small distortion. In other words, the autoencoder learns a compressed (lower dimensional) version of the inputs. In that respect, the autoencoder is similar to PCA and performs clustering.

Loosely speaking, the outputs of the first hidden layer neurons are trained to recognize some specific features, as linear combinations of the original image features, such as edges, in different positions and orientations. This is what is meant by saying that new features are automatically learned/extracted by deep neural networks. The same process is repeated at the output of the second hidden layer, which takes as input the output of the first hidden layer. The outcome of doing it is that the previously extracted features, say, edges, are aggregated into more complex features, say silhouettes of objects. Supervised learning is only then used to train the weights between the last hidden and the output layer in order to assign labels to the input images.

Instead of describing Neocognitron or convolutional neural network of LeCun for which many excellent online resources exist, we describe below a network called IRNN (Image Recognition Neural Network), which was inspired by the works of Hubel and Wiesel and Fukushima [2]. In the IRNN the hidden layers perform explicit clustering operations of the (sub) images for the purpose of extracting key features at each level of hierarchical processing. IRNN consists of an input layer, an output layer, and one or more hidden layers, as shown in Fig. 11. The Sensory layer extracts local features from the images. The role of the hidden layer(s) is to aggregate local features to generate higher level semi-global features. The output layer, in a supervised mode, associates the semi-global features with the known labels. Notice that IRNN operates like a semi-supervised convolutional DNN. It uses windowing, which is based on a biological observation that a neuron connected to the sensory system receives inputs from only a portion of the sensory neurons.

**Fig. 11** IRNN's architecture: unsupervised part consists of the sensory and feature aggregating layers while the associative part is supervised

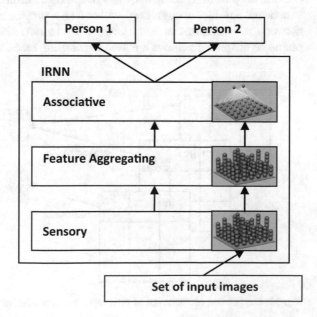

Figure 11 shows stacks of neurons represented by small balls. How they are generated and what they represent is explained in Fig. 12. We see there three (hashed) subimages/windows of the three input images, which are clustered using a novel image similarity measure [2]. If the first two subimages are similar (as shown) they are clustered together in neuron $n_1$. Since subimage 3 was found quite different from the subimages 1 and 2, it creates its own cluster, so the second neuron, $n_2$, is generated. The weights $w_1$ and $w_2$ are initially set to the first subimage pixel values vector but are later updated to represent cluster center (thus representing an "average" subimage). At the end of scanning of entire images the result might be as the one shown in Fig. 13. Notice, that at the center more neurons (clusters) were created to represent image details, such as nose, eyes and mouth, while at the periphery where background was about the same in all images only single neurons/clusters were needed.

The same process is repeated on the outputs of the sensory layer to aggregate the local features into more complex semi-global features. Clustering of subimages at each level is performed with all other layers disabled. Finally, the associative layer associates the images with recognition codes—like Person 1 or Person 2—using winner-takes-all learning rule. Note that the number of clusters/neurons in the IRNN is not predetermined by the user: it depends only on a similarity between the subimages. Characteristic feature of the IRNN is that if new data become available the already trained network can be used in two ways. If the new data points are labeled then additional training continues with the new data. However, if new data are not labeled only the output layer needs to be trained.

The networks described so far, including convolutional DNN, used only simple, non spiking, neuron models as their basic processing units. But is it possible to perform deep learning using networks of spiking neurons? Shin et al. [27] used such a network for face recognition, without any preprocessing of the images. The network self-organizes at each level of its hierarchical processing. Even at the output layer spiking neurons are used for labeling faces, in contrast to more popular

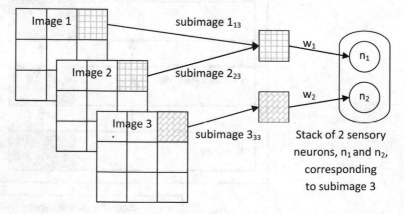

**Fig. 12** Explanation of clustering of subimages into a number of clusters/neurons

| 1 | 1 | 1 | 1 | 1 | 1 | 1 | 1 |
|---|---|---|---|---|---|---|---|
| 1 | 1 | 2 | 2 | 3 | 1 | 1 | 1 |
| 1 | 2 | 2 | 3 | 4 | 2 | 2 | 1 |
| 1 | 2 | 4 | 5 | 3 | 2 | 1 | 1 |
| 1 | 1 | 3 | 6 | 5 | 3 | 1 | 1 |
| 1 | 2 | 4 | 4 | 5 | 3 | 1 | 1 |
| 1 | 1 | 2 | 2 | 2 | 1 | 1 | 1 |
| 1 | 1 | 1 | 1 | 1 | 1 | 1 | 1 |

**Fig. 13** A hypothetical result of clustering of a set of registered face images

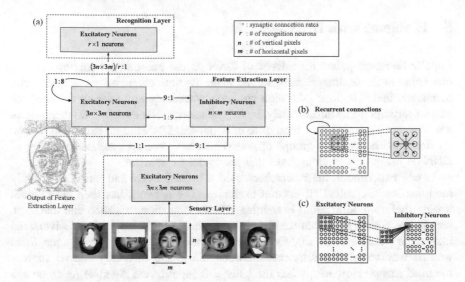

**Fig. 14** Architecture of the network of spiking neurons, **a** High-level block diagram. **b** Recurrent synaptic connections between the excitatory neurons in the feature extraction layer. **c** Synaptic connections between the excitatory neurons in the sensory/feature extraction layer and the inhibitory neurons in the feature extraction layer (taken from Shin et al. [27] paper)

use of supervised methods such as backpropagation; similar approach was later used in Cao, et al. [1]. Specifically, the spiking neuron model used was McGregor's with SAPR and STDP learning rules for self-organization of the neurons; self-organization in essence being a clustering operation. Figure 14 shows architecture of this a network. The sensory layer serves as relay of the input image but increases the input dimension, from an image of size n × m to an image of size 3n × 3m. The only hidden layer, the Feature Extraction layer, is composed of excitatory and inhibitory neurons, while keeping their ratio close to the one observed in human brain. Notice that no supervised training is performed by this layer. Instead, multiple submitting of the input images is required until there is a negligible change in the result of self-organization. It was shown that using SAPR rule gave better results in recognizing face images than using STDP. The Recognition layer also uses spiking neurons: the neuron that spikes the most for a known input face image "recognizes" the person. The network performed particularly well on rotated and partially occluded images. In short, the network uses raw images for input, extracts key features without any training between the sensory and feature extracting layers. This is in contrast to using an autoencoder that can be seen as a supervised training method where the label is the compressed pattern of the input pattern.

## 5 Problems with DNN Learning

Popular literature paints the advent of DNN as the panacea for solving difficult problems, such as image recognition, hand written character recognition, etc. Moreover, that it is done with high confidence/accuracy and without the need for human participation. Unfortunately, history of science tells us that new technologies are often accompanied by a high dose of hype, and DNN are no exception to this. As described below, two groups of researchers have shown spectacular failing of DNN on image recognitions tasks that are trivial for humans.

In one experiment, the researchers used a trained DNN and ran it on slightly modified images, called adversarial examples. The network has seen the original images (before modification) in training. The modification was such that that there was no perceptible to the human eye difference between the original and adversarial image; the latter had only slightly different statistical properties. For example, there was no way to tell the difference between the image of a dog and its slightly modified image. However, when the latter was input to the AlexNet (open source implementation of a convolutional neural network), which was trained on the original image of the dog, it failed to recognize it [32].

In another work, the researchers took the opposite approach. Namely, they modified (using genetic algorithms) the image used in training in such a way that it had no resemblance whatsoever to the original image. For example, an image looking like a TV static noise was not only recognized by LeNet (part of Caffe software package) say, as peacock, but also was very certain (accuracy of 99.6%) about its recognition decision [22].

# 6 Conclusions

The described above DNN shortcomings do not outweigh their many advantages. However, lots of research is needed to answer the question of why they failed in those experiments. I think it is increasingly more important for the computational researchers to team up with neuroscientists to come up with better algorithms for image recognition so that the algorithms cannot be so easily fooled [15]. That may require, in the first place, more work by the neuroscientists to better understand processes used by the brain in recognition tasks.

The easy fooling of DNN in some recognition tasks, which are easily recognized by humans, poses a very serious cybersecurity risk. Modern society heavily relies on machine learning techniques, like DNN, for performing many everyday tasks such as medical diagnosis, self-driving cars, investing financial assets, and even in a legal system. Since the researchers have shown that it is relatively easy to come up with adversarial examples, the automated systems we so much now depend on can produce possibly disastrous results. It is thus increasingly important for researchers to add safety features to deep learning algorithms they are developing, something that software engineers have been doing for a long time to assure safety of their code. To start with, researchers should routinely use in training adversarial examples, in addition to original ones, to make their systems more secure.

# References

1. Cao, Y., Chen, Y., Khosla, D.: Spiking deep convolutional neural networks for energy-efficient object recognition. Int. J. Comput. Vis. (2014)
2. Cios, K.J., Shin, I.: Image recognition neural network: IRNN. Neurocomputing 7(2), 159–185 (1995)
3. Fukushima, K.: Neocognitron: a self organizing neural network model for a mechanism for pattern recognition unaffected by shift in position. Biol. Cybern. 36, 193–202 (1980)
4. Hebb DO. 1949. The Organization of Behavior. Wiley
5. Hinton, G.E., Osindero, S., Teh, Y.: A fast learning algorithm for deep belief nets. Neural Comput. 18, 1527–1554 (2006)
6. Hinton, G.E., Salakhutdinov, R.: Reducing the dimensionality of data with neural networks. Science 313(5786), 504–507 (2006)
7. Hodgkin, A.L., Huxley, A.F.: A quantitative description of membrane current and its application to conduction and excitation in nerve. J. Physiol. 177, 500–544 (1952)
8. Hubel, D.H., Wiesel, T.N.: Receptive fields, binocular interaction and functional architecture in the cat's visual cortex. J. Physiol. 160, 106–154 (1962)
9. Izhikevich, E.M.: Simple model of spiking neurons. IEEE Trans. Neural Networks 14, 1569–1572 (2003)
10. Izhikevich, E.M.: Polychronization: computation with spikes. Neural Comput. 18(2), 245–282 (2006)
11. Kandel E.R., et al.: Principles of Neural Science, 5th edn. McGraw-Hill (2013)
12. Kohonen, T.: Self-organized formation of topologically correct feature maps. Biol. Cybern. 43, 59–69 (1982)

13. Konorski J.: Conditioned Reflexes and Neuron Organization. Cambridge University Press, Cambridge (267 pp.); Reprinted with a supplementary chapter in 1968 by Hafner Publ. Co., New York (1948)

14. Krizhevsky, A., Sutskever, I., Hinton, G.: ImageNet classification with deep convolutional neural networks. In: NIPS' 2012 (2012)

15. Lim, H.K., Keniston, L.P., Cios, K.J.: Modeling of multisensory convergence with a network of spiking neurons: a reverse engineering approach. IEEE Trans. Biomed. Eng. **58**(7), 1940–1949 (2011)

16. LeCun, Y., Bottou, L., Bengio, Y., Haffner, P.: Gradient-based learning applied to document recognition. In: Proceedings of IEEE, vol. 86, no. 11, pp. 2278–2324 (1998)

17. LeCun, Y., Kavukcuoglu, K., Farabet, C.: Convolutional networks and applications in vision. In: Proceedings of 2010 IEEE International Symposium on Circuits and Systems (ISCAS), pp. 253–256. IEEE (2010)

18. LeCun, Y., Bengio, Y., Hinton, G.: Deep learning. Nature **521**, 436–444 (2015)

19. McCulloch, W.S., Pitts, W.H.: A logical calculus of the ideas immanent in nervous activity. Bull. Math. Biophys. **5**, 115–133 (1943)

20. MacGregor, R.J.: Neural and Brain Modeling. Academic Press (1987)

21. Minsky, M., Papert, S.: An Introduction to Computational Geometry. MIT Press (1969)

22. Nguyen, A., Yosinski, J., Clune, J.: Deep Neural Networks are Easily Fooled: High Confidence Predictions for Unrecognizable Images (2014). arXiv:1412.1897v2 [cs.CV] 18 Dec 2014

23. Robbins, H., Monro, S.: A stochastic approximation method. Ann. Math. Stat. **22**(3), 400–407 (1951)

24. Rumelhart, D.E., Hinton, G.E., Williams, R.J.: Learning representations by back-propagating errors. Nature **323**, 533–536 (1986)

25. Rosenblatt, F.: The Perceptron—a perceiving and recognizing automaton. Report 85-460-1, Cornell Aeronautical Laboratory (1957)

26. Sala, D.M., Cios, K.J.: Solving graph algorithms with networks of spiking neurons. IEEE Trans. Neural Netw. **10**(4), 953–957 (1999)

27. Shin, J.H., Smith, D., Swiercz, W., Staley, K., Rickard, T., Montero, J., Kurgan, L., Cios, K. J.: Recognition of partially occluded and rotated images with a network of spiking neurons. IEEE Trans. Neural Netw. **21**(11), 1697–1708 (2010)

28. Song, S., Miller, K.D., Abbot, L.F.: Competitive Hebbian learning through spike timing-dependent synaptic plasticity. Nat. Neurosci. **3**(9) (2000)

29. Strack, B., Jacobs, K., Cios, K.J.: Biological restraint on the Izhikevich neuron model essential for seizure modeling. In: Proceedings of 6th International IEEE EMBS Conference on Neural Engineering, San Diego, 6–8 Nov, pp. 395–398 (2013)

30. Strack, B., Jacobs, K., Cios, K.J.: Simulating vertical and horizontal inhibition with short term dynamics in a multi-column multi-layer model of Neocortex. Int. J. Neural Syst. **24**(5), 1440002 [19 pp.] (2014)

31. Swiercz, W., Cios, K.J., Staley, K., et al.: New synaptic plasticity rule for networks of spiking neurons. IEEE Trans. Neural Netw. **17**(1), 94–105 (2006)

32. Szeged, C., Zaremba, W., Sutskever, I., Bruna, J., Erhan, D., Goodfellow, D., Fergus, R.: Intriguing properties of neural networks. In: International Conference on Learning Representations (2014)

33. Werbos, P.: Beyond regression: new tools for prediction and analysis in the behavioral sciences. Ph.D. thesis, Harvard University (1974)

# Part III
# Intelligent Technologies
# in Systems Modeling

# Techniques for Construction and Integration of Rule Bases

Grzegorz J. Nalepa

**Abstract** This chapter discusses issues in the practical integration approaches for intelligent rule-based systems. In it selected issues that need to be addressed for performing integration of rule based systems are identified and discussed. These include high level modeling techniques for rule bases, integration architectures for rule-based systems, and rule interoperability challenges. In the chapter a short review of different rule types and languages used to express them is given. Moreover, important issues regarding construction of complex rule bases are introduced. Furthermore, the execution issues of rule bases are considered, with the emphasis on addressing the structure identified during modeling. Finally, main approaches to integration and interoperability of rule-based systems are given.

## 1 Introduction

Intelligent systems that use rules for capturing and executing knowledge have been a widely used technology for several decades. Rule-based shells are a commonly referred technology supporting the execution of such systems. Originally developed for rule-based expert systems [22, 32, 39], shells are software frameworks that support knowledge engineers by providing a rule language for encoding the rule base and a generic inference engine for execution. CLIPS (C Language Integrated Production System) [22, 64] is probably the one best known. Currently, the CLIPS rule language is a multi paradigm programming language that provides support for rule-based, object-oriented and procedural programming. The wide spread and acceptance of CLIPS resulted in the development of Jess [19]. While the differences in the language were minimal, Jess was entirely written in Java which improved its integration capabilities. Today construction of rule-based systems (RBS) is a well-studied field with important handbooks available [22, 32].

G.J. Nalepa (✉)
AGH University of Science and Technology, al. A. Mickiewicza 30,
30-059 Krakow, Poland
e-mail: gjn@agh.edu.pl

© Springer International Publishing AG 2018
A.E. Gawęda et al. (eds.), *Advances in Data Analysis with Computational Intelligence Methods*, Studies in Computational Intelligence 738,
https://doi.org/10.1007/978-3-319-67946-4_8

The development of intelligent systems in last decades shows that rule-based systems (RBS) are still a technology of great potential and many applications [24]. However, it is also clear that rules, while very useful, need to be integrated with other paradigms. This integration concerns not only other models of data and knowledge processing, but also software development and implementation paradigms. Practical integration approaches for intelligent systems are and probably will be an area of active research. This also gives motivation for this chapter. Its objective is the identification of selected issues that need to be addressed for performing integration of rule based systems. These issues include: (1) high level modeling techniques for rule bases, (2) integration architectures for rule-based systems, and (3) and rule interoperability. They will be discussed in the reminder of the chapter.

It is important to note, that this chapter has mostly the knowledge engineering perspective on knowledge representation [7, 29]. It means we assume that the rulebase is interactively developed by knowledge engineers using knowledge acquired from human experts. While it was the first and original perspective in knowledge-based systems, for several decades there have been number of advanced methods for an automatic construction of rule sets from data [16, 69]. Today, many classic machine learning [17] algorithms are available to build and optimize rule sets and decision trees [66], that in general correspond to rule-based knowledge. These methods are commonly used in data mining [27] systems, including recent works in learning (mining) from data streams, e.g. [67]. While the perspective on rules and their applications is slightly different (see the discussion on rule types in the next section), some of problems identified in this chapter remain the challenge. This includes handling large rules sets through structuring, integration of rule-based components, as well as rule interoperability issues.

The structure of the chapter is as follows. In Sect. 2 a short review of different rule types and languages used to express them is given. Then in Sect. 3 important issues regarding construction of complex rule bases are introduced, including modeling, structuring, and analysis. Section 4 is devoted to the execution issues of rule bases; the emphasis is put on addressing the structure identified during modeling. Next, in Sect. 5 main approaches to integration of rule-based systems are given. The presentation of problems ends in Sect. 6 where important tools for rule interoperability are presented. The chapter is summarized in Sect. 7.

## 2 Expressing Rules with Rule Languages

Rules are often simply considered conditional statements, that are evaluated or executed to make decision. Rules can also express constraints or regularities. There might be different sources of rules, and thus diverse ways to construct them. They can be provided based on knowledge possessed by human experts and then acquired and properly represented in the knowledge engineering process. Another common case is the automatic construction of rules and rule sets based on the available data. In computational intelligence [33] number of data mining techniques are used for this purpose, e.g. [16].

To evaluate and execute rules a special mechanism has to be implemented. In the common case, its role is to check to conditional part of the rule and if required run the decision/derivation part. The basic form of rules as well as the construction of the execution mechanism may seem straightforward. However, in practice proper formulation of rules, and rule sets turns out to be quite challenging. Not only, rules express different kinds of knowledge acquired from experts or harvested from the available data, but also the use of this knowledge can differ. Example tasks include (but are not limited to) identification/classification, or decision/control. Therefore, number of rule types are identified in the literature.

## 2.1 Types of Rules

Considering logical aspects of inference with rules a basic distinction could be on *deductive* and *abductive* (derivation) rules (used in forward and backwards chaining respectively). Moreover, concepts of *facts* (rules with no condition) and *constraint* rules (defining certain conditions that must hold) [7] are introduced. In [75] an interesting classification is proposed. It is oriented on rule exchange and follows OMG MDA [44]. On the "computation independent" level three general types of rules are identified: integrity, derivation, and reaction. An extended classification is provided by the RuleML organization [58]: integrity, derivation, reaction, production, and transformation rules. Furthermore, in business rules approach [26] a high level BR classification scheme is considered with: terms, facts, and rules. Then the following types of rules are identified: mandatory constraint, guideline, action enabler, computation, inference. Finally, in machine learning [17] and data mining [27] *association rules*, expressing certain correlation between features (attributes) are considered, as well as classification rules. However, the later can be simply interpreted as derivations.

## 2.2 Rule Languages

In general, rules need to be encoded in some kind of notation for computer processing, with the use of rule language. First of all, a rule language can be a certain well-defined notation for encoding and storing rules. In such a case only its syntax has to be defined. Such a language can be oriented on rule execution, thus being close (in terms of its goals) to general programming languages. Examples of such languages are CLIPS, Jess, or Drools. Another objective might be rule interchange and translation. In such a case the language should offer a richer syntax, that allows for expressing different types of rules (perhaps not all of them would be present in every rule base). Examples of such languages are RuleML and RIF. In the case of these languages the semantics of rules is also considered, although not always fully defined.

Formalized languages are an important class of rule languages. Both syntax and semantics of such languages is formally defined. In most of the cases such languages serve not only to represent rules, but are more general knowledge representation languages [29]. Examples include F-Logic [36], or more recently Description Logics [2]. In our research we proposed a dedicated formalized language for rules based on attributive logic [53] called XTT2 [54]. With formalized languages the inference is well-defined, and interchange much simplified. The design issue can also be better addressed. A limitation can be a lower flexibility and expressiveness compared to solutions like CLIPS. This is due to the fact that "programming rule languages" have often vague (undefined) semantics. A clear benefit of formalized solution is also the possibility of formal model checking and verification.

From the perspective of this chapter an important and large group of rule languages are *attributive languages*. Knowledge representation based on attributes is very common and intuitive, as it is related to technical ways of presentation. In such a case the behavior of a physical system is described by providing the values of system variables. This kind of logic approach is used in various applications e.g. relational database tables [13], attributive decision tables and trees [37, 60], and attributive RBS [40]. In order to define characteristics of the system one selects some specific set of attributes and assigns them some values. This way of describing an object and system properties is both simple and intuitive. Such languages provide a number of features, making them efficient tool for practical representation and manipulation of knowledge. After [40], these features can be as follows: introducing variables (the same attribute can take different values and there is no need to introduce new propositional symbols), specification of constraints (using relations between attribute values it is possible to specify constraints), and parametrization (attributes are parameters to be instantiated). Thanks to these advantages, the attributive logic is more expressive than the propositional logic.

## 3   From Construction of Rule Sets to Design of Rule Bases

Basic discussion of rule-based systems is commonly focused on building single rules, or constructing relatively small sets of rules. This is justified in simple cases, or studies of rule extraction algorithms. However, in engineering practice the size, structure, properties, and quality of such rule sets is very important. These challenges give motivation to consider proper building of rule bases as a dedicated design process. In such a process certain activities, often ordered as phases can be identified:

- rule base modeling—may include selection of a certain representation mechanism that can simply the design. Very often these are visual forms of rule sets, such as decision tables or decision trees.
- structure identification—where relation between groups of rules are identified.
- analysis—allows for assuring the quality of the rule base.

We discuss these issues next.

## 3.1 Rulebase Modeling

Rule base modeling is an evolutionary process. Due to the large amount of rules or complex dependencies, the modeled knowledge may not reflect the acquired knowledge in an appropriate way. Therefore, in order to make the modeling process more efficient a number of visual methods are developed. The visual (or semi-visual) languages facilitate modeling phase making it more transparent for the knowledge engineer. As the number of rules identified in the system is increasing, it may be difficult to model and manage them. Thus, in complex systems having rule sets consisting of thousands of rules, various forms of rule set representation are used. Such forms as tables or trees are logically equivalent to a set of single rules, but they are easier to understand. Moreover decision diagrams are also used [40]. They represent the decision making process in a the form of graph, so a more general structure than a tree). In some cases basic decision tables, or diagrams can also be built automatically from computational intelligence models [11].

Decision tables are used to group similar rules. Rules grouped into a table usually correspond to the *canonical set* of rules, i.e. a rule set satisfying the assumptions that [40]: all rules use the same propositional symbols in the same order, and rules differ only with respect to using the negation symbol before the propositional symbol. A set of rules which is not in the canonical form, can always be transformed to an equivalent canonical set. Typically, such canonical sets of rules are used for creating decision tables. In the basic binary decision table each rule is specified in a single row, in which the first $n$ columns specify conditions under which specific conclusion is fulfilled.

To enhance the expressive power and knowledge representation capabilities Attributive Logic can be used, for Attribute-Value Pair Table (AV-Pair Table) or [40] for Attributive Decision Table (AD-Table). A row of an AD-Table represents a rule, expressed as follows:

$$r_i : (p_1 = v_{i1}) \land (p_2 = v_{i2}) \land \cdots \land (p_n = v_{in}) \rightarrow h_1 = w_{i1} \land h_2 = w_{i2} \land \cdots \land h_m = w_{im}$$

Conditions of such a rule can take several values from a specified domain. Moreover, this approach can be extended in order to allow for specifying an attribute value as an interval or a subset of the domain [40]. An example of such a table determining a rented car category based on the driver age and driving license holding period is presented in Table 1.

Decision trees allow for organizing rules in a hierarchical manner. As they show the dependencies between conditions and decisions, this clarifies the thinking about the consequences of certain decisions being made [25]. A decision tree has a flowchart-like structure in which a node represent an attribute and branches from such a node represent the attribute values. The end nodes (leaves) represent the final decision values. Such a form of knowledge representation allows for clear presentation of the decision process. Unfortunately, decision trees become much more complex if each attribute has a large number of different values because of the

**Table 1** An example of decision table

| Driver age | Driving license holding period | Rented car category |
|---|---|---|
| <18 | Any | None |
| <21 | <2 | A |
| <21 | >=2 | {A, B} |
| >=21 | <2 | {A, B, C} |
| >=21 | >=2 | {A, B, C, D} |

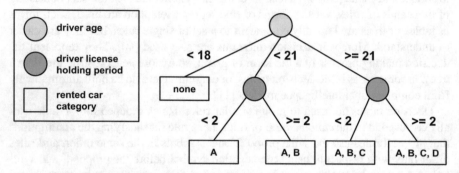

**Fig. 1** An example of a decision tree corresponding to the decision table

redundancy of nodes. An example of a decision tree corresponding to the decision table from Table 1 is presented in Fig. 1.

Both trees and tables are also useful from practical point of view, as they provide a visual representation of the rule base. We assume after [40] that the transformation between rules, tables, and trees is always possible (having some syntactic restrictions). These visual models help during the design process of rule bases.

## 3.2 Structure

Large rule sets should more commonly be referred to as "rule bases" as rules in such a set are most often interrelated. Relations of rules in a rule base can be expressed explicitly in rule bodies. Examples include rules for decision control, where execution of certain rules explicitly calls another rules. Moreover, there are cases of rewriting systems, where some rules can be modified by others. Furthermore, rule sets can be explicitly partitioned into groups operating in or regarding given situations. There might also occur implicit relations between rules. Probably the most common and important case is when rules that share the same attributes. Even if such rules are not grouped together they could be related to similar situations, or objects. What makes such cases even more complicated are logical relations between rules.

This can result in contradicting or excluding which can lead to unexpected operation of the system as a whole. There are different solutions to address these issues.

A simple solution, common in rule-based shells, is the introduction of modularization of the rule base. CLIPS offers functionality for organizing rules into so-called modules. They allow for the restriction of access to their elements from other modules, and can be compared to global and local scoping in other programming languages. In CLIPS each module has its own pattern-matching network for rules and its own agenda (rules to be fired). Jess provides a similar module mechanism that helps to manage large rule bases. Modules provide a control mechanism: rules in a module will fire only when this module has the focus, and only one module can have focus at a time. In general, although any Jess rule can be activated at any time, only rules in a module having focus will fire. In certain cases this mechanism can improve management of the rule base., as the large set of rules can be partitioned into smaller ones. It also has positive impact on the performance of the inference process as not all of the rules need to be analyzed. A similar approach was employed in the Drools system [8]. However, as Drools moved away from CLIPS-like inference in large rule bases to a dedicated process engine, it will be described in the subsequent section on inference.

Another approach is to introduce the structure into the model of the knowledge base during the design. Use of visual representation methods such as decision tables can simplify grouping of rules sharing the same attributes [72]. A decision tree can also be used to represent a group of rules but emphasizing the inference process. There exist hybrid representations such as XTT2 (eXtended Tabular Trees) that combine tables with trees [54]. While tables group rules with the same attributes, a high level inference network allows to control the inference process. The tables can be connected during the design process to denote relations between groups of rules. These connections can be further used by the inference engine to optimize the inference process. For more detail see [46]. In [50] a complete design and integration approach for formalized rule-based systems was introduced. It is called *Semantic Knowledge Engineering* (SKE) as it put emphasis on the proper interpretation of rule based knowledge as well as on its integration with other software engineering paradigms. In the subsequent parts of this chapter we will briefly discuss how inference control and integration are handled in SKE.

## 3.3  Analysis

Rule-based systems are widely used in areas where high performance and reliability are important. In some cases a failure of such system may have serious consequences. Therefore, it is crucial to ensure that the system will work correctly in every possible situation.

Verification and validation were discussed by many authors, including [1, 55, 57, 61]. Verification concerns proving correctness of the set of rules in terms of some verifiable characteristics [4]. In fact features such as consistency, completeness, and various features of correctness may be efficiently checked only by using formal methods [41]. In turn, validation is related to checking if the system provides correct answers to specific inputs. In other words, validation consists in assuring that the system is sound and fits the user requirements.

According to [76] verification and validation procedures of the system can be understood as one of the following: anomaly detection, formal verification, parallel use, rule base visualization to aid review, code review, testing. That study shows, that verification of the rule based systems is dominated by testing and code review. This approach highly depends on human skills, since incorrectly written test may produce wrong results. Formal verification and anomaly detection are not so widely used despite the fact that those methods usually have strong logical foundations and in most cases exceed testing and debugging approach.

According to comparison of existing verification tools in [71] one can draw a conclusion, that the main reason why formal verification is not widely used among expert system developers is that it requires *formal knowledge representation*. In fact, most of these tools are usually based on propositional or predicate logic. Melodia [10] uses propositional logic and flat rule base. Clint [15], Cover [63], use predicate logic and a flat rule base. Moreover, Indepth [43] introduces a hierarchical representation, Covadis [65] uses simple production rules language with a flat rule base. However, common expert system shells such as CLIPS, Jess or Drools do not provide formal knowledge representation, so it is not possible to apply formal methods to these tools. Although there are some analysis tools that are dedicated to aforementioned shells like CrsvClips [14], Drools Verifier [12], their aim is not to provide formal verification, but to offer a framework for writing tests.

Verification of knowledge in RBS is typically considered the last stage of the design procedure [39]. It is assumed to be performed on a complete, specified knowledge base [62, 73], as such it is costly and difficult. In the SKE approach we advocate for approaches that introduce a formalized description of rule base. Thanks to them formal verification of the rule base is possible during its design. A verification framework for XTT2 knowledge bases called HalVA was introduced in [47]. It allows for verification of formal properties such as determinism, or local completeness, as well a redundancy (subsumption). The toolset works on the XTT2 table level and is integrated as a module of the HeaRT [48] inference engine, that can be called during the design of the rulebase. Its full description can be found in [50].

## 4 Execution of Rule Bases

### 4.1 Inference in Rule Bases

In RBS the execution of rule base is related to the automated inference process performed by an inference engine. It uses specific algorithms to analyze the contents of the rule base, identify rules that can be fired, and fires them. It is generally assumed that the engine and algorithm are independent from the encoded knowledge and allow for processing knowledge from any domain. Important aspects that determine operations of the inference engine include the inference mode and tasks. *Inference mode* defines how the knowledge contained in the rule base is processed [32]. Forward chaining is the *data-driven* (or bottom-up) reasoning. This mode of reasoning starts from the existing knowledge stored as facts and continues until no further conclusions can be drawn. *Backward chaining* is a reverse process to forward chaining and is called *goal-driven* reasoning. In this mode the system has a goal (a hypothetical solution) and the inference engine attempts to find the evidence to prove it with the help of the facts stored within fact base. Both inference modes can be applied to different kinds of problems. However, according to [22] forward chaining is a natural way to design expert systems for analysis and interpretation. An *inference task* is a scenario of using rules that is performed by inference engine working in a given inference mode. Thus, it is important to distinguish between inference modes, like forward and backward chaining, and inference tasks. Therefore, a given inference task can be performed in different inference modes. Examples include *final consequence* that determines the evaluation of a given set of rules in order to infer all possible conclusions based on the existing facts and the facts drawn during inference. *consequence reduction* forces an inference engine to answer question if a given hypothesis can be proved to be true according to existing facts. Executing this task, an inference engine tries to find such sequence of rules that allows for expressing the hypothesis by means of the existing facts.

A typical forward chaining inference process performed in RBS is an iterative process consisting of the following steps: match, conflict set resolution, action, return. Among these four steps, the first step is the bottleneck of inference because it requires to match facts stored within fact base to rules in order to check if a given rule has satisfied conditions. An important and now commonly used algorithm is called Rete [18]. This algorithm allows for avoiding the Naïve approach and makes the match step much more efficient. Knowledge compilation is used where each knowledge base is compiled and the set of all rules is transformed into so-called discrimination network that represents all the rules in the form of directed and acyclic graph. Each node of this graph corresponds to single condition of a certain rule. The second idea is to store the information concerning facts satisfying certain condition within corresponding node. Thanks to that, operations performed during this step are limited to monitor only changes (adding or removing) made in the fact base. When such change is observed, it is passed through the network in order to identify rules having satisfied their conditions. As many unnecessary rules can be fired, we shall refer to the inference scheme as a *blind* one.

State saving mechanism implemented in Rete is not very efficient. The structure of the network is often redundant, and the number of elements stored in memory may be combinatorially explosive. To address these problems an improved algorithm called TREAT was proposed by Miranker [45]. The conflict set is explicitly retained across production system cycles which allows for advancements over Rete. Both Rete and TREAT offer static networks. The structures of the networks are defined arbitrary by the design engineer and look mostly the same for all kinds of knowledge bases. This often leads to the creation of networks that are not optimal for some knowledge bases. To address this problem a new discrimination network algorithm called Gator [28] was proposed. It is based on Rete, but additionally implements mechanisms for optimizing network structure according to specific knowledge base characteristic.

## 4.2 Improving Inference in Structured Rule Bases

Modularization of knowledge base helps managing rules, and improves efficiency of rule-based system execution. The structuring of the rule base can be used during the inference process. CLIPS modules allow for restriction of access to their elements from other modules, and can be compared to global and local scoping in other programming languages. In CLIPS each module has its own pattern-matching network for its rules and its own agenda. When a *run* command is given, the agenda of the module which is the current focus is executed. Rule execution continues until another module becomes the current focus, no rules are left on the agenda, or the return function is used from the RHS of a rule. Whenever a module that was focused on runs out of rules on its agenda, the current focus is removed from the focus stack and the next module on the focus stack becomes the current focus. Before a rule executes, the current module is changed to the module in which the executing rule is defined. The current focus can be dynamically switched. A similar mechanism is present in Jess.

The Drools platform introduced a RuleFlow tool. It is a workflow and process engine that allows advanced integration of processes and rules. It provides a graphical interface for processes and rules modeling. Drools have built-in a functionality to define the structure of the rulebase which can determine the order of the rules evaluation and execution. The rules can be grouped in a ruleflow-groups which defines the subset of rules that are evaluated and executed. The ruleflow-groups have a graphical representation as the nodes on the *ruleflow* diagram. The ruleflow-groups are connected with the links what determines the order of its evaluation. A *ruleflow* diagram is a graphical description of a sequence of steps that the rule engine needs to take, where the order is important.

More recently, Drools moved from a dedicated flow control engine into the integration of rule-based reasoning system with a complete Business Process Management systems in Drools 5. In this case rule-based subsystems can be called arbitrarily by a high-level flow control mechanism. In this case it is a Business Process engine jBPM. This approach to controlling the rule-based inference will be described in the section regarding integration of RBS.

## 4.3 Inference Control in SKE

In the SKE approach the rule base is composed of extended decision tables in the XTT2 notation. Any table can have input links (inputs) as well as output links (outputs). Links are related to the possible inference order. Tables to which no connections point are referred to as input (or start) tables. Tables with no connections pointing to other tables are referred to as output tables. All the other tables (ones having both input and output links) are referred to as middle tables.

We consider a network of tables connected according to the following principles: there is at least one input table, there is at least one output table, there is zero or more middle tables, and all the tables are interconnected. The aim is to choose the inference order. The basic principle is that before firing a table, all the immediately preceding tables must have already been fired. The structure of the network imposes a partial order with respect to the order of table firing. Firing the table involves processing in a sequence all the rules in the table. In [46] several dedicated algorithms for inference control were described. This approach is only suitable for relatively small knowledge bases, where the manual analysis is possible. Therefore, more complex modes are considered, including DDI (Data-Driven Inference), TDI (Token-Driven Inference), and GDI (Goal-Driven Inference).

The *Data-Driven Inference* algorithm identifies start tables, and puts all the tables that are linked to the initial ones in the XTT2 network into a FIFO queue. When there are no more tables to be added to the queue, the algorithm fires selected tables in the order they are popped from the queue. The forward-chaining strategy is suitable for simple tree-like inference structures. However, it has limitations in a general case, because it cannot determine tables having multiple dependents. The *Token-Driven Inference* approach is based on monitoring the partial inference order defined by the network structure with tokens assigned to tables. A table can be fired only when there is a token at each input. Intuitively, a token is a flag signaling that the necessary data generated by the preceding table is ready for use. The *Goal-Driven inference* approach works backwards with respect to selecting the tables necessary for a specific task, and then fires the tables forward so as to achieve the goal. One or more output tables are identified as the ones that can generate the desired goal values and are put into a LIFO queue. As a consequence, only the tables that lead to the desired solution are fired, and no rules are fired without purpose. All of the mentioned inference modes are implemented as a part of a dedicated inference engine for SKE called HeaRT [48].

## 5  Integration of Rule-Based Systems

Historically, rule-based systems were considered as stand alone. This meant such a systems was an independent software component (sometimes integrated in a hardware systems). As such, it was fully responsible to process input data, perform processing, and then appropriate decision making and ultimately produce output data, or carry out control actions. Therefore, with time, in classic RBS systems such as CLIPS, number of additional libraries were created to support such an environment. Today however, such an approach seems redundant and it is rather rare. RBS are considered software components, that have to be integrated in a larger software environment using some well-defined software engineering approaches [70]. Therefore, here we give a short account of main architectures to integrate rule-based systems with a larger software environment.

The already mentioned, classic approach with standalone systems can be considered a *homogeneous* one. As in such a case the RBS should be able to provide not just the decision making, but also vital part of interfaces on the software runtime level. An important aspect is in fact related to the rule language level. In this case, the rule language should be powerful enough to program all of these features, as it is the only language available for the system designer. This results is the design of expressive rule languages like in the case of CLIPS with additional programming libraries, or language extensions such as COOL [23].

An alternative approach is to restrict the role of the RBS only to decision making. In this case, the remaining functionality is delegated to another systems or components. The RBS only needs to posses interfaces allowing for such lower-level integration. It also operates as intelligent middleware, not a stand-alone system. Therefore, such an architecture can be simply referred to as *heterogeneous* one.

## 5.1 Heterogeneous Integration

The rule-based component can be integrated with a larger software system using common software design patterns [20]. An example of such an approach was previously proposed in [50]. It is related to bridging knowledge engineering with software engineering [70]. Historically, when the software systems became more complex, the engineering process became more and more declarative in order to model the systems in a more comprehensive way. It made the design stage independent of programming languages, which resulted in a number of approaches. One of the best examples is the MDA (Model-Driven Architecture) approach [44]. Since there is no direct "bridge" between declarative design and sequential implementation, a substantial work is needed to turn a design into a running application. This problem is often referred to as a *semantic gap* between a design and its implementation [42]. It is worth noting that while the conceptual design can sometimes be partially formally analyzed, the full formal analysis is impossible in most cases [52]. However, there is no way to assure that even a fully formally correct model would translate to a correct code in a programming language. Moreover, if an application is automatically generated from a designed conceptual model, then any changes in the generated code have to be synchronized with the design. Another issue is the common lack of separation between core software logic, interfaces, and presentation layers.

Some of the methodologies e.g. the MDA, and the design approaches e.g. the MVC (Model-View-Controller) [9] try to address this issue. The main goal is to avoid semantic gaps, mainly the gap between the design and the implementation. In order to do so, the following elements should be developed: a rich and expressive design method, a high-level runtime environment, and an effective design process. Methodologies which embody all of these elements should eventually shorten the development time, improve software quality, and transform the "implementation" into the runtime-integration and introduce the so-called "executable design".

Using these ideas the *heterogeneous integration* of a RBS may be considered on several levels:

- *runtime level*: the application is composed of the rule-based model run by the inference engine integrated with the external interfaces.
- *service level*: the rule-based core is exposed to external applications using a network-based protocol. This allows for a SOA (Service-Oriented Architecture)-like integration [3] where the rule-based logic is designed and deployed using an embedded inference engine.
- *design level*: integration considers a scenario, where the application has a clearly identified logic-related part, which is designed using visual design method for rules (such as decision table, or decision trees), and then translated to a domain-specific representation.
- *rule language level*: in this case rule expressions can be mixed with another programming language, and both syntax and semantics are mixed. However, this allows for an easy integration of rule-based code with rich features of another programming environment (e.g. Java).

We will now discuss how these are integrated in the SKE.

## 5.2   Integration in the SKE Approach

The SKE approach provides a heterogeneous solution through a clear *separation of core business logic*. Eventually it can shorten the development time by transforming the "implementation" into the runtime-integration with rule-based model. The approach introduces a strong separation of the core application logic from the interfaces. In fact, it is assumed that the MVC-like software design pattern is used. The intelligent application is decomposed into a Model that captures the logic, a View that corresponds to different interfaces, and a Controller that links these two. The SKE architecture provides means for the design and implementation of software logic and the integration of this logic with the presentation layer. The emphasis is on a rich and formally designed and analyzed knowledge-based model. It is important to observe that as opposed to standard software engineering approaches there are no differences in the semantics of design methods. Thanks to the XTT-based logic core, the knowledge base is represented using a formalized knowledge model. This allows for using formal analysis of the model and avoiding common evaluation problems.

**Fig. 2** Heterogeneous
system architecture [50]

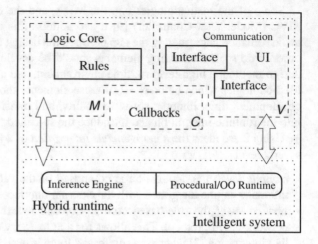

Such an analysis can be provided at the design stage, which allows for a gradual refinement of the designed system (Fig. 2).

The concept of the model considered here is based on certain concepts related to *classic control theory* and dynamic system modeling. The primary assumption is that the rule-based model is a model of a dynamic system having a certain *state*. The state is described using attributes that represent important properties of the system. A statement that an attribute has a given value can be interpreted as a fact in terms of classic expert systems. The concept of the state is similar to the one in dynamic systems and state-machines. The current state of the system is considered as a complete set of values of all the attributes at the instant of time. The *dynamics* of the system (transitions between states) is modeled with the use of *rules* described by a logical representation. The conditional part of a rule is an expression related to the state to be matched. The decision part includes statements that modify a system state in case the rule is fired (the proper decision) and actions that do not change attribute values, thus the state. This is a *declarative* model.

In general, the values of the XTT attributes in the model can be modified by an independent external system (or user). This case concerns attributes representing some process variables, which are taken into account in the inference process, but depend only on the environment and external systems. As such, the variables cannot be directly changed by the XTT system. Values of those variables are obtained as a result of some measurement or observation process, and are assumed to be put into the inference system via a *blackboard* communication method [30].

To connect the internal system memory with the external environment, *callbacks* are used. These are dedicated functions related to system attributes. Callbacks are invoked to get and send attribute values from and to the environment. The values of internal attributes can only be modified by the inference process itself. In such a case, values of attributes are obtained at certain stages of reasoning as the result of the operations performed in the decision part of XTT rules.

The *integration of a heterogeneous system* is considered mainly at the at the run-time, service and design levels.

- *runtime level*: the application is composed of the rule-based model run by the inference engine integrated with the external interfaces using the callback mechanisms. The model is run by the XTT inference engine that uses callbacks to communicate with front-ends implemented with other languages e.g. Java.
- *service level*: the rule-based core is exposed to external applications using a network-based protocol. This allows for a SOA-like integration [3] where the XTT logic is designed and deployed using the dedicated inference engine. The state of the XTT-based system can be modified with the use of callbacks triggered by attribute changes.
- *Design level*: integration considers a scenario, where the application has a clearly identified logic-related part, which is designed using the XTT method, and then translated to a domain-specific representation. As an example, in [51] the translation from XTT to the UML notation is discussed.

An important issue is the integration of rules and processes that will be discussed next.

## 5.3 Integration of Rules and Business Process System

Drools 5 platform includes several integrated modules including *Drools Expert* and *jBPM* [68]. The former is a business rules execution engine. It implements an extended version of Rete, called ReteOO. Improvements include the object-oriented type system that allows for tight integration with Java. In fact the whole platform is implemented in JavaEE. JBPM is a full-fledged process execution engine. It executes business process models encoded in the BPMN notation [59]. This notation includes dedicated syntactic constructs, so-called rule tasks. Thanks to them it is possible to delegate the execution of details of business process logic to a rule-bases system. Practically it can be any system implemented in Drools. However, from the design transparency perspective a reasonable approach is to connect only restricted well-defined subsystems, or even single modules (tables).

Following the previously defined levels, such a scenario for integration is mainly runtime-oriented. While proper design tools are currently not available for Drools, with some extensions this integration can also be reflected on the design level. Preliminary work in this direction was presented in [38], where a web design frameworks for business process with rules where presented. Drools also allows for service-level integration, as the whole runtime environment is web-enabled. It supports the orchestration of web services using rules. Execution of such solutions is supported by the runtime environment.

# 6 Rule Interoperability

Having complete, possibly verified and validated system, it is desirable to ensure method for sharing knowledge with other systems, representations and tools. Furthermore, with the increasing number of rules application areas, the number of different rule representations is also growing. The differences between these representations cause that the rule-based knowledge cannot be easily shared among different rule bases. Usually, the naive translation methods do not take rule base semantics into account what leads the semantics mismatch before and after translation. This problem is called *rule interoperability problem* and it has been known since classic expert systems [22]. Today it returns because of novel nature of rules applications in business technologies. In the context of this problem, a lot of research has been conducted. The goal is to facilitate the process of interoperability between representations by providing intermediate and formalized format for knowledge translation.

In general, the methodology of interoperability must take two aspects into account: syntax that is used for knowledge encoding and semantics. On each of these two levels some problems can be identified, including ambiguous semantics, different expressiveness, and syntactic power.

Over the time, many different methods and approaches to the knowledge interoperability problem were developed. Some of them are general-purpose i.e. aim at providing framework for translation between many different representations. Historically, first of such approaches that were developed was called Knowledge Interchange Framework. Due to the difficulty of maintaining of such approaches, there is very few technologies use this framework. This is why, modern methods providing wide support for many different representations, are usually divided into so-called dialects. Each dialect has a well-defined semantics and thus is intended to translation of rules expressed in some well-defined representations. *Rule Interchange Framework* consists of several dialects providing support for example for production rules. Apart from the methods supporting many different representations, more specialized approaches are also developed. Many of the existing technologies are dedicated for a certain set of representations that share similar assumptions and thus have similar semantics. *Rule Markup Language* is an example of such technologies that is dedicated for representations used mainly within Semantic Web. In turn, *Production Rule Representation* allows for expressing production rules that perform actions and thus allow for changing system state.

*Knowledge Interchange Framework* (KIF) [21] constitutes one of the first implementation of formal knowledge interoperability approach that uses unified intermediate representation model providing declarative semantics. KIF was intended to be a formal language for the translation of knowledge among disparate computer programs providing possibility of precise definition of knowledge semantics. It was not limited only to rules but supports also other representation techniques like frames, graphs, natural language, etc. It is important to note that KIF was not intended as a primary language for interaction with human users (though it can be used for this purpose). Different programs could interact with their users in whatever forms that are most appropriate to their applications. The formal definition (specification) of KIF provides very complex meta-model consisting of large number of classes. Moreover, its complexity led to very weak tool support and currently there is no tools that support KIF even partially.

*Rule Interchange Format* (RIF) [34, 35] is a result of research conducted by Rule Interchange Format Working Group. This group was established by the World Wide Web Consortium (W3C) in 2005 to create a standard for exchanging rules among rule systems, in particular among web rule engines. Although originally envisioned by many as a *rule layer* for the Semantic Web, in reality the design of RIF is based on the observation that there are many rule languages in existence, and what is needed to exchange rules between them. In RIF rule systems fall into three main categories: *first-order*, *logic-programming*, and *action rules*. These paradigms share little in terms of syntax and semantics. Moreover, there are large differences between systems even within the same paradigm. The approach taken by the group was to design a family of languages, called dialects with rigorously specified syntax and semantics of different rule systems.

*Production Rule Representation* (PRR) [58] is an OMG standard for production rule representation, that addresses the need for a representation of production rules in UML models (i.e. business rule modeling as part of a modeling process). It adopts the rule classification scheme supplied by the RuleML Initiative and supports only production rules. It provides the MOF-based metamodel and profile that are composed of a core structure referred to as PRR Core and a non-normative abstract OCL-based syntax for the expressions, defined as an extended PRR Core metamodel referred to as PRR OCL [58].

*Rule Markup Language* (RuleML) [5, 6] is defined by the RuleML Initiative.[1] This initiative aims at developing an open, vendor neutral XML/RDF-based rule language allowing for exchange of rules between various systems including: distributed software components on the web, heterogeneous client-server systems found within large corporations, etc. RuleML is intended to be used in Semantic Web and this is why it offers XML-based language syntax for rules. In turn, the abstract syntax of this language is specified by means of a version of Extended BNF, similar to EBNF notation used for XML. RuleML provides an underlying formalism which precisely defines semantics of the language. This formalism is based on the partial logic [31] and provides a formal meaning for RuleML knowledge bases written in the abstract syntax. The foundation for the kernel of RuleML is the Datalog (constructor-function-free) sublanguage of Horn logic. Its expressiveness allows for expressing both forward (bottom-up) and backward (top-down) rules in XML. It also supports different kind of rules: derivation rules, transformation rules, reaction rules and production rules. The formal model of RuleML is comprehensively described in [74].

Interoperability problems exist not only with rule bases design by knowledge engineers but also with rule sets built with data mining tools. This short review of existing tools for translating rules bases gives a certain insight to the main problems encountered when translating one rule set into another. Clearly a proper formaliza-

---

[1] See http://www.ruleml.org.

tion of both syntax and semantics of rules can be useful in such translation. However, few methods and tools support such formalization, which turns out to be quite tedious. Finally, collaborative tools for knowledge engineering, such as semantic wikis can improve the way rule-based knowledge is created, and shared [49].

# 7 Concluding Remarks

The objective of this chapter was to emphasize and discuss selected important challenges in integration of rule-based systems. They include modeling techniques for structures rule bases, integration architectures using software engineering paradigms, as well as rule interoperability issues. These challenges exist both in rule bases developed by knowledge engineers using expert knowledge, and in cases where rule sets are built by data mining approaches [56], in computational intelligence paradigm [33]. We gave examples of selected tools, and techniques to address these challenges.

# References

1. Andert, E.P.: Integrated knowledge-based system design and validation for solving problems in uncertain environments. Int. J. Man-Mach. Stud. **36**(2), 357–373 (1992). http://www.reviews. com/reviewer/quickreview/frameset_toplevel.cfm?bib_id=144453
2. Baader, F., Calvanese, D., McGuinness, D.L., Nardi, D., Patel-Schneider, P.F. (eds.): The Description Logic Handbook: Theory, Implementation, and Applications. Cambridge University Press (2003)
3. Bieberstein, N., Bose, S., Fiammante, M., Jones, K., Shah, R.: Service-Oriented Architecture (SOA) Compass: Business Value, Planning, and Enterprise Roadmap. IBM Press (2006)
4. Boehm, B.W.: Verifying and validating software requirements and design specifications. IEEE Softw. **1**(1), 75–88 (1984)
5. Boley, H., Paschke, A., Shafiq, O.: RuleML 1.0: the overarching specification of web rules. In: M. Dean, J. Hall, A. Rotolo, S. Tabet (eds.) Semantic Web Rules—International Symposium, RuleML 2010, Washington, DC, USA, 21–23 Oct 2010. Proceedings. Lecture Notes in Computer Science, vol. 6403, pp. 162–178. Springer (2010). doi:10.1007/978-3-642-16289-3
6. Boley, H., Tabet, S., Wagner, G.: Design rationale for RuleML: a markup language for semantic web rules. In: Cruz, I.F., Decker, S., Euzenat, J., McGuinness, D.L. (eds.) Proceedings of SWWS'01, The First Semantic Web Working Symposium, Stanford University, California, USA, 30 July–1 Aug 2001, pp. 381–401 (2001). http://www.semanticweb.org/SWWS/program/full/paper20.pdf
7. Brachman, R., Levesque, H.: Knowledge Representation and Reasoning, 1st edn. Morgan Kaufmann (2004)
8. Browne, P.: JBoss Drools Business Rules. Packt Publishing (2009)
9. Burbeck, S.: Applications programming in Smalltalk-80(TM): How to use Model-View-Controller (MVC). Department of Computer Science, University of Illinois, Urbana-Champaign, Technical report (1992)

10. Charles, E., Dubois, O.: Melodia: logical methods for checking knowledge bases. In: Ayel, M., Laurent, J.P. (eds.) Validation, Verification and Test of Knowledge-Based Systems, pp. 95–105. Wiley, New York (1991). http://portal.acm.org/citation.cfm?id=130251.130258
11. Chorowski, J., Zurada, J.M.: Extracting rules from neural networks as decision diagrams. IEEE Trans. Neural Netw. **22**(12), 2435–2446 (2011). doi:10.1109/TNN.2011.2106163
12. Community, J.: Drools verifier. http://community.jboss.org/wiki/DroolsVerifier (2009)
13. Connolly, T., Begg, C., Strechan, A.: Database Systems, A Practical Approach to Design, Implementation, and Management, 2nd edn. Addison-Wesley (1999)
14. Culbert, S.: Expert system verifications and validation. In: Proceedings of First AAAI Workshop on V,V & Testing, Aug 1988
15. De Raedt, L., Sablon, G., Bruynooghe, M.: Using interactive concept-learning for knowledge base validation and verification. In: Ayel, M., Laurent, J. (eds.) Validation, Verification and Testing of Knowledge Based Systems, pp. 177–190. Wiley (1991)
16. Duch, W., Setiono, R., Zurada, J.M.: Computational intelligence methods for rule-based data understanding. In: Proceedings of the IEEE, pp. 771–805 (2004)
17. Flach, P.: Machine Learning: The Art and Science of Algorithms That Make Sense of Data. Cambridge University Press, New York (2012)
18. Forgy, C.: Rete: a fast algorithm for the many patterns/many objects match problem. Artif. Intell. **19**(1), 17–37 (1982)
19. Friedman-Hill, E.: Jess in Action, Rule Based Systems in Java. Manning (2003)
20. Gamma, E., Helm, R., Johnson, R., Vlissides, J.: Design Patterns, 1st edn. Addison-Wesley Pub Co. (1995)
21. Genesereth, M.R., Fikes, R.E.: Knowledge Interchange Format Version 3.0 Reference Manual (1992)
22. Giarratano, J., Riley, G.: Expert Systems. Principles and Programming, 4th edn. Thomson Course Technology, Boston, MA, United States (2005). ISBN 0-534-38447-1
23. Giarratano, J.C., Riley, G.D.: Expert Systems. Thomson (2005)
24. Giurca, A., Gašević, D., Taveter, K. (eds.): Handbook of Research on Emerging Rule-Based Languages and Technologies: Open Solutions and Approaches. Information Science Reference, Hershey, New York (2009)
25. Graham, I.: Business Rules Management and Service Oriented Architecture. Wiley (2006)
26. von Halle, B.: Business Rules Applied: Building Better Systems Using the Business Rules Approach. Wiley (2001)
27. Han, J., Kamber, M.: Data Mining: Concepts and Techniques. Morgan Kaufmann Publisher (2000)
28. Hanson, E.N., Hasan, M.S.: Gator: An Optimized Discrimination Network for Active Database Rule Condition Testing. Technical Report 93-036, CIS Department University of Florida (1993)
29. van Harmelen, F., Lifschitz, V., Porter, B. (eds.): Handbook of Knowledge Representation. Elsevier Science (2007)
30. Hayes-Roth, B.: A blackboard architecture for control. Artif. Intell. **26**(3), 251–321 (1985)
31. Herre, H., Jaspars, J.O.M., Wagner, G.: Partial logics with two kinds of negation as a foundation for knowledge-based reasoning. Centrum voor Wiskunde en Informatica (CWI) **158**, 35 (1995)
32. Jackson, P.: Introduction to Expert Systems, 3rd edn. Addison-Wesley (1999). ISBN 0-201-87686-8
33. Kacprzyk, J., Pedrycz, W. (eds.): Springer Handbook of Computational Intelligence. Springer (2015). doi:10.1007/978-3-662-43505-2
34. Kifer, M.: Rule interchange format: the framework. In: Calvanese, D., Lausen, G. (eds.) Web Reasoning and Rule Systems, Second International Conference, RR 2008, Karlsruhe, Germany, 31 Oct–1 Nov 2008. Proceedings. Lecture Notes in Computer Science, vol. 5341, pp. 1–11. Springer (2008). doi:10.1007/978-3-540-88737-9_1
35. Kifer, M., Boley, H.: RIF overview. W3C working draft, W3C. http://www.w3.org/TR/rif-overview (2009)

36. Kifer, M., Lausen, G., Wu, J.: Logical foundations of object-oriented and frame-based languages. J. ACM **42**(4), 741–843 (1995). doi:10.1145/210332.210335
37. Klösgen, W., Żytkow, J.M. (eds.): Handbook of Data Mining and Knowledge Discovery. Oxford University Press, New York (2002)
38. Kluza, K., Kaczor, K., Nalepa, G.J.: Enriching business processes with rules using the Oryx BPMN editor. In: Rutkowski, L., et al. (eds.) Artificial Intelligence and Soft Computing: 11th International Conference, ICAISC 2012: Zakopane, Poland, 29 Apr–3 May 2012. Lecture Notes in Artificial Intelligence, vol. 7268, pp. 573–581. Springer (2012). http://www.springerlink.com/content/u654r0m56882np77/
39. Liebowitz, J. (ed.): The Handbook of Applied Expert Systems. CRC Press, Boca Raton (1998)
40. Ligęza, A.: Logical Foundations for Rule-Based Systems. Springer, Berlin, Heidelberg (2006)
41. Ligęza, A., Nalepa, G.J.: Rules verification and validation. In: Giurca, A., Gašević, D., Taveter, K. (eds.) Handbook of Research on Emerging Rule-Based Languages and Technologies: Open Solutions and Approaches, pp. 273–301. IGI Global, Hershey, New York (2009)
42. Mellor, S.J., Balcer, M.J.: Executable UML: A Foundation for Model Driven Architecture, 1st edn. Addison-Wesley Professional (2002)
43. Meseguer, P.: Incremental verification of rule-based expert systems. In: Proceedings of the 10th European conference on Artificial intelligence, ECAI '92, pp. 840–844. Wiley, New York, NY, USA (1992). http://portal.acm.org/citation.cfm?id=145448.147581
44. Miller, J., Mukerji, J.: MDA Guide Version 1.0.1. OMG (2003)
45. Miranker, D.P.: TREAT: A Better Match Algorithm for AI Production Systems; Long Version. Technical Report 87-58, University of Texas (1987)
46. Nalepa, G., Bobek, S., Ligęza, A., Kaczor, K.: Algorithms for rule inference in modularized rule bases. In: Bassiliades, N., Governatori, G., Paschke, A. (eds.) Rule-Based Reasoning, Programming, and Applications. Lecture Notes in Computer Science, vol. 6826, pp. 305–312. Springer, Berlin, Heidelberg (2011)
47. Nalepa, G., Bobek, S., Ligęza, A., Kaczor, K.: HalVA—rule analysis framework for XTT2 rules. In: Bassiliades, N., Governatori, G., Paschke, A. (eds.) Rule-Based Reasoning, Programming, and Applications. Lecture Notes in Computer Science, vol. 6826, pp. 337–344. Springer, Berlin, Heidelberg (2011). http://www.springerlink.com/content/c276374nh9682jm6/
48. Nalepa, G.J.: Architecture of the HeaRT hybrid rule engine. In: Rutkowski, L., et al. (eds.) Artificial Intelligence and Soft Computing: 10th International Conference, ICAISC 2010: Zakopane, Poland, 13–17 June 2010, Pt. II. Lecture Notes in Artificial Intelligence, vol. 6114, pp. 598–605. Springer (2010)
49. Nalepa, G.J.: Loki—semantic wiki with logical knowledge representation. In: Nguyen, N.T. (ed.) Transactions on Computational Collective Intelligence III. Lecture Notes in Computer Science, vol. 6560, pp. 96–114. Springer (2011). http://www.springerlink.com/content/y91w134g03344376/
50. Nalepa, G.J.: Semantic Knowledge Engineering. A Rule-Based Approach. Wydawnictwa AGH, Kraków (2011)
51. Nalepa, G.J., Kluza, K.: UML representation for rule-based application models with XTT2-based business rules. Int. J. Softw. Eng. Knowl. Eng. (IJSEKE) **22**(4), 485–524 (2012). doi:10.1142/S021819401250012X, http://www.worldscientific.com
52. Nalepa, G.J., Ligęza, A.: Conceptual modelling and automated implementation of rule-based systems. In: Software Engineering: Evolution and Emerging Technologies. Frontiers in Artificial Intelligence and Applications, vol. 130, pp. 330–340. IOS Press, Amsterdam (2005)
53. Nalepa, G.J., Ligęza, A.: HeKatE methodology, hybrid engineering of intelligent systems. Int. J. Appl. Math. Comput. Sci. **20**(1), 35–53 (2010)
54. Nalepa, G.J., Ligęza, A., Kaczor, K.: Formalization and modeling of rules using the XTT2 method. Int. J. Artif. Intell. Tools **20**(6), 1107–1125 (2011)
55. Nazareth, D.L.: Issues in the verification of knowledge in rule-based systems. Int. J. Man-Mach. Stud. **30**(3), 255–271 (1989). http://www.reviews.com/reviewer/quickreview/frameset_toplevel.cfm?bib_id=69244

56. Nguyen, M.N., Zurada, J.M., Rajapakse, J.C.: Toward better understanding of protein secondary structure: Extracting prediction rules. IEEE/ACM Trans. Comput. Biol. Bioinform. **8**(3), 858–864 (2011). doi:10.1109/TCBB.2010.16
57. Nguyen, T.A., Perkins, W.A., Laffey, T.J., Pecora, D.: Checking an expert systems knowledge base for consistency and completeness. In: IJCAI, pp. 375–378 (1985). http://dli.iiit.ac.in/ijcai/IJCAI-85-VOL1/PDF/070.pdf
58. OMG: Production Rule Representation (OMG PRR) version 1.0 specification. Technical Report formal/2009-12-01, Object Management Group (2009). http://www.omg.org/spec/PRR/1.0
59. OMG: Business Process Model and Notation (BPMN): Version 2.0 specification. Technical Report formal/2011-01-03, Object Management Group (2011)
60. Pawlak, Z.: Rough Sets. Theoretical Aspects of Reasoning about Data. Kluwer Academic Publishers, Dordrecht/Boston/London (1991)
61. Preece, A.D.: A new approach to detecting missing knowledge in expert system rule bases. Int. J. Man-Mach. Stud. **38**(4), 661–688 (1993). http://users.cs.cf.ac.uk/A.D.Preece/publications/download/ijhcs1993.pdf
62. Preece, A.D.: A new approach to detecting missing knowledge in expert system rule bases. Int. J. Man-Mach. Stud. **38**, 161–181 (1993)
63. Preece, A.D., Shinghal, R., Batarekh, A.: Principles and practice in verifying rule-based systems. Knowl. Eng. Rev. **7**(02), 115–141 (1992). doi:10.1017/S026988890000624X
64. Riley, G.: CLIPS—A Tool for Building Expert Systems. http://clipsrules.sourceforge.net (2008)
65. Rousset, M.C.: On the consistency of knowledge bases: the COVADIS system. In: ECAI, pp. 79–84 (1988)
66. Rutkowski, L., Jaworski, M., Pietruczuk, L., Duda, P.: The CART decision tree for mining data streams. Inf. Sci. **266**, 1–15 (2014). doi:10.1016/j.ins.2013.12.060
67. Rutkowski, L., Jaworski, M., Pietruczuk, L., Duda, P.: Decision trees for mining data streams based on the Gaussian approximation. IEEE Trans. Knowl. Data Eng. **26**(1), 108–119 (2014). doi:10.1109/TKDE.2013.34
68. Salatino, M.: jBPM Developer Guide. Packt Publishing Ltd (2009)
69. Setiono, R., Leow, W.K., Zurada, J.M.: Extraction of rules from artificial neural networks for nonlinear regression. IEEE Trans. Neural Netw. **13**(3), 564–577 (2002). doi:10.1109/TNN.2002.1000125
70. Sommerville, I.: Software Engineering, 7th edn. International Computer Science. Pearson Education Limited (2004)
71. Tsai, W.T., Vishnuvajjala, R., Zhang, D.: Verification and validation of knowledge-based systems. IEEE Trans. Knowl. Data Eng. **11**, 202–212 (1999). doi:10.1109/69.755629
72. Vanthienen, J., Dries, E., Keppens, J.: Clustering knowledge in tabular knowledge bases. In: ICTAI, pp. 88–95 (1996)
73. Vermesan, A.I., Coenen, F. (eds.): Validation and Verification of Knowledge Based Systems. Theory, Tools and Practice. Kluwer Academic Publisher, Boston (1999)
74. Wagner, G., Antoniou, G., Tabet, S., Boley, H.: The abstract syntax of RuleML—towards a general web rule language framework. In: Web Intelligence, pp. 628–631. IEEE Computer Society (2004). doi:10.1109/WI.2004.134, http://doi.ieeecomputersociety.org
75. Wagner, G., Damásio, C.V., Antoniou, G.: Towards a general web rule language. Int. J. Web Eng. Technol. **2**(2/3), 181–206 (2005). doi:10.1504/IJWET.2005.008483
76. Zacharias, V.: Development and verification of rule based systems—a survey of developers. In: Proceedings of the International Symposium on Rule Representation, Interchange and Reasoning on the Web, RuleML '08, pp. 6–16. Springer, Berlin, Heidelberg (2008). doi:10.1007/978-3-540-88808-6_4

# New Aspects of Interpretability of Fuzzy Systems for Nonlinear Modeling

Krystian Łapa, Krzysztof Cpałka and Leszek Rutkowski

**Abstract** Fuzzy systems are well suited for nonlinear modeling. They can be effectively used if their structure and structure parameters are properly chosen. Moreover, it should be ensured that system rules are clear and interpretable. In this paper we propose a new algorithm for automatic learning and new interpretability criteria of fuzzy systems. Interpretability criteria are related to all aspects of those systems, not only their fuzzy sets and rules. Therefore, proposed criteria also concern parameterized triangular norms, discretization points and weights of importance from the rule base. As of the present time similar solutions have not been discussed in the literature. The proposed criteria are taken into account in the learning process, which is carried out with the use of a new learning algorithm. It was created by combining the genetic and the firework algorithms (this particular combination makes it possible to automatically choose not only system parameters but also its structure). It is an important advantage as most of the learning algorithms can only select system parameters when their structure has been specified by the designer. Proposed solutions were tested using typical simulation problems of nonlinear modeling.

K. Łapa · K. Cpałka · L. Rutkowski (✉)
Institute of Computational Intelligence, Czestochowa University of Technology,
Al. Armii Krajowej 36, 42-200 Częstochowa, Poland
e-mail: leszek.rutkowski@iisi.pcz.pl

K. Łapa
e-mail: krystian.lapa@iisi.pcz.pl

K. Cpałka
e-mail: krzysztof.cpalka@iisi.pcz.pl

L. Rutkowski
Information Technology Institute, Academy of Social Sciences,
Ul. Sienkiewicza 9, 90-113 Łódź, Poland

© Springer International Publishing AG 2018
A.E. Gawęda et al. (eds.), *Advances in Data Analysis with Computational Intelligence Methods*, Studies in Computational Intelligence 738,
https://doi.org/10.1007/978-3-319-67946-4_9

# 1  Introduction

Modeling is creation of simplified models of objects. It provides predictability, increases safety, reduces running costs, provides the ability to control, explains the principles of operation, etc. Most modeling problems are nonlinear [7]. Fuzzy systems work within this application range. This paper deals with interpretability of fuzzy systems which can be used in nonlinear modeling.

## 1.1  Model Representation

The model is expected, among others, to work accurately and perform readable (interpretable) operations. This last feature allows one to acquire knowledge about how a given object works. There are three different representations of a model related to the possibility of its interpretation: white-box, gray-box and black-box. White-box modeling uses phenomenological model, which is the mathematical description [13, 43, 61]. White-box models are readable but often simplified. Simplifying assumptions usually refer to characteristics idealization and linearization, skipping the saturation phenomenon, friction, etc. In black-box methods behavior of the object is modeled on the basis of the cause-and-effect relationships [35, 63, 88, 103]. Black-box models are accurate, but mostly not interpretable. From the point of view of interpretability, gray-box models are the most important methods. They offer a satisfactory compromise between accuracy and interpretability. Gray-box models often base on computational intelligence methods, e.g. decision trees [6, 51, 64], fuzzy systems [17, 30, 42, 44, 72, 76, 78, 94, 99–102], neural networks [23, 25, 41] etc.

In this work we discuss fuzzy systems since they use clear and intuitive fuzzy rules [72]. These rules take the form of IF … THEN …, which provides good opportunities for interpretation. Fuzzy systems can be used in direct or indirect nonlinear modeling [55, 70]. In indirect modeling they can, for example, explain the idea of switching component models in the methods of sectoral non-linearity [45] or model coefficients of so-called matrices of state variables [65], thus explaining elementary dependences which occur in the object. They can also model derogations of state variable matrix coefficients from their linear counterparts while explaining the source of nonlinearity [8]. Generally, fuzzy systems are used not only for non-linear modeling but also for classification. In particular, they are used e.g. in medical diagnostics, economics, controlling, forecasting, biometrics, databases, natural language processing, image processing, and many others. In each of these applications interpretation of the knowledge stored in the system is of great importance.

## 1.2 Interpretability of Fuzzy Systems

In the past researchers mainly paid attention to the accuracy of fuzzy systems while ignoring issues of their interpretability. However, in the 1990s they started to notice the fact that a large number of rules or fuzzy sets in those rules is not conducive to the readability of the rule base. Then, they began to form solutions which related to the term of "interpretability". Nowadays fuzzy system designers are trying to reach an acceptable compromise between accuracy and interpretability [31, 40, 54, 81, 83, 90, 95]. It is more difficult in the case of modeling than in the case of classification, because any change in the system structure developed to improve rules readability may result in deterioration of accuracy.

In the literature a number of papers on the subject of interpretability of fuzzy systems can be found. Their authors have proposed among the others:

- Solutions aimed at reducing the number of fuzzy rules [1, 4, 31, 33, 40, 50, 54], reducing the number of fuzzy sets [34] and aimed at reducing the number of system inputs [4, 91, 93]. Limitations were also related to the number of antecedences in fuzzy rules. The optimal number was most often set to Miller number, which equals $7 \pm 2$ [2, 4, 40, 66]. Miller number was designated in 1956 by George Miller and it represents the maximum pieces of information that can be directly distinguished by a human [59]. The use of restrictions in a system structure was often associated with a reduction of redundant elements and merging of similar ones [15, 34, 36, 38, 46, 67, 84].
- Solutions related to correct notation of fuzzy rules [4, 50], correct activation of fuzzy rules [2, 26, 54], distinguishability and interdependence of fuzzy sets (e.g. their overlapping) [57, 58, 66] as well as solutions on issues such as: complementarity, fitting in with data, etc. [9, 28, 31].
- Solutions related to fuzzy systems construction aimed at interpretability. In the papers [15, 18, 19, 22, 68, 72–75, 79, 80, 86, 87] the use of additional weights of importance of the rules, antecedences, consequences and system inputs was proposed. In the paper [82] a dynamic structure of connections between fuzzy sets and rules was considered. It was proposed, among others, in order to reduce system complexity and to simplify the rule-based notation. In the papers [15, 27] parameterized triangular norms were used (as precise aggregation operators) in order to increase accuracy, and in the paper [27] the authors reviewed parameterized triangular norms in terms of their suitability for the construction of fuzzy systems. In the paper [15] the authors used an extended (precise) defuzzification mechanism in which the number of discretization points does not have to be equal to the number of rules. This was suggested, among others, in order to increase accuracy of the system with a fixed number of rules and to provide opportunities to reduce the complexity of rules.

## 1.3 Attempts at Systematizing Solutions for Interpretability of Rule-Based Systems

The literature abounds in numerous attempts to systematize solutions for inter-pretability (e.g. [32, 83, 85]). The systematics presented in [32] deserves a spe-cial attention. Its authors have proposed division of solutions for interpretability into four groups-quadrants: (a) Quadrant concerning solutions aimed at reducing com-plexity at fuzzy rules level (it takes into account, among others, the number of fuzzy rules, the number of antecedences in each rule and using Miller number), (b) quad-rant concerning solutions aimed at reducing the complexity at the fuzzy partitioning level (it takes into account, among others, the number of fuzzy sets associated with various inputs and outputs and the number of inputs), (c) quadrant concerning solu-tions aimed at increasing semantic readability at the fuzzy rules' level (it takes into account, among others, the consistency of the rules, activation level of rules and readability of rule-based notation), and (d) quadrant concerning solutions aimed at increasing semantic readability at the fuzzy partitioning level (it takes into account, among others, a coverage degree of the input data by fuzzy sets, normalization of fuzzy sets, distinctness of fuzzy sets and fuzzy sets complementarity).

An interesting semantics has also been proposed in [3], which can complement the semantics proposed in [32]. In this semantic interpretability criteria were divided in terms of their readability of the knowledge accumulated in the system and a different importance was symbolically assigned to the criteria. There are: (a) very important criteria (for the complexity of fuzzy rules and notation readability), (b) important criteria (for the semantics of rules and fuzzy sets, including the criteria for order-ing fuzzy sets, semantic phrases used, sharing of fuzzy sets by a number of rules, etc.), and (c) the least important criteria (for the normalization of fuzzy sets, their complementarity, coverage of input data area, etc.).

## 1.4 Solutions Proposed in This Paper

The solutions proposed in this paper can be summarized as follows (Fig. 1): (a) In this paper the issue of interpretability has been treated comprehensively in terms of semantics considered in Sect. 1.3. Moreover, the proposed interpretability criteria apply to all aspects of fuzzy system designing. In particular, they include, among oth-ers, interpretability of fuzzy sets and rules, parameterized triangular norms, weights of importance and discretization points. This approach gives a broader look at the issue of interpretability, going beyond the concept of interpretability conception of a fuzzy set and fuzzy rule (most often discussed in the literature). (b) In this paper we propose a new hybrid algorithm for selection of the structure and parameters of a fuzzy system, constructed on the basis of the genetic [24, 29, 96, 98] and the fire-work [89] algorithms. This algorithm uses the idea of the genetic algorithm based on the biological evolution of species for the selection of the system structure (3)

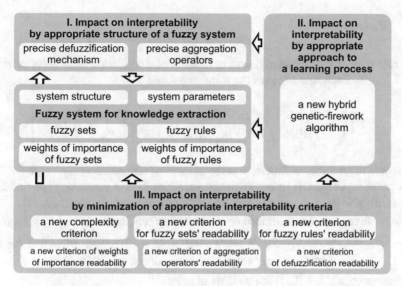

**Fig. 1** Summary of the solutions proposed in this work

and the idea of the firework algorithm based on the behavior of exploding fireworks for the selection of system structure parameters (3). The vast majority of algorithms presented in the literature can select system parameters only when its structure has been indicated by the designer (selected earlier by trial and error). Moreover, our algorithm takes into account all the interpretability criteria considered in this paper and it belongs to the methods based on populations [11, 49, 52, 53, 60, 62, 72].

It can be noted that the solutions proposed in this paper relate to interpretability directly and indirectly. The direct approach formulates appropriate criteria (Sect. 4) and uses them in the automatic process of fuzzy system selection. The indirect approach uses precise aggregation and inference operators in system design. They were proposed in our previous work [15, 16, 77] and called flexible. Their use allows for achieving good accuracy with a simpler system structure. Therefore, it makes a good starting point for direct impact on interpretability of a rule-based system. The use of the learning algorithm also affects the indirect impact on interpretability. It creates a good opportunity to find an appropriate trade-off between interpretability and accuracy.

## 2 Description of a Neuro-Fuzzy System for Non-linear Modeling

Further on in this paper a typical multi-input, multi-output Mamdani-type flexible fuzzy system will be considered [15, 16, 71, 72]. This system performs mapping $\mathbf{X} \to \mathbf{Y}$, where $\mathbf{X} \subset \mathbf{R}^n$ and $\mathbf{Y} \subset \mathbf{R}^m$.

## 2.1 Rule Base

The rule base of the considered system consists of a collection of $N$ fuzzy rules $R^k$, $k = 1, \ldots, N$. Each rule $R^k$ takes the following form:

$$R^k : \left[ \left( \begin{array}{l} \text{IF } \left( \bar{x}_1 \text{ is} A_1^k \right) \Big| w_{k,1}^A \text{ AND} \ldots \text{AND} \left( \bar{x}_n \text{ is} A_n^k \right) \Big| w_{k,n}^A \\ \text{THEN } \left( y_1 \text{ is} B_1^k \right) | w_{1,k}^B, \ldots, \left( y_m \text{ is} B_m^k \right) | w_{m,k}^B \end{array} \right) \Big| w_k^{\text{rule}} \right], \quad (1)$$

where $n$ is the number of inputs, $m$ is the number of outputs, $\bar{\mathbf{x}} = \left[ \bar{x}_1, \ldots, \bar{x}_n \right] \in \mathbf{X}$ is a vector of input signals (input linguistic variables for the singleton type fuzzification used), $\mathbf{y} = \left[ y_1, \ldots, y_m \right] \in \mathbf{Y}$ is a vector of output linguistic variables, $A_1^k, \ldots, A_n^k$ are input fuzzy sets characterized by membership functions $\mu_{A_i^k} \left( x_i \right)$ $(i = 1, \ldots, n)$, $B_1^k, \ldots, B_m^k$ are output fuzzy sets characterized by membership functions $\mu_{B_j^k} \left( y_j \right)$ $(j = 1, \ldots, m)$, $w_{k,i}^A \in [0, 1]$ are weights of antecedents, $w_{j,k}^B \in [0, 1]$ are weights of consequences, $w_k^{\text{rule}} \in [0, 1]$ are weights of rules.

Fuzzy sets $A_i^k$ and $B_j^k$ are fuzzy values of linguistic variables representing values such as e.g. *'very low'*, *'low'*, *'medium low'*, *'medium'*, *'medium high'*, *'high'*, *'very high'*, *'near* [value]', etc. Later in this paper we consider the system, in which membership functions $\mu_{A_i^k} \left( x_i \right)$ and $\mu_{B_j^k} \left( y_j \right)$ of fuzzy sets $A_i^k$ and $B_j^k$ are Gaussian functions, represented as follows:

$$\mu(x) = \exp \left( -\left( \frac{x - \bar{x}}{\sigma} \right)^2 \right). \quad (2)$$

Selection of a membership function allowed us to give more detailed information in Sects. 3.1 and 4.2. The Gaussian function reflects well the industrial, natural, medical and social processes; however, our solutions may be related to any other membership function.

The flexibility of the system (3) is a result of using: (a) weights in the rule base, (b) precise aggregation operators of antecedences and rules (Sect. 2.3), (c) precise inference operators (Sect. 2.3), and (d) a precise defuzzification process (Sect. 2.2).

## 2.2 Defuzzification Process

Defuzzification is used to determine output signals of fuzzy system $\bar{y}_j$ for given input signals. This is carried out as follows:

$$\bar{y}_j = \frac{\sum_{r=1}^{R_j} \bar{y}_{j,r}^{\text{def}} \cdot \overset{*}{\underset{k=1}{\overset{N}{S}}} \left\{ \overset{*}{\overset{\leftrightarrow}{T}} \left\{ \begin{array}{c} \tau_k(\bar{\mathbf{x}}), \mu_{B_j^k}\left(\bar{y}_{j,r}^{\text{def}}\right); \\ 1, w_{j,k}^B, p^{\text{imp}} \\ w_k^{\text{rule}}, p^{\text{agr}} \end{array} \right\}; \right\}}{\sum_{r=1}^{R_j} \overset{*}{\underset{k=1}{\overset{N}{S}}} \left\{ \overset{*}{\overset{\leftrightarrow}{T}} \left\{ \begin{array}{c} \tau_k(\bar{\mathbf{x}}), \mu_{B_j^k}\left(\bar{y}_{j,r}^{\text{def}}\right); \\ 1, w_{j,k}^B, p^{\text{imp}} \\ w_k^{\text{rule}}, p^{\text{agr}} \end{array} \right\}; \right\}}, \tag{3}$$

where $\tau_k(\bar{\mathbf{x}})$ is the activation level of the rule $k$. It is determined for the input signals vector $\bar{\mathbf{x}}$ and defined as follows:

$$\tau_k(\bar{\mathbf{x}}) = \overset{*}{\underset{i=1}{\overset{n}{\overset{\leftrightarrow}{T}}}} \left\{ \mu_{A_i^k}(\bar{x}_i); w_{k,i}^A, p^\tau \right\}, \tag{4}$$

and $\overset{*}{\overset{\leftrightarrow}{T}}\{\cdot\}$ and $\overset{*}{\overset{\leftrightarrow}{S}}\{\cdot\}$ are Dombi parameterized triangular norms with weights of arguments (Sect. 2.3), $p^\tau$ is a shape parameter of t-norm used for aggregation of antecedences, $p^{\text{imp}}$ is a shape parameter of t-norm used for inference, $p^{\text{agr}}$ is a shape parameter of t-conorm used for aggregation of inferences from rules, and $\bar{y}_{j,r}^{\text{def}}$ ($r = 1, \ldots, R_j$) are discretization points.

Discretization points are points in space $\mathbf{Y}$, which are related to the defuzzification and they are independent of the rule base (1). In these points discretization of output fuzzy sets and fuzzy sets obtained in response to the input signals of the system $\bar{\mathbf{x}}$ is performed. The most frequently used defuzzification methods (Center of area, Center of gravity, Fuzzy mean, Weighted fuzzy mean, Quality method, etc., [48, 72]) associate the number of discretization points with the number of output fuzzy sets (rules). In the system considered in this paper the number of discretization points $R_j$ for any output $j$ does not have to equal the number of rules $N$. This creates good opportunities for increasing the interpretability and accuracy of the fuzzy system. This issue was discussed in detail in our previous papers [15, 16]. In these papers detailed information on derivation of the formula (3) and linking it with the rule base of the form (1) can also be found.

## 2.3 Aggregation and Inference Operators

In this section parameterized Dombi-type triangular norms with weights of arguments, used in Eqs. (3) and (4), are considered. Their use contributes indirectly to an increase of the interpretability of the system (3). This is due to high working precision of these operators, which allows for achieving the expected accuracy of the system (3) with a smaller number of rules $N$.

Parameterized Dombi-type triangular norms with weights of arguments have the following form:

$$
\begin{cases}
\overset{*}{\vec{T}}\{\mathbf{a};\mathbf{w},p\} = \overset{*}{\underset{i=1}{\vec{T}}}\left\{\begin{array}{c}a_i;\\w_i,p\end{array}\right\} = \left(1 + \left(\sum_{i=1}^{n}\left(\frac{w_i\cdot(1-a_i)}{1-w_i\cdot(1-a_i)}\right)^p\right)^{\frac{1}{p}}\right)^{-1} \\
\overset{*}{\vec{S}}\{\mathbf{a};\mathbf{w},p\} = \overset{*}{\underset{i=1}{\vec{S}}}\left\{\begin{array}{c}a_i;\\w_i,p\end{array}\right\} = 1 - \left(1 + \left(\sum_{i=1}^{n}\left(\frac{w_i\cdot a_i}{1-w_i\cdot a_i}\right)^p\right)^{\frac{1}{p}}\right)^{-1},
\end{cases}
\tag{5}
$$

where $p \in [0, \infty)$ and parameters $w_1, \dots, w_n \in [0, 1]$ are weights of arguments $a_1, \dots, a_n \in [0, 1]$. Operators of the form (5) were formed from the combination of two types of triangular norms. The first type comprises parameterized Dombi-type triangular norms (marked with the symbol "÷"). Their way of working depends on the value of the parameter $p$. By changing value of the parameter $p$ it is possible to achieve similar behavior to typical non-parametric norms, such as min/max norms, Hamacher norms, Łukasiewicz norms, algebraic norms, etc. Apart from the Dombi-type norms, in the literature many other types of parameterized triangular norms (e.g. Frank, Dubois and Prade, Schweizer and Skalar, Weber, Yager, Yu, etc.) can be found. The second type of triangular norms used in the construction of operators of the form (5) are standard triangular norms with weights of arguments (marked with the symbol "$*$") [15, 16]. They can be described as follows:

$$
\begin{cases}
T^*\{\mathbf{a};\mathbf{w}\} = \underset{i=1}{\overset{n}{T^*}}\{a_i;w_i\} = \underset{i=1}{\overset{n}{T}}\{1 - w_i\cdot(1-a_i)\} \\
S^*\{\mathbf{a};\mathbf{w}\} = \underset{i=1}{\overset{n}{S^*}}\{a_i;w_i\} = \underset{i=1}{\overset{n}{S}}\{a_i\cdot w_i\}.
\end{cases}
\tag{6}
$$

Examples of triangular norms with weights of arguments are standard algebraic norms:

$$
\begin{cases}
T^*\{\mathbf{a};\mathbf{w}\} = \prod_{i=1}^{n}\left(1 + (a_i - 1)\cdot w_i\right) \\
S^*\{\mathbf{a};\mathbf{w}\} = 1 - \prod_{i=1}^{n}\left(1 - a_i\cdot w_i\right).
\end{cases}
\tag{7}
$$

The idea of operation of triangular norms with weights of arguments (especially the idea of reducing arguments with weights equal to 0) can be summarized as follows: $T^*\{a_1, a_2; w_1, 0\} = a_1$, $T^*\{a_1, a_2; 1, 1\} = T\{a_1, a_2\}$, $S^*\{a_1, a_2; w_1, 0\} = a_1$ i $S^*\{a_1, a_2; 1, 1\} = S\{a_1, a_2\}$. More detailed information on the operators of the forms (5)–(7) can be found in our previous papers [15, 16].

## 3   Description of a New Fuzzy System Learning Algorithm

The proposed hybrid genetic-firework algorithm aims at selection of the structure and parameters of the fuzzy system (3) (Fig. 1). The purpose of the algorithm is also to minimize interpretability criteria presented in Sect. 4.

The proposed algorithm belongs to so-called population-based algorithms. They provide a method for solving optimization problems. They can be defined as search procedures based on the mechanisms of natural selection and inheritance and they use the evolutionary principle of survival of the fittest individuals. What differs population algorithms from traditional optimization methods, among others, is that they (a) do not process task parameters directly, but their encoded form, (b) do not conduct a search starting from a single point, but from a population of points, (c) use only the objective function and not its derivatives, and (d) use probabilistic, not deterministic selection rules. Owing it to the above mentioned features, population algorithms have the advantage over other optimization techniques such as analytical, inspection, random methods, etc. [72].

## 3.1 Encoding of Potential Solutions

Encoding of population of potential solutions used in the algorithm refers to the Pittsburgh approach [37]. A single individual of the population ($\mathbf{X}_{ch}$) is therefore an object that encodes the complete structure of the fuzzy system (3) ($\mathbf{X}_{ch}^{str}$) and its parameters ($\mathbf{X}_{ch}^{par}$):

$$\mathbf{X}_{ch} = \left\{ \mathbf{X}_{ch}^{str}, \mathbf{X}_{ch}^{par} \right\}. \tag{8}$$

Part $\mathbf{X}_{ch}^{str}$ of the individual $\mathbf{X}_{ch}$ encodes the whole structure of the fuzzy system (3) in a binary form, which has the following form:

$$\mathbf{X}_{ch}^{str} = \left\{ \begin{array}{c} x_1, \ldots, x_n, \\ A_1^1, \ldots, A_n^1, \ldots, A_1^{Nmax}, \ldots, A_n^{Nmax}, \\ B_1^1, \ldots, B_m^1, \ldots, B_1^{Nmax}, \ldots, B_m^{Nmax}, \\ \text{rule}_1, \ldots, \text{rule}_{Nmax}, \\ \bar{y}_{1,1}^{def}, \ldots, \bar{y}_{1,Rmax}^{def}, \ldots, \bar{y}_{m,1}^{def}, \ldots, \bar{y}_{m,Rmax}^{def} \end{array} \right\} = \left\{ X_{ch,1}^{str}, \ldots, X_{ch,L^{str}}^{str} \right\}, \tag{9}$$

where $ch = 1, \ldots, Npop$ is the index of an individual in a population, $Npop$ is the number of individuals in a population, $Nmax$ is the maximum (allowed) number of rules in the system (3) (selected individually for the considered problem), $Rmax$ is the maximum (allowed) number of discretization points in the system (3) (also selected individually for the considered problem) and $L^{str}$ is the number of the individual components $\mathbf{X}_{ch}^{str}$ (referred to as genes from now on), which is determined as follows:

$$L^{str} = Nmax \cdot (n + m + 1) + n + Rmax \cdot m. \tag{10}$$

In the encoding procedure of $\mathbf{X}_{ch}^{str}$ it is assumed that each individual of the population encodes the maximum number of rules $Nmax$ indicated by the user and number of discretization points $Rmax$. The algorithm searches the real number of the system (3) rules in the range $N \in [1, Nmax]$ and the real number of discretization points in the range $R_j \in [1, Rmax]$ $(j = 1, \dots, m)$. Therefore, it is a different approach than in the conventional methods of learning, in which the user (mostly using the trial-and-error method) had to clearly indicate $N$ and $R_j$.

The principle adopted in the encoding procedure $\mathbf{X}_{ch}^{\mathrm{str}}$ is such that the gene with value 0 of the individual $\mathbf{X}_{ch}^{\mathrm{str}}$ excludes the associated element from the target system structure (3) and vice versa. This element can be: a rule (rule$_k$, $k = 1, \dots, Nmax$), an antecedence ($A_i^k$, $i = 1, \dots, n$, $k = 1, \dots, Nmax$), a consequence ($B_j^k$, $j = 1, \dots, m$, $k = 1, \dots, Nmax$), an input ($\bar{x}_i$, $i = 1, \dots, n$) and a discretization point ($\bar{y}^r$, $r = 1, \dots, Rmax$).

Part $\mathbf{X}_{ch}^{\mathrm{par}}$ of the individual $\mathbf{X}_{ch}$ encodes the real parameters of the fuzzy system and it has the following form:

$$\mathbf{X}_{ch}^{\mathrm{par}} = \left\{ \begin{matrix} \bar{x}_{1,1}^A, \sigma_{1,1}^A, \dots, \bar{x}_{n,1}^A, \sigma_{n,1}^A, \dots \\ \bar{x}_{1,Nmax}^A, \sigma_{1,Nmax}^A, \dots, \bar{x}_{n,Nmax}^A, \sigma_{n,Nmax}^A, \\ \bar{y}_{1,1}^B, \sigma_{1,1}^B, \dots, \bar{y}_{m,1}^B, \sigma_{m,1}^B, \dots \\ \bar{y}_{1,Nmax}^B, \sigma_{1,Nmax}^B, \dots, \bar{y}_{m,Nmax}^B, \sigma_{m,Nmax}^B, \\ w_{1,1}^A, \dots, w_{1,n}^A, \dots, w_{Nmax,1}^A, \dots, w_{Nmax,n}^A, \\ w_{1,1}^B, \dots, w_{m,1}^B, \dots, w_{1,Nmax}^B, \dots, w_{m,Nmax}^B, \\ w_1^{\mathrm{rule}}, \dots, w_{Nmax}^{\mathrm{rule}}, \\ p^\tau, p^{\mathrm{imp}}, p^{\mathrm{agr}}, \\ \bar{y}_{1,1}^{\mathrm{def}}, \dots, \bar{y}_{1,Rmax}^{\mathrm{def}}, \dots, \bar{y}_{m,1}^{\mathrm{def}}, \dots, \bar{y}_{m,Rmax}^{\mathrm{def}} \end{matrix} \right\} = \left\{ X_{ch,1}^{\mathrm{par}}, \dots, X_{ch,L^{\mathrm{par}}}^{\mathrm{par}} \right\}, \quad (11)$$

where $\left\{ \bar{x}_{i,k}^A, \sigma_{i,k}^A \right\}$ are membership function parameters (2) of input fuzzy sets $A_1^k, \dots, A_n^k$, $\left\{ \bar{y}_{j,k}^B, \sigma_{j,k}^B \right\}$ are membership function parameters (2) of output fuzzy sets $B_1^k, \dots, B_m^k$ and $L^{\mathrm{par}}$ is the number of components of individual $\mathbf{X}_{ch}^{\mathrm{par}}$, determined as follows:

$$L^{\mathrm{par}} = Nmax \cdot (3 \cdot n + 3 \cdot m + 1) + Rmax \cdot m + 3. \quad (12)$$

In the encoding procedure of $\mathbf{X}_{ch}^{\mathrm{par}}$ it is assumed that only genes $\mathbf{X}_{ch}^{\mathrm{par}}$, whose counterparts in $\mathbf{X}_{ch}^{\mathrm{str}}$ are equal to 1, are considered in the construction of the system (3). Moreover, analyzing $\mathbf{X}_{ch}^{\mathrm{str}}$ of the form (9) the actual number of inputs encoded in the individual $\mathbf{X}_{ch}$ can be easily indicated:

$$n_{ch} = \sum_{i=1}^n \mathbf{X}_{ch}^{\mathrm{str}} \{x_i\}, \quad (13)$$

where $\mathbf{X}_{ch}^{\mathrm{str}} \{x_i\}$ is the parameter of the individual $\mathbf{X}_{ch}^{\mathrm{str}}$ associated with the input $x_i$. The adoption of this notation greatly facilitated, among others, notation of interpretability criteria considered in Sect. 4. Similarly to $n_{ch}$, the actual number of rules $N_{ch}$, the actual number of discretization points $R_{j,ch}$ for any input $j$, the number of input fuzzy sets $nant_{ch}$ and the number of output fuzzy sets $ncon_{ch}$ can be determined. They are taken into account in the target structure of the fuzzy system (3) encoded in the individual $\mathbf{X}_{ch}$. On the basis of the notation used in (13), function noifs $(\cdot)$, which allows us to determine the number of fuzzy sets for the input $i$, can also be defined as follows:

$$\text{noifs}\,(i) = \sum_{k=1}^{N_{ch}} \mathbf{X}_{ch}^{\text{str}}\,\{\text{rule}_k\} \cdot \mathbf{X}_{ch}^{\text{str}}\,\{A_i^k\}. \tag{14}$$

Function noofs $(\cdot)$, which allows us to determine the number of fuzzy sets for the output $j$, can be defined analogously. Functions noifs $(\cdot)$ and noofs $(\cdot)$ are used in Sect. 4.2.

## 3.2 Evaluation of Potential Solutions

As already mentioned, each individual in the population $(\mathbf{X}_{ch})$ encodes parameters $\mathbf{X}_{ch}^{\text{par}}$ (formula (11)) and structure $\mathbf{X}_{ch}^{\text{str}}$ (formula (9)) of a single system (3). The purpose of the algorithm is to minimize the value of the evaluation function specified for the individual $\mathbf{X}_{ch}$ in the following way:

$$\text{ff}\,(\mathbf{X}_{ch}) = T^* \left\{ \begin{array}{c} \text{ffacc}\,(\mathbf{X}_{ch})\,,\text{ffint}\,(\mathbf{X}_{ch})\,; \\ w_{\text{ffacc}},\,w_{\text{ffint}} \end{array} \right\}, \tag{15}$$

where component $\text{ffacc}\,(\mathbf{X}_{ch})$ specifies the accuracy of the system (3), component $\text{ffint}\,(\mathbf{X}_{ch})$ specifies interpretability of the system (3) according to the adopted interpretability criteria, $T^*\,\{\cdot\}$ is a weighted algebraic triangular norm of the form (7), $w_{\text{ffacc}} \in [0, 1]$ represents weight of the component $\text{ffacc}\,(\mathbf{X}_{ch})$ and $w_{\text{ffint}} \in [0, 1]$ represents weight of the component $\text{ffint}\,(\mathbf{X}_{ch})$. Values of weights $w_{\text{ffacc}}$ and $w_{\text{ffint}}$ result from expectations of the user regarding the ratio between the accuracy of the system (3) and its interpretability.

Component $\text{ffacc}\,(\mathbf{X}_{ch})$ is determined as follows:

$$\text{ffacc}\,(\mathbf{X}) = \frac{1}{m} \sum_{j=1}^{m} \frac{\frac{1}{Z}\sum_{z=1}^{Z} \left| d_{z,j} - \bar{y}_{z,j} \right|}{\max\limits_{z=1,\dots,Z}\{d_{z,j}\} - \min\limits_{z=1,\dots,Z}\{d_{z,j}\}}, \tag{16}$$

where $Z$ is the number of sets of a learning sequence, $d_{z,j}$ is the desired output value of output $j$ for input vector $z$ $(z = 1, \dots, Z)$, $\bar{y}_{z,j}$ is the real output value $j$ calculated by the system for the input vector $\bar{x}_z$. Equation (16) takes into account the normalization of errors at different outputs of the system (3) in order to eliminate significant differences between them.

Component $\text{ffint}\,(\mathbf{X}_{ch})$ takes into account interpretability criteria proposed in Sect. 4. Their aggregation is realized as follows:

$$\text{ffint}\left(\mathbf{X}_{ch}\right) = T^* \left\{ \begin{matrix} \text{ffint}_A\left(\mathbf{X}_{ch}\right), \text{ffint}_B\left(\mathbf{X}_{ch}\right), ...; \\ w_{\text{ffintA}}, w_{\text{ffintB}}, ... \end{matrix} \right\}, \tag{17}$$

where $\text{ffint}_A\left(\cdot\right)$, $\text{ffint}_B\left(\cdot\right)$, ... are functions representing considered interpretability criteria defined in Sect. 4, $w_{\text{ffintA}} \in [0, 1]$, $w_{\text{ffintB}} \in [0, 1]$,... are weights of importance of function $\text{ffint}_A\left(\cdot\right)$, $\text{ffint}_B\left(\cdot\right)$,... and $T^*\left\{\cdot\right\}$ is weighted algebraic triangular norm of the form (7). The values of weights in Eq. (17) can be selected on the basis of suggestions given in the paper [3], which was done in our simulations.

In most applications, the objective adopted in the design phase is to obtain a single fuzzy system. It is expected that the system will be characterized by good accuracy and interpretability. But if it was necessary to obtain a set of solutions with different proportions of accuracy-interpretability (in terms of Eq. (15)), then possibilities offered by the methods based on Pareto fronts [21] could be used instead of criteria aggregation.

### 3.3 Processing of Potential Solutions

The hybrid genetic-firework algorithm under consideration works according to the steps shown in Fig. 2.

**Step 1. Initiation of population**

In this step the population of individuals $\mathbf{X}_{ch}$, $ch = 1, \dots, Npop$ is initialized. These individuals are interpreted as fireworks. Fireworks are defined as locations of their "explosion". Each gene $X_{ch,g}^{\text{str}}$ of these individuals (affecting the form of the fuzzy system structure (3)) is initially drawn from the set $\{0, 1\}$ and each gene $X_{ch,g}^{\text{par}}$ (determining the values of the structure parameters) is initially drawn taking into account any possible limitations on its value.

**Step 2. Evaluation of individuals**

In this step evaluation of $Npop$ individuals (referred to as fireworks) belonging to the population is performed by using the evaluation function $\text{ff}\left(\mathbf{X}_{ch}\right)$ of the form (15).

**Step 3. Generation of sparks**

The idea of this step is to generate sparks from the fireworks in order to exploit the search space. In this step only individuals $\mathbf{X}_{ch}^{\text{par}}$ which encode parameters are modified. Individuals $\mathbf{X}_{ch}^{\text{str}}$ encoding structure are not modified-a change of those individuals is performed in the next step. If individuals encoding structure are modified with the same intensity as the ones encoding parameters, the algorithm has a small chance to find a satisfactory solution. The approach proposed in our algorithm indicates a set of structures (resulting from the population) and then looks for a solutions in their surrounding (with close/similar values of parameters). In turn, a mechanism for generating random sparks (described in the next step) allows us to search for new structures of the system (3).

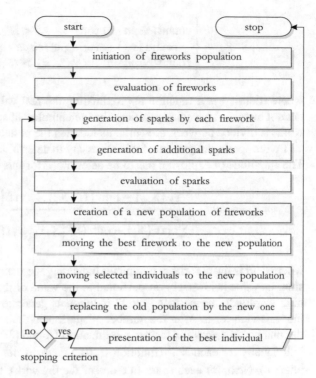

**Fig. 2** Block schema of the hybrid genetic-firework algorithm

The operation of sparks generation discussed in this step involves determination of the number of sparks for each firework. This number is dependent on the value of the firework fitness function (15). If the value of the adaptation function of a firework is smaller (in minimization problem), the number of its sparks is greater. The number of sparks for the firework $\mathbf{X}_{ch}$ is determined as follows:

$$s_{ch} = Nspa \cdot \frac{\max\left\{\mathrm{ff}\left(\mathbf{X}_1\right), \dots, \mathrm{ff}\left(\mathbf{X}_{Npop}\right)\right\} - \mathrm{ff}\left(\mathbf{X}_{ch}\right) + \xi}{\sum\limits_{i=1}^{Npop}\left(\max\left\{\mathrm{ff}\left(\mathbf{X}_1\right), \dots, \mathrm{ff}\left(\mathbf{X}_{Npop}\right)\right\} - \mathrm{ff}\left(\mathbf{X}_i\right)\right) + \xi}, \tag{18}$$

where $Nspa$ is a parameter of the algorithm which controls the number of created sparks and $\xi$ is a small real number which prevents dividing by 0. The algorithm assumes that the target number of sparks for each firework (denoted as $\hat{s}_{ch}$) has to be within the range $\left[b \cdot Nspa, a \cdot Nspa\right]$, while parameters $a$ and $b$ of the algorithm should meet the assumption $a < b < 1$. Value $a \cdot Nspa$ is the minimum number of generated sparks (value $a$ has to be relatively low), value $b \cdot Nspa$ is the maximum number of generated sparks (value $b$ has to be greater than $a$, but not greater than 1). In this way, the worst fireworks in a population always receive at least $Nspa \cdot a$ sparks and the best fireworks get maximum $Nspa \cdot b$ sparks. It is calculated as follows:

$$\hat{s}_{ch} = \begin{cases} \text{round}\,(Nspa \cdot a) \text{ for} & s_{ch} < Nspa \cdot a \\ \text{round}\,(s_{ch}) \quad \text{ for } s_{ch} \in (Nspa \cdot a, Nspa \cdot b) \,, \\ \text{round}\,(Nspa \cdot b) \text{ for} & s_{ch} > Nspa \cdot b \end{cases} \tag{19}$$

where $\text{round}\,(\cdot)$ is a function approximating the real value of the argument to the nearest integer. After this operation, the total number of sparks generated by all fireworks is divided between fireworks included in the population.

Before sparks are generated it is necessary to determine the area of their location. The amplitude of explosion has to be determined for this purpose:

$$amp_{ch} = \overline{amp} \cdot \frac{\text{ff}\,(\mathbf{X}_{ch}) - \min\left\{\text{ff}\,(\mathbf{X}_1), \dots, \text{ff}\,(\mathbf{X}_{Npop})\right\} + \xi}{\sum\limits_{i=1}^{Npop} \left(\text{ff}\,(\mathbf{X}_i) - \min\left\{\text{ff}\,(\mathbf{X}_1), \dots, \text{ff}\,(\mathbf{X}_{Npop})\right\}\right) + \xi}, \tag{20}$$

where $\overline{amp}$ is the algorithm parameter indicating the maximum amplitude of explosion. Its value is inversely proportional to the value of the evaluation function. The high amplitude means that "good" individuals generate sparks in their surroundings and vice versa. After the number of sparks and their amplitude are determined, the number of genes for further modification (round $\left(L^{par} \cdot U_r\,(0, 1)\right)$ is calculated individually for each $\mathbf{X}_{ch}^{par}$ (function $U_r\,(0, 1)$ returns the real value of random unit interval). Next, for each spark (a clone of the firework), the round $\left(L^{par} \cdot U_r\,(0, 1)\right)$ randomly chosen genes $\mathbf{X}_{ch}^{par}$ are modified as follows:

$$X_{ch,g}^{par} := X_{ch,g}^{par} + amp_{ch} \cdot U_r\,(-1, 1), \tag{21}$$

where $U_r\,(-1, 1)$ is a random number of the range $[-1, 1]$. Update of the individual genes $\mathbf{X}_{ch}^{par}$, according to the relation (21), is called "generation of sparks". Created sparks are evaluated using the defined evaluation function (15).

### Step 4. Generation of additional sparks

Generating additional sparks in order to explore the search space involves random selection of $Nsparnd$ fireworks from the set of $Npop$ fireworks. Then, the part encoding parameters $\mathbf{X}_{ch}^{par}$ and the part encoding structure $\mathbf{X}_{ch}^{str}$ are modified for each of the selected $Nsparnd$ fireworks. Modification of the part $\mathbf{X}_{ch}^{par}$ encoding parameters starts with determination of the directions of sparks propagation. This is similar to the previous step but the procedure for updating selected genes of the individual $\mathbf{X}_{ch}^{par}$ is different. The revision is performed as follows:

$$X_{ch,g}^{par} := X_{ch,g}^{par} \cdot U_g\,(1, 1), \tag{22}$$

where $U_g\,(1, 1)$ is a real number drawn from the Gaussian distribution (normal distribution). Modification of selected individuals $\mathbf{X}_{ch}^{str}$ encoding the structure takes place using a mutation operator (known from the genetic algorithm). For each gene of modified individuals $\mathbf{X}_{ch}^{str}$ a number from the range $[0, 1]$ is drawn. If it is lower than so-called probability of mutation $p_m \in (0, 1)$ (which is a parameter of the algorithm), then the value of the gene is changed to its opposite value (i.e. from 0 to 1 or vice versa). Created sparks are evaluated using the defined fitness function (15).

**Step 5. Creation of a new population of individuals**

The new population of individuals is joined by the currently best firework (having the lowest value of the minimized fitness function of the form (15)) and $Npop - 1$ individuals selected from the sparks generated in the last two steps and other fireworks are chosen. The selection of $Npop - 1$ individuals is done using the roulette wheel method. Selection probability of the individual $\mathbf{X}_{ch}$ is determined as follows:

$$p\left(\mathbf{X}_{ch}\right) = \frac{\sum_{ch2=1}^{Npop+Nspa} \left\|\mathbf{X}_{ch} - \mathbf{X}_{ch2}\right\|}{ff\left(\mathbf{X}_{ch}\right)}, \tag{23}$$

where $\|\cdot\|$ is an adopted method of calculating the distance (e.g. Euclidean, Manhattan type, etc.).

**Step 6. Replacement of the population**

In this step the old population of individuals is replaced by the population generated in the previous step. In the new population all individuals are treated as fireworks. Stopping criterion check is also performed. In turn, the stopping criterion may take into account achievement of the threshold value of the evaluation function by the best individual from the population or performance of the maximum allowed number of the algorithm steps. If this condition is not met, the algorithm goes back to step 3.

## 4 New Interpretability Criteria of a Fuzzy System for Nonlinear Modeling

In this section new interpretability criteria of the form (3) which can be used in nonlinear modeling issues are described. Those criteria are general measures of interpretability, whose values are in the range $[0, 1]$. Due to that, they can be used in the function of the form (17) and minimized. The purpose of this minimization is to reduce the complexity of the system (3) (especially the rules of the form (1)) and to increase its interpretability.

The advantages of interpretability measures proposed in this section can be summarized as follows:

- They were designed taking into account both semantics discussed in Sect. 1.3, in particular all the quadrants considered in the paper [32].

- They were adapted to the general specifics of the rule base. They also refer to the aspects of the system (3) flexibility (presented in Sect. 2) and are used to evaluate readability of weights of importance, parameters of triangular norms and discretization points of the defuzzification mechanism. Apart from a single paper (this paper appeared in the field of discretization points interpretability enforcement [56]) no similar issues have been considered in the literature so far.

An additional advantage of the proposed criteria is drawing attention to the fact that interpretability of fuzzy systems is an issue that can be considered more comprehensively, without focusing only on the "fuzzy set" or "fuzzy rule" notions (Fig. 1).

## *4.1 Complexity Evaluation Criterion*

This criterion allows us to evaluate complexity of the fuzzy system (3). It bases on the genes analysis of the individual $\mathbf{X}_{ch}^{str}$ encoding the structure of the form (9). It takes into account the number of fuzzy rules of the form (1), antecedences of rules, consequences of rules, inputs and discretization points.

The method of operation of the considered criterion is shown in Fig. 3 and it is expressed as follows:

$$
\mathrm{ffint}_A\left(\mathbf{X}_{ch}\right) = \frac{\left(\begin{array}{c} \sum\limits_{i=1}^{n} \mathbf{X}_{ch}^{str}\left\{x_i\right\} \cdot \sum\limits_{k=1}^{Nmax} \mathbf{X}_{ch}^{str}\left\{\mathrm{rule}_k\right\} \cdot \mathbf{X}_{ch}^{str}\left\{A_i^k\right\}+ \\ +\sum\limits_{j=1}^{m}\sum\limits_{k=1}^{Nmax} \mathbf{X}_{ch}^{str}\left\{\mathrm{rule}_k\right\} \cdot \mathbf{X}_{ch}^{str}\left\{B_j^k\right\}+ \\ +\sum\limits_{j=1}^{m}\sum\limits_{r=1}^{Rmax} \mathbf{X}_{ch}^{str}\left\{\bar{y}_{j,r}^{def}\right\} \end{array}\right)}{N_{ch}\cdot\left(n_{ch}+m\right)+m\cdot Rmax}, \tag{24}
$$

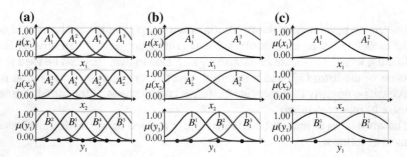

**Fig. 3** Three exemplary cases obtained for criterion (24): **a** negative, **b** intermediate, **c** preferred (low complexity of the system (3), low criterion value). Discretization points are denoted as black circles

where $\mathbf{X}_{ch}^{str}\{x_i\}$ are parameters of $\mathbf{X}_{ch}^{str}$ associated with the input $x_i$, etc. A detailed explanation of the adopted notation is given in the context of the formula (13).

## 4.2 Fuzzy Sets Readability Evaluation Criterion

In this section two criteria related to readability of fuzzy sets are proposed. The first one concerns the position of fuzzy sets while the other refers to the consistency of their shape. The criteria have been adapted to the membership function (2) considered in this paper, but they can be easily adapted to some other membership functions.

### Fuzzy sets position evaluation criterion

The criterion under consideration makes it possible to evaluate the correctness of the input and output fuzzy sets' distribution. Incorrect distribution of fuzzy sets results from their overlapping and their remoteness. The distribution of fuzzy sets can be evaluated, among others, by analyzing the intersections of adjacent fuzzy sets. The considered criterion takes into account two points of intersection for each pair of adjacent fuzzy sets. The use of the first intersection point allows us to assess the distance between fuzzy sets while the use of the other allows us to assess overlapping.

The method of operation of the considered criterion is shown in Fig. 4 and it is expressed as follows:

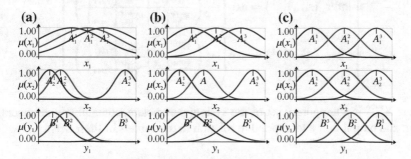

**Fig. 4** Three exemplary cases obtained for criterion (25): **a** negative, **b** intermediate, **c** preferred (correct distribution of fuzzy sets, low criterion value)

$$\text{ffint}_B\left(\mathbf{X}_{ch}\right) =$$

$$= \frac{1}{2\cdot(n_{ch}+m)}\left(\begin{array}{l} \displaystyle\sum_{i=1}^{n}\sum_{k=1}^{\text{noifs}(i)-1}\left(\dfrac{2\cdot\left|c_{\text{intB}}-\text{inter}_1\left(A_i^k,A_i^{k+1}\right)\right|+}{\text{+inter}_2\left(A_i^k,A_i^{k+1}\right)}{\displaystyle\sum_{i=1}^{n}\text{noifs}(i)-1}\right)+ \\[2em] \displaystyle+\sum_{j=1}^{m}\sum_{k=1}^{\text{noofs}(j)-1}\left(\dfrac{2\cdot\left|c_{\text{intB}}-\text{inter}_1\left(B_j^k,B_j^{k+1}\right)\right|+}{\text{+inter}_2\left(B_j^k,B_j^{k+1}\right)}{\displaystyle\sum_{j=1}^{m}\text{noofs}(j)-1}\right) \end{array}\right), \qquad (25)$$

where $c_{\text{intB}}$ determines the desired value of the membership function at the intersection point between two adjacent fuzzy sets (in the simulations we adopted value 0.5), noifs (i) is a function of the form (14) which determines the number of active fuzzy sets for $i$-th input, noofs (j) is a function which analogously determines the number of active fuzzy sets for $j$-th output, $\text{inter}_1(\cdot)$ and $\text{inter}_2(\cdot)$ are the functions which determine the values of two intersection points of fuzzy sets. In the case where the sets are expressed by the Gaussian function of the form (2), then functions $\text{inter}_1(\cdot)$ and $\text{inter}_2(\cdot)$ take the following form:

$$\left\{\begin{array}{l} \text{inter}_1\left(A_i^k,A_i^{k+1}\right) = \exp\left\{-\dfrac{1}{2}\left(\dfrac{\mathbf{X}_{ch}^{\text{supp}}\left\{\bar{x}_{i,k}^A\right\}+\mathbf{X}_{ch}^{\text{supp}}\left\{\bar{x}_{i,k+1}^A\right\}}{\mathbf{X}_{ch}^{\text{supp}}\left\{\sigma_{i,k}^A\right\}+\mathbf{X}_{ch}^{\text{supp}}\left\{\sigma_{i,k+1}^A\right\}}\right)^2\right\} \\[2em] \text{inter}_2\left(A_i^k,A_i^{k+1}\right) = \exp\left\{-\dfrac{1}{2}\left(\dfrac{\mathbf{X}_{ch}^{\text{supp}}\left\{\bar{x}_{i,k}^A\right\}+\mathbf{X}_{ch}^{\text{supp}}\left\{\bar{x}_{i,k+1}^A\right\}}{\mathbf{X}_{ch}^{\text{supp}}\left\{\sigma_{i,k}^A\right\}-\mathbf{X}_{ch}^{\text{supp}}\left\{\sigma_{i,k+1}^A\right\}}\right)^2\right\} \end{array}\right., \qquad (26)$$

where $\mathbf{X}_{ch}^{\text{supp}}$ is a temporary set of the system parameters, containing parameters of input and output fuzzy sets sorted in relation to the centers of these sets:

$$\mathbf{X}_{ch}^{\text{supp}} = \left\{\begin{array}{l} \bar{x}_{1,1}^A,\sigma_{1,1}^A,\bar{x}_{1,2}^A,\sigma_{1,2}^A,\ldots, \\ \bar{x}_{n_{ch},1}^A,\sigma_{n_{ch},1}^A,\bar{x}_{n_{ch},2}^A,\sigma_{n_{ch},2}^A,\ldots, \\ \bar{y}_{1,1}^B,\sigma_{1,1}^B,\bar{y}_{2,N_{ch}}^B,\sigma_{2,N_{ch}}^B,\ldots, \\ \bar{y}_{m,N_{ch}}^B,\sigma_{m,N_{ch}}^B,\bar{y}_{m,N_{ch}}^B,\sigma_{m,N_{ch}}^B,\ldots \end{array}\right\}. \qquad (27)$$

**Criterion for assessing similarity of fuzzy sets width**

The considered criterion is a cohesion measure of the widths of input and output fuzzy sets. It is of a great importance for semantic readability of the fuzzy system (3) rule base, because it facilitates the understanding of the rule-based notation (1).

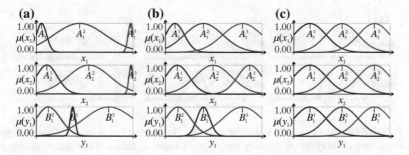

**Fig. 5** Three exemplary cases obtained for criterion (28): **a** negative, **b** intermediate, **c** preferred (minor differences between the widths of fuzzy sets, low criterion value)

The method of operation of this criterion is shown in Fig. 5 and it is expressed as follows:

$$\text{ffint}_C\left(\mathbf{X}_{ch}\right) = \frac{\left(\begin{array}{c} \sum\limits_{i=1}^{n} \mathbf{X}_{ch}^{\text{str}}\{x_i\} \cdot \sum\limits_{k=1}^{Nmax} \mathbf{X}_{ch}^{\text{str}}\{\text{rule}_k\} \cdot \text{shx}\left(\mathbf{X}_{ch}, i, k\right) + \\ + \sum\limits_{j=1}^{m} \cdot \sum\limits_{k=1}^{Nmax} \mathbf{X}_{ch}^{\text{str}}\{\text{rule}_k\} \cdot \text{shy}\left(\mathbf{X}_{ch}, j, k\right) \end{array}\right)}{n_{ch} + m}, \tag{28}$$

where $\text{shx}\left(\mathbf{X}_{ch}, i, k\right)$ and $\text{shy}\left(\mathbf{X}_{ch}, j, k\right)$ are functions used to determine the proportion between the widths of fuzzy sets, working analogously for input and output fuzzy sets. Function $\text{shx}\left(\mathbf{X}_{ch}, i, k\right)$ is defined as follows:

$$\begin{aligned} \text{shx}\left(\mathbf{X}_{ch}, i, k\right) = \\ = 1 - \frac{\min\left(\mathbf{X}_{ch}^{\text{par}}\left\{\sigma_{i,k}^A\right\}, \frac{1}{N_{ch}} \sum\limits_{l=1}^{Nmax}\left(\mathbf{X}_{ch}^{\text{str}}\{\text{rule}_l\} \cdot \mathbf{X}_{ch}^{\text{par}}\left\{\sigma_{i,l}^A\right\}\right)\right)}{\max\left(\mathbf{X}_{ch}^{\text{par}}\left\{\sigma_{i,k}^A\right\}, \frac{1}{N_{ch}} \sum\limits_{l=1}^{Nmax}\left(\mathbf{X}_{ch}^{\text{str}}\{\text{rule}_l\} \cdot \mathbf{X}_{ch}^{\text{par}}\left\{\sigma_{i,l}^A\right\}\right)\right)}, \end{aligned} \tag{29}$$

where $\mathbf{X}_{ch}^{\text{par}}\left\{\sigma_{i,k}^A\right\}$ is a gene of the individual $\mathbf{X}_{ch}^{\text{par}}$ associated with the parameter $\sigma_{i,k}^A$. A function $\text{shy}\left(\mathbf{X}_{ch}, j, k\right)$ related to the outputs can be determined in a similar way.

## 4.3 Fuzzy Rules Readability Evaluation Criteria

In this section two criteria related to the readability of fuzzy rules are presented: the criterion considering uniformity of covering data points with input fuzzy sets and the criterion limiting the number of simultaneously activated fuzzy rules.

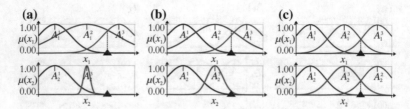

**Fig. 6** Three exemplary cases obtained for criterion (30): **a** negative, **b** intermediate, **c** preferred (good matching of fuzzy sets to the data, low criterion value). Location of the signals coming from the learning sequence sample is denoted as a triangle

## Criterion for assessing coverage of the data space by input fuzzy sets

The considered criterion allows one to evaluate matching of input fuzzy sets to the input data. For properly positioned input fuzzy sets associated with the input $i$, the sum of memberships determined for the signal given on the input $i$ is equal to 1. This assumption is valid for all system inputs (3) and it is evaluated in the context of the whole learning sequence $\{\bar{\mathbf{x}}_z, \mathbf{d}_z\}$ ($z = 1, \ldots, Z$).

The method of operation of the considered criterion is shown in Fig. 6 and it is expressed as follows:

$$\text{ffint}_D\left(\mathbf{X}_{ch}\right) = \frac{\sum_{z=1}^{Z}\sum_{i=1}^{n}\left( \cdot \max\left\{ \left| 1 - \sum_{k=1}^{N\max} \mathbf{X}_{ch}^{\text{str}}\{\text{rule}_k\} \cdot \mu_{A_i^k}\left(\bar{x}_{z,i}\right) \right|, 1 \right\} \right)}{Z \cdot n_{ch}}. \quad (30)$$

## Criterion for assessing fuzzy rules activity

The considered criterion makes it possible to evaluate activation level of the rules in the system (3). The proper rule activation level (1) is achieved when for each set from the learning sequence activation of a single rule from the rule base occurs and activation of the other rules is minimal. Activation level of the rule $k$ of the system (3) is expressed using Eq. (4).

The method of operation of the considered criterion is shown in Fig. 7 and it is expressed as follows:

$$\text{ffint}_E\left(\mathbf{X}_{ch}\right) = 1 - \frac{1}{Z}\sum_{z=1}^{Z} \frac{\left( \max_{k=1,\ldots,Nmax}\left\{\mathbf{X}_{ch}^{\text{str}}\{\text{rule}_k\} \cdot \tau_k\left(\bar{\mathbf{x}}_z\right)\right\} \right)^2}{\sum_{k=1}^{Nmax} \mathbf{X}_{ch}^{\text{str}}\{\text{rule}_k\} \cdot \tau_k\left(\bar{\mathbf{x}}_z\right)}. \quad (31)$$

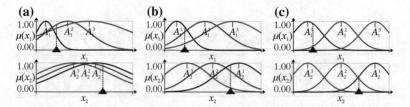

**Fig. 7** Three exemplary cases obtained for criterion (31): **a** negative, **b** intermediate, **c** preferred (a small number of fuzzy rules activated at the same time, low criterion value). Location of signals coming from the learning sequence sample is denoted as a triangle

## 4.4 Criterion for Assessing the Readability of Weight Values in the Fuzzy Rule Base

The use of weights in the rules base of the fuzzy system has many advantages: (a) it increases the flexibility of the problem description, (b) it allows for the introduction of a hierarchy of importance in the rules base, and (c) it increases the accuracy of the system [15, 39, 86]. However, weights in the rule base may sometimes affect readability of fuzzy systems [69]. In the system of the form (3), weights have specified interpretation in the context of rules of the form (1) and dedicated aggregation operators of the forms (5) and (6), used for their processing. Therefore, it seems that this usage of weights positively affects the readability of the system (3).

The considered criterion allows us to evaluate readability of weight values from the rule base of the form (1) of the system (3). Preferred values of weights are those which are close to the values of the set $\{0.0, 0.5, 1.0\}$. Then, they can be easily labeled as: "not important", "important" and "very important".

The method of operation of the considered criterion is shown in Fig. 8 and it is expressed as follows:

$$
\text{ffint}_F\left(\mathbf{X}_{ch}\right) =
$$
$$
= 1 - \frac{\sum\limits_{k=1}^{Nmax} \mathbf{X}_{ch}^{\text{str}}\{rule_k\} \cdot \left( \begin{array}{l} \sum\limits_{i=1}^{n} \mathbf{X}_{ch}^{\text{str}}\{x_i\} \cdot \mathbf{X}_{ch}^{\text{str}}\{A_i^k\} \cdot \mu_w\left(w_{i,k}^A\right) + \\ + \sum\limits_{j=1}^{m} \mathbf{X}_{ch}^{\text{str}}\{B_j^k\} \cdot \mu_w\left(w_{i,k}^B\right) + \\ + \sum\limits_{k=1}^{Nmax} \mu_w\left(w_k^{\text{rule}}\right) \end{array} \right)}{\sum\limits_{k=1}^{Nmax} \mathbf{X}_{ch}^{\text{str}}\{rule_k\} \cdot \left( \begin{array}{l} \sum\limits_{i=1}^{n} \mathbf{X}_{ch}^{\text{str}}\{x_i\} \cdot \mathbf{X}_{ch}^{\text{str}}\{A_i^k\} + \\ + \sum\limits_{j=1}^{m} \mathbf{X}_{ch}^{\text{str}}\{B_j^k\} + 1 \end{array} \right)},
$$

$$(32)$$

where $\mu_w(\cdot)$ is a function "promoting" values 0.0, 0.5 and 1.0, expressed as follows:

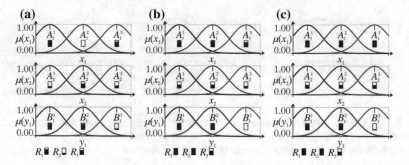

**Fig. 8** Three exemplary cases obtained for criterion (32): **a** negative, **b** intermediate, **c** preferred (good readability of weights, low criterion value). Weights' values are denoted as rectangles. "Black" rectangles indicate value of weight equal to 1.0 and "white" ones indicate value of weight equal to 0.0

$$\mu_w(x) = \begin{cases} \frac{a-x}{a} & \text{for } x \in [0, a] \\ \frac{x-a}{b-a} & \text{for } x \in (a, b] \\ \frac{c-x}{c-b} & \text{for } x \in (b, c] \\ \frac{x-c}{1-c} & \text{for } x \in (c, 1] \end{cases}, \tag{33}$$

where $a = 0.25$, $b = 0.50$, and $c = 0.75$.

## 4.5 Criterion for Assessing the Readability of Triangular Norms

The considered criterion allows one to evaluate readability of the shape parameter of parametrized triangular norms with weights of arguments of the form (5) described in Sect. 2.3. In the system (3) these norms are used for aggregation of antecedences (they have a parameter $p^\tau$), generation of inferences from the rules (they have a parameter $p^{\text{imp}}$) and aggregation of inferences from the rules (they have a parameter $p^{\text{agr}}$). The precision of their operation can usually achieve a better accuracy of the system [15, 27, 72]. This allows for a better use of abilities of the rule base (1) without substantially increasing the number of rules. Moreover, application of the norms of the form (5) facilitates the selection of aggregation operators for the system (3), which takes place automatically by changing the shape parameter and not by the trial-and-error method.

The readable parameter of the norms (5) is the one for which norms (5) approximate the shape of the typical, nonparametric triangular norms. This is because it

**Table 1** A set of values of the parameter of Dombi-type parametrized triangular norms of the form (5), for which their shape approximates the shape of typical and non-parametric triangular norms

| Parameter value | Non-parametrized norm | Similarity |
|---|---|---|
| $p = 0.00$ | Drastic | Full |
| $p = 0.43$ | Algebraic | High |
| $p = 0.71$ | Łukasiewicz | High |
| $p = 1.00$ | Hamacher | Full |
| $p \to \infty$ | Min/max | Full |

is assumed that operation of nonparametric norms (e.g. minimum/maximum operator) is more intuitive. Table 1 contains a set of selected values of the parameter of Dombi-type parametrized triangular norms of the form (5), for which their shape approximates the shape of typical and non-parametric triangular norms [47]. The data presented in the table were generated for the two-argument norms with an accuracy of 0.01.

The method of operation of the considered criterion is shown in Fig. 9 and it is expressed as follows:

$$\text{ffint}_G \left( \mathbf{X}_{ch} \right) = \frac{1}{3} \left( \mu_p \left( \mathbf{X}_{ch}^{par} \{ p^\tau \} \right) + \mu_p \left( \mathbf{X}_{ch}^{par} \{ p^{imp} \} \right) + \mu_p \left( \mathbf{X}_{ch}^{par} \{ p^{agr} \} \right) \right), \quad (34)$$

where $\mu_p(\cdot)$ is a function "promoting" values presented in Table 1. The formula of function $\mu_p(x)$ is as follows:

$$\mu_p(x) = \begin{cases} \frac{a-x}{a} & \text{for } x \in [0, a] \\ \frac{x-a}{b-a} & \text{for } x \in (a, b] \\ \frac{c-x}{c-b} & \text{for } x \in (b, c] \\ \frac{x-c}{d-c} & \text{for } x \in (c, d] \\ \frac{d-x}{e-x} & \text{for } x \in (d, e] \\ \frac{x-d}{f-e} & \text{for } x \in (e, f] \\ \frac{g-x}{g-f} & \text{for } x \in (f, g] \\ \frac{x-g}{h-g} & \text{for } x \in (g, h] , \end{cases} \quad (35)$$

where $a = 0.21$, $b = 0.43$, $c = 0.57$, $d = 0.71$, $e = 0.85$, $f = 1.00$, $g = 1.20$, $h = 10.00$.

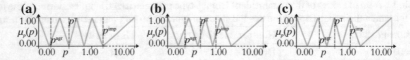

**Fig. 9** Three exemplary cases obtained for criterion (34): **a** negative, **b** intermediate, **c** preferred (good readability of parameters of Dombi-type norm, low criterion value)

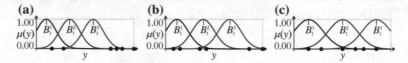

**Fig. 10** Three exemplary cases obtained for criterion (36): **a** negative, **b** intermediate, **c** preferred (properly distributed discretization points, low criterion value). Discretization points are denoted as black circles

## 4.6 Criterion for Assessing the Defuzzification Mechanism

The considered criterion allows us to evaluate distribution of discretization points of the system (3) described in Sect. 2.2. Points distributed correctly are the ones which are placed close to the centers of output fuzzy sets and in their borders (for the proposed criterion this is controlled by the parameter $c_{\text{intH}}$). Properly spaced discretization points increase precision of the defuzzification operator, which can help to increase accuracy of the system (3). This is due to the fact that a greater number of discretization points increases the importance of the shape of fuzzy sets, which makes it possible to achieve better accuracy of the system without increasing the number of rules. The other advantages of the defuzzification operator used are described in Sect. 2.2.

The method of operation of the considered criterion is shown in Fig. 10 and it is expressed as follows:

$$
\text{ffint}_H \left( \mathbf{X}_{ch} \right) = \\
= \frac{\sum\limits_{j=1}^{m} \sum\limits_{k=1}^{Nmax} \left( \begin{array}{c} \mathbf{X}_{ch}^{\text{str}} \left\{ B_j^k \right\} \cdot \\ \cdot \left( \begin{array}{c} 1 - \max \left\{ \mu_{B_j^k} \left( \bar{y}_{j,1}^{\text{def}} \right), ..., \mu_{B_j^k} \left( \bar{y}_{j,R_j}^{\text{def}} \right) \right\} + \\ + \left| c_{\text{intH}} - \max \left\{ \mu_{B_j^k} \left( \bar{y}_{j,1}^{\text{def}} \right), ..., \mu_{B_j^k} \left( \bar{y}_{j,R_j}^{\text{def}} \right) \right\} \right| \end{array} \right) \end{array} \right)}{2 \cdot \sum\limits_{j=1}^{m} \sum\limits_{k=1}^{Nmax} \left( \mathbf{X}_{ch}^{\text{str}} \left\{ B_j^k \right\} \right)}, \tag{36}
$$

where $c_{\text{intH}} \in [0, 1]$ is a parameter specifying the desired value of membership function in the "redundant" discretization points.

**Table 2** Simulation problems discussed

| No. | Test set name | Number of input attributes | Number of output attributes | Number of sets | Problem label |
|-----|---------------|----------------------------|-----------------------------|----------------|---------------|
| 1. | Airfoil self-noise [12] | 5 | 1 | 1503 | ASN |
| 2. | Box & Jenkins gas furnace [10] | 6 | 1 | 290 | BJG |
| 3. | Energy efficient [92] | 8 | 2 | 768 | EE |
| 4. | Concrete slump [97] | 7 | 3 | 103 | CS |
| 5. | Van der Pol oscillator [5] | 2 | 2 | 1000 | VPO |
| 6. | Brusselator [14] | 2 | 2 | 1000 | BR |

# 5 Simulations

The set of issues examined in the simulations is shown in Table 2. The purpose of the simulations was to obtain systems of the forms (3) characterized by the lowest values of elements of the forms (16) and (17).

The method of conducting simulations and interpreting the results can be summarized as follows:

- In the simulations we used the fuzzy system of the form (3). We used the new hybrid genetic-firework algorithm described in Sect. 3 to select its structure and parameters. In this process the new interpretability criteria described in Sect. 4 were taken into account.
- The simulations were performed for seven different variants of weights of the evaluation function (15): from the one focused on accuracy (W1) to the one focused on interpretability (W7). The set of these variants is shown in Table 3.
- The simulations were performed taking into account all the criteria described in Sect. 4. The function of the form (17) was used for aggregation of these criteria. They have the following weight values: $w_{ffintA} = 0.5$, $w_{ffintB} = 1.0$, $w_{ffintC} = 0.5$, $w_{ffintD} = 0.5$, $w_{ffintE} = 0.2$, $w_{ffintF} = 0.2$, $w_{ffintG} = 0.2$, $w_{ffintH} = 0.2$. These values refer to the semantics presented in [3].
- Each simulation (for each variant W1...W7) was repeated 100 times, each time drawing a population of individuals of the form (8). The obtained results were averaged and they are presented in Table 7 and in Fig. 17. Due to the varying complexity of the considered simulation problems, Fig. 17 is indicative.

**Table 3** A set of variants of the weights of the evaluation function (15)

| Variant | $w_{ffacc}$ | $w_{ffint}$ | Description |
|---------|-------------|-------------|-------------|
| W1 | 1.00 | 0.10 | Focused on high accuracy |
| W2 | 0.85 | 0.25 | Focused on accuracy |
| W3 | 0.70 | 0.40 | Intermediate between W2 and W4 |
| W4 | 0.55 | 0.55 | Taking into account the compromise between interpretability and accuracy |
| W5 | 0.40 | 0.70 | Intermediate between W4 and W6 ˙ |
| W6 | 0.25 | 0.85 | Focused on interpretability |
| W7 | 0.10 | 1.00 | Focused on good interpretability |

**Table 4** A set of parameters of the hybrid genetic-firework algorithm

| Description | Notation | Value |
|-------------|----------|-------|
| Number of iterations | $Niter$ | 1000 |
| Number of fireworks | $Npop$ | 10 |
| Parameter controlling the number of sparks | $Nspa$ | 100 |
| Number of additional sparks | $Nsparnd$ | 10 |
| Parameter limiting the minimum number of sparks | $a$ | 0.02 |
| Parameter limiting the maximum number of sparks | $b$ | 0.40 |
| Maximum amplitude of explosion | $\overline{amp}$ | 0.50 |

**Table 5** A set of parameters of the fuzzy system of the form (3) described in Sect. 2

| Description | Notation | Value |
|-------------|----------|-------|
| Maximum number of rules | $Nmax$ | 7 |
| Maximum number of discretization points | $Rmax$ | 21 |
| Minimum value of Dombi-norm parameters | $\underline{p}$ | 0.00 |
| Maximum value of Dombi-norm parameters | $\bar{p}$ | 10.00 |
| Expected intersection point of fuzzy sets | $c_{intB}$ | 0.5 |
| Parameter concerning distribution of discretization points | $c_{intH}$ | 0.5 |

- A set of parameters of the hybrid genetic-firework algorithm is presented in Table 4 and a set of parameters of the fuzzy system of the form (3) is presented in Table 5. Moreover, the Eq. (23) uses the Euclidean measure.

The remarks on the way of interpretation of fuzzy rules of the form (1) obtained in simulations can be summarized as follows:

- The fuzzy rules are presented in Table 6 and in Figs. 11, 12, 13, 14, 15 and 16. Each fuzzy set and each fuzzy rule has a weight of importance represented in the Figs. 11, 12, 13, 14, 15 and 16 by the rectangle. Filling of the rectangle depends on the weight value: full filling means that weight value is 1.0 and an empty rectangle

**Table 6** Summary with examples of fuzzy rules in the form of (1) of the fuzzy system (3) for variant W6

| | | |
|---|---|---|
| **ASN** | $R^1$: IF $\begin{pmatrix} frequency_{[1]} \text{ is } medium\,\|i \text{ AND} \\ angle_{[2]} \text{ is } high\,\|n \text{ AND} \\ chord\ length_{[3]} \text{ is } medium\,\|i \text{ AND} \\ fs\ velocity_{[4]} \text{ is } low\,\|v \end{pmatrix}$ THEN $\left( pressure_{[1]} is low\,\|i \right)$ $\|i$ | |
| | $R^2$: IF $\begin{pmatrix} frequency_{[1]} is low\,\|i \text{ AND} \\ angle_{[2]} is low\,\|i \text{ AND} \\ chord\ length_{[3]} is low\,\|n \text{ AND} \\ fs\ velocity_{[4]} is high\,\|i \text{ AND} \\ displacement_{[5]} is low\,\|i \end{pmatrix}$ THEN $\left( pressure_{[1]} is high\,\|v \right)$ $\|i$ | |
| | $R^3$: IF $\begin{pmatrix} frequency_{[1]} is high\,\|i \text{ AND} \\ chord\ length_{[3]} is high\,\|i \text{ AND} \\ displacement_{[4]} is high\,\|i \end{pmatrix}$ THEN $\left( pressure_{[1]} is medium\,\|i \right)$ $\|i$ | |
| **BJG** | $R^1$: IF $\begin{pmatrix} gas\ flow(t-2)_{[2]} \text{ is } near\ 55.55\,\|i \text{ AND} \\ gas\ flow(t-5)_{[5]} \text{ is } near\ -1.61\,\|i \text{ AND} \\ gas\ flow(t-6)_{[6]} \text{ is } near\ -1.73\,\|i \end{pmatrix}$ THEN $\left( CO_{2[1]} \text{ is } high\,\|i \right)$ $\|i$ | |
| | $R^2$: IF $\left( gas\ flow(t-1)_{[1]} \text{ is } low\,\|i \right)$ THEN $\left( CO_{2[1]} \text{ is } low\,\|v \right)$ $\|i$ | |
| | $R^3$: IF $\left( gas\ flow(t-1)_{[1]} \text{ is } high\,\|i \right)$ THEN $\left( CO_{2[1]} is medium\,\|v \right)$ $\|i$ | |
| **EE** | $R^1$: IF $\begin{pmatrix} compactness_{[1]} \text{ is } high\,\|i \text{ AND} \\ surface\ area_{[2]} is low\,\|i \text{ AND} \\ roof\ area_{[4]} is medium\,\|i \text{ AND} \\ height_{[5]} \text{ is } near\ 5.35\,\|i \text{ AND} \\ orientation_{[6]} \text{ is } low\,\|v \text{ AND} \\ glazing\ area_{[7]} \text{ is } low\,\|i \end{pmatrix}$ THEN $\begin{pmatrix} heating_{[1]} \text{ is } low\,\|i \text{ AND} \\ cooling_{[2]} \text{ is } high\,\|i \end{pmatrix}$ $\|i$ | |
| | $R^2$: IF $\begin{pmatrix} roof\ area_{[4]} \text{ is } low\,\|i \text{ AND} \\ glazing\ area_{[7]} \text{ is } high\,\|i \end{pmatrix}$ THEN $\begin{pmatrix} heating_{[1]} \text{ is } high\,\|i \text{ AND} \\ cooling_{[2]} \text{ is } medium\,\|i \end{pmatrix}$ $\|i$ | |
| | $R^3$: IF $\begin{pmatrix} compactness_{[1]} \text{ is } low\,\|i \text{ AND} \\ surface\ area_{[2]} \text{ is } high\,\|i \text{ AND} \\ wall\ area_{[3]} \text{ is } near\ 343.93\,\|i \text{ AND} \\ roof\ area_{[4]} \text{ is } high\,\|i \text{ AND} \\ orientation_{[6]} \text{ is } high\,\|i \end{pmatrix}$ THEN $\left( cooling_{[2]} \text{ is } low\,\|i \right)$ $\|i$ | |
| **CS** | $R^1$: IF $\begin{pmatrix} cement_{[1]} \text{ is } low\,\|i \text{ AND} \\ coarse\ aggr._{[6]} \text{ is } near\ 741.26\,\|i \text{ AND} \\ fine\ aggr._{[7]} \text{ is } near\ 696.57\,\|i \end{pmatrix}$ THEN $\begin{pmatrix} slump_{[1]} \text{ is } low\,\|i \text{ AND} \\ flow_{[2]} \text{ is } high\,\|i \text{ AND} \\ strength_{[3]} \text{ is } low\,\|i \end{pmatrix}$ $\|i$ | |
| | $R^2$: IF $\begin{pmatrix} cement_{[1]} \text{ is } high\,\|i \text{ AND} \\ slag_{[2]} \text{ is } near\ 123.55\,\|i \text{ AND} \\ fly\ ash_{[3]} \text{ is } near\ 108.30\,\|i \text{ AND} \\ water_{[4]} \text{ is } near\ 191.53\,\|i \end{pmatrix}$ THEN $\begin{pmatrix} slump_{[1]} \text{ is } high\,\|i \text{ AND} \\ mpa_{[3]} \text{ is } medium\,\|i \end{pmatrix}$ $\|i$ | |
| | $R^3$: IF $\left( sp_{[5]} \text{ is } near\ 7.06\,\|v \right)$ THEN $\begin{pmatrix} flow_{[2]} \text{ is } low\,\|i \text{ AND} \\ strength_{[3]} \text{ is } high\,\|i \end{pmatrix}$ $\|i$ | |
| **VPO** | $R^1$: IF $\begin{pmatrix} x(t)_{[1]} \text{ is } low\,\|i \text{ AND} \\ y(t)_{[2]} \text{ is } high\,\|i \end{pmatrix}$ THEN $\left( x(t+1)_{[1]} \text{ is } low\,\|i \right)$ $\|i$ | |
| | $R^2$: IF $\left( x(t)_{[1]} \text{ is } high\,\|v \right)$ THEN $\begin{pmatrix} x(t+1)_{[1]} \text{ is } high\,\|i \text{ AND} \\ y(t+1)_{[2]} \text{ is } high\,\|i \end{pmatrix}$ $\|i$ | |
| | $R^3$: IF $\begin{pmatrix} x(t)_{[1]} \text{ is } medium\,\|i \text{ AND} \\ y(t)_{[2]} \text{ is } low\,\|i \end{pmatrix}$ THEN $\begin{pmatrix} x(t+1)_{[1]} \text{ is } medium\,\|i \text{ AND} \\ y(t+1)_{[2]} \text{ is } low\,\|i \end{pmatrix}$ $\|i$ | |
| **BR** | $R^1$: IF $\left( y(t)_{[2]} \text{ is } near\ -0.02\,\|i \right)$ THEN $\begin{pmatrix} x(t+1)_{[1]} \text{ is } low\,\|i \text{ AND} \\ y(t+1)_{[2]} \text{ is } low\,\|i \end{pmatrix}$ $\|i$ | |
| | $R^2$: IF $\left( x(t)_{[1]} \text{ is } high\,\|i \right)$ THEN $\begin{pmatrix} x(t+1)_{[1]} \text{ is } high\,\|v \text{ AND} \\ y(t+1)_{[2]} \text{ is } high\,\|i \end{pmatrix}$ $\|i$ | |
| | $R^3$: IF $\left( x(t)_{[1]} \text{ is } low\,\|i \right)$ THEN $\begin{pmatrix} x(t+1)_{[1]} \text{ is } medium\,\|v \text{ AND} \\ y(t+1)_{[2]} \text{ is } medium\,\|v \end{pmatrix}$ $\|i$ | |

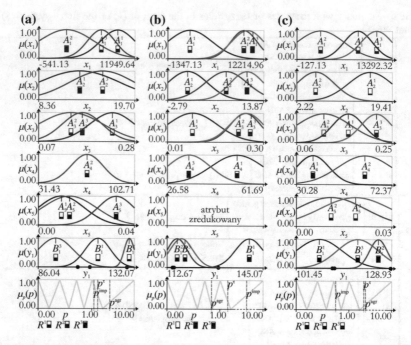

**Fig. 11** Exemplary representation of the fuzzy system rules (3) for the ASN problem and variants: **a** W2, **b** W4, **c** W6

means that the weight is 0.0. Graphic representation of fuzzy rules also takes into account the values of parameters of Dombi-type norm of the form (5).

- Weight symbols in notation of rules of the form (1) were replaced by linguistic labels (Table 6). They are: '$v$' when weight value is greater than 0.75 (*very important*), '$i$' when weight value is in the range [0.25, 0.75] (*important*), '$n$' when weight value is less than 0.25 (*not important*).
- Names of input fuzzy sets $A_i^k$ and output fuzzy sets $B_j^k$ in notation of rules of the form (1) were replaced by the following linguistic labels: '*very low*', '*low*', '*medium low*', '*medium*', '*medium high*', '*high*', '*very high*' (Table 6). Fuzzy sets, which were reduced in the system, were not included in the notation of rules (1). Sometimes in the literature these sets are described as '*don't care*' sets [72]. If the fuzzy system has only one fuzzy set assigned to a specific input or output, its label is set to '*near* [value]'.
- Names of inputs and outputs in notation of the rules of the form (1) were replaced by linguistic labels taken from the description of the described simulation problems (Table 6). Moreover, these names were extended by input or output index placed in square brackets (e.g. '*frequency*$_{[1]}$'). It makes it possible to clearly associate general notation of fuzzy sets (i.e. $A_i^k$, $B_j^k$), presented in Figs. 11, 12, 13, 14, 15 and 16, with linguistic labels used in rules notation presented in Table 6.

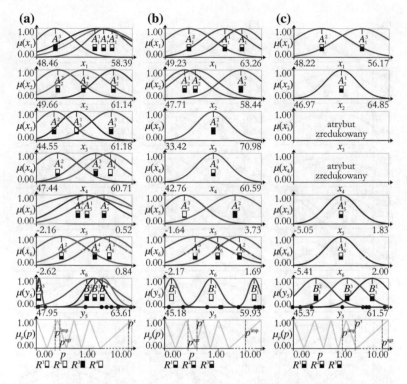

**Fig. 12** Exemplary representation of the fuzzy system rules (3) for the BJG problem and variants:
**a** W2, **b** W4, **c** W6

The conclusions from the simulations can be summarized as follows:

- The fuzzy sets for variant W2 (column a in Figs. 11, 12, 13, 14, 15 and 16) are characterized by low readability. However, the systems related to these sets work with high accuracy. The accuracy is similar to the one obtained in variant W1 focused on accuracy. It is also comparable to the results obtained using methods of other authors which focused on accuracy [20, 65, 92].
- The fuzzy sets for variant W4 (column b in Figs. 11, 12, 13, 14, 15 and 16) have good interpretability. Number of rules for this variant is in the range from 3 to 4 with a good accuracy of the system (Fig. 17). This is a good basis for interpretation of these rules.
- The fuzzy sets for variant W6 (column c in Figs. 11, 12, 13, 14, 15 and 16), have very good interpretability. In these cases reduction of system outputs often occurs, and the number of rules is usually equal to 3. Moreover, the system accuracy is acceptable (Fig. 17).

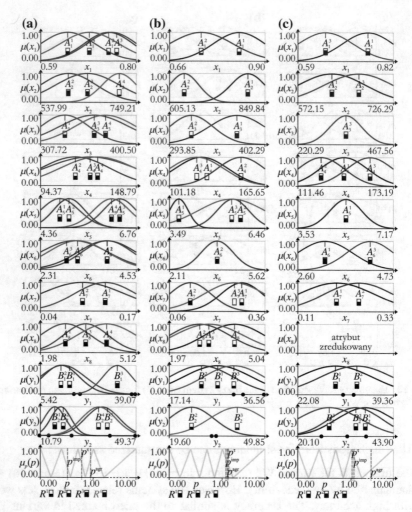

**Fig. 13** Exemplary representation of the fuzzy system rules (3) for the EE problem and variants:
**a** W2, **b** W4, **c** W6

- The results for intermediate variants W3 and W5 and extreme variants W1 and W7 are shown in Table 7 and in Fig. 17. They show dependence between the system accuracy (3) and its interpretability. The results are (as expected) differential. This is also reflected in Figs. 11, 12, 13, 14, 15 and 16.

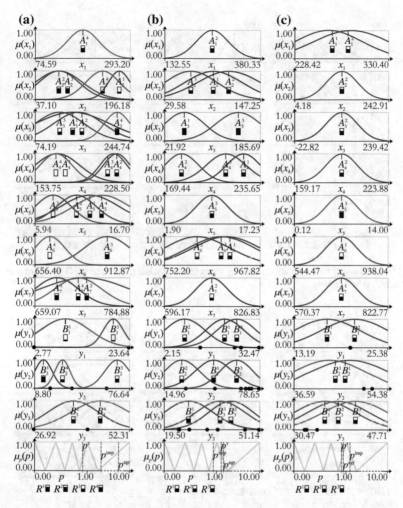

**Fig. 14** Exemplary representation of fuzzy rules of the system (3) for the CS problem and variants: **a** W2, **b** W4, **c** W6

- The average number of rules $N_{ch}$, antecedences $nant_{ch}$, consequences $ncon_{ch}$, inputs $n_{ch}$ and discretization points $R_{j,ch}$ was different for different simulation variants (Fig. 17). Values of these components decrease for cases characterized by greater interpretability.

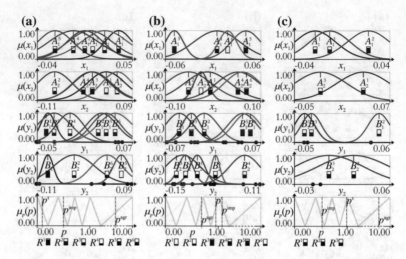

**Fig. 15** Exemplary representation of fuzzy rules of the system (3) for the VPO problem and variants: **a** W2, **b** W4, **c** W6

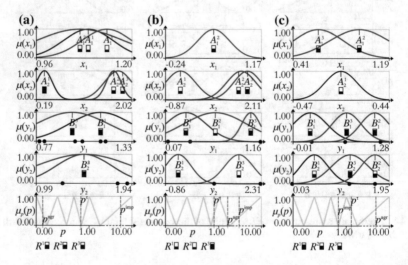

**Fig. 16** Exemplary representation of fuzzy rules of the system (3) for the BR problem and variants: **a** W2, **b** W4, **c** W6

- The results obtained for the AGF algorithm are in all aspects better than the ones obtained for the AGS algorithm (genetic algorithm cooperating with evolutionary strategy). The AGS algorithm was tested as a primary algorithm in order to compare obtained results.

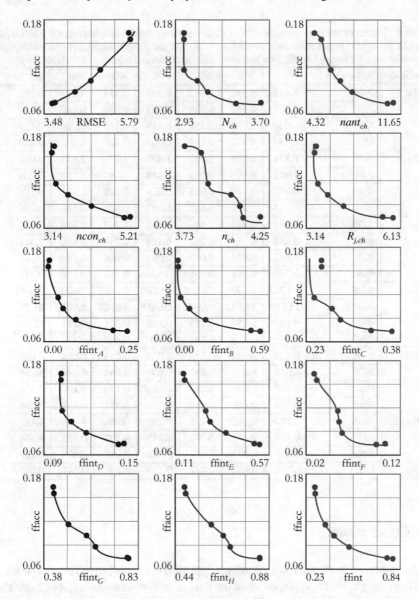

**Fig. 17** Graphical presentation of the components of the evaluation function of the forms of (15) and (17) averaged in the context of all the problems of simulation and performed 100 times. These values were referred to the component ffacc (·) defined by Eq. (16)

**Table 7** Values of the evaluation function (15) components and RMSE, averaged for 100 repetitions of the hybrid genetic-firework algorithm (AGF) and the genetic algorithm cooperating with the evolutionary strategy (AGS). The best results in the context of considered variants are in bold

| Problem | Algorithm | W1 | W2 | W3 | W4 | W5 | W6 | W7 |
|---------|-----------|-----|-----|-----|-----|-----|-----|-----|
| ASN | AGS | 5.635 | 5.358 | 8.192 | 8.246 | 8.314 | 8.391 | 8.286 |
| | AGF | **4.777** | **4.553** | **6.930** | **6.903** | **7.059** | **6.960** | **6.999** |
| | ffacc | 0.098 | 0.094 | 0.150 | 0.149 | 0.150 | 0.149 | 0.153 |
| | ffint | 0.736 | 0.791 | 0.272 | 0.245 | 0.270 | 0.245 | 0.260 |
| BJG | AGS | 1.329 | 1.275 | 1.203 | 1.430 | 1.950 | 3.828 | 4.015 |
| | AGF | **1.062** | **0.993** | **0.966** | **1.148** | **1.557** | **3.257** | **3.245** |
| | ffacc | 0.052 | 0.048 | 0.048 | 0.054 | 0.078 | 0.181 | 0.180 |
| | ffint | 0.723 | 0.787 | 0.653 | 0.587 | 0.434 | 0.280 | 0.278 |
| EE | AGS | 6.541 | 7.074 | 7.330 | 7.433 | 7.787 | 8.312 | 8.101 |
| | AGF | **5.624** | **5.950** | **6.268** | **6.445** | **6.725** | **7.060** | **7.441** |
| | ffacc | 0.082 | 0.084 | 0.090 | 0.093 | 0.096 | 0.102 | 0.110 |
| | ffint | 0.705 | 0.781 | 0.436 | 0.444 | 0.421 | 0.402 | 0.377 |
| CS | AGS | 20.833 | 19.767 | 22.103 | 25.400 | 25.793 | 26.783 | 26.448 |
| | AGF | **16.668** | **16.082** | **17.800** | **20.458** | **20.900** | **22.225** | **21.288** |
| | ffacc | 0.158 | 0.157 | 0.178 | 0.198 | 0.201 | 0.211 | 0.207 |
| | ffint | 0.763 | 0.750 | 0.492 | 0.370 | 0.377 | 0.352 | 0.346 |
| VPO | AGS | 0.033 | 0.034 | 0.045 | 0.054 | 0.053 | 0.054 | 0.053 |
| | AGF | **0.027** | **0.028** | **0.037** | **0.044** | **0.044** | **0.044** | **0.045** |
| | ffacc | 0.065 | 0.062 | 0.083 | 0.151 | 0.152 | 0.150 | 0.159 |
| | ffint | 0.784 | 0.860 | 0.529 | 0.272 | 0.251 | 0.248 | 0.274 |
| BR | AGS | 0.187 | 0.182 | 0.195 | 0.259 | 0.348 | 0.441 | 0.435 |
| | AGF | **0.152** | **0.146** | **0.167** | **0.219** | **0.302** | **0.353** | **0.364** |
| | ffacc | 0.033 | 0.033 | 0.041 | 0.052 | 0.087 | 0.105 | 0.118 |
| | ffint | 0.796 | 0.855 | 0.471 | 0.379 | 0.356 | 0.287 | 0.280 |

## 6 Conclusions

In this paper we have proposed a complex approach to the design of fuzzy systems. It has been developed for applications in the field of nonlinear modeling, but it can also be used in classification issues. The new aspects of the proposed approach include: **(a)** the hybrid genetic-firework algorithm and **(b)** the interpretability criteria of fuzzy systems.

The genetic-firework algorithm was created by combining the genetic and firework algorithms. Not only can the algorithm select parameters of the fuzzy system rules but, owing it to the particular combination of the two algorithms, it can also select its structure. All this creates a significant advantage of the proposed algorithm.

The proposed new interpretability criteria of fuzzy systems are related to all components of fuzzy systems: fuzzy sets, fuzzy rules, weights of importance of rules, weights of importance of the rules antecedences, weights of importance of rules consequences, discretization points of the system, shape parameters of used aggregation and inference operators. Therefore, these criteria do not focus only on fuzzy sets and rules, as it is often the case in the solutions proposed by other authors.

In the simulations we have obtained systems characterized not only by a good accuracy but also a suitable readability in terms of the proposed criteria. Therefore, the solutions proposed in this paper (the algorithm and the criteria) allow one to use the abilities of the fuzzy system more comprehensively and at the same time to receive good results.

In our future papers on interpretability of fuzzy systems we are planning to, among others, develop a new hybrid algorithm searching Pareto fronts (associated with accuracy and interpretability), generalize our discussion for any membership function and parametrized triangular norms of different type. We find the pre-obtained results encouraging so as to continue our research studies in this particular direction.

**Acknowledgements** The project was financed by the National Science Centre (Poland) on the basis of the decision number DEC-2012/05/B/ST7/02138.

# References

1. Alcalá, R., Ducange, P., Herrera, F., Lazzerini, B., Marcelloni, F.: A multi-objective evolutionary approach to concurrently learn rule and data base soft linguistic fuzzy rule-based systems. IEEE Trans. Fuzzy Syst. **17**, 1106–1122 (2009)
2. Alonso, J.M., Magdalena, L., Cordón, O.: Embedding HILK in a three-objective evolutionary algorithm with the aim of modeling highly interpretable fuzzy rule-based classifiers. In: 4th International Workshop on Genetic and Evolving Fuzzy Systems (GEFS2010), pp. 15–20 (2010)
3. Alonso, J.M.: Modeling highly interpretable fuzzy systems. Eur. Centre Soft Comput. (2010)
4. Alonso, J.M., Magdalena, L.: HILK++: an interpretability-guided fuzzy modeling methodology for learning readable and comprehensible fuzzy rule-based classifiers. Soft Comput. **15**(10), 1959–1980 (2011)
5. Althoff, M., Stursberg, O., Buss, M.: Reachability analysis of nonlinear systems with uncertain parameters using conservative linearization. In: Proceedings of the 47th IEEE Conference on Decision and Control, pp. 4042–4048 (2008)
6. Amor, N.B., Salem, B., Zied, E.: Naive Bayes vs decision trees in intrusion detection systems. In: Proceedings of the 2004 ACM Symposium on Applied Computing (2004)
7. Andrieu, C., Doucet, A.: Particle filtering for partially observed Gaussian state space models. JR Stat. Soc. B **64**(4), 827–836 (2002)
8. Bartczuk, Ł., Przybył, A., Koprinkova-Hristova, P.: New method for non-linear correction modelling of dynamic objects with genetic programming. In: Artificial Intelligence and Soft Computing. Lecture Notes in Computer Science, vol. 9120, pp. 318–329 (2015)
9. Botta, A., Lazzerini, B., Marcelloni, F., Stefanescu, D.C.: Context adaptation of fuzzy systems through a multi-objective evolutionary approach based on a novel interpretability index. Soft Comput. **13**, 437–449 (2009)
10. Box, G., Jenkins, G.: Time Series Analysis: Forecasting and Control. Holden-Day, San Francisco (1970)

11. Brasileiro, Í., Santos, I., Soares, A., Rablo, R., Mazullo, F.: Ant colony optimization applied to the problem of choosing the best combination among M combinations of shortest paths in transparent optical networks. J. Artif. Intell. Soft Comput. Res. 6(4), 231–242 (2016)
12. Brooks, T.F., Pope, D.S., Marcolini, A.M.: Airfoil self-noise and prediction. Technical report, NASA RP-1218 (1989)
13. Chen, K.: Global modeling of different vehicles. IEEE Veh. Technol. Mag. 4(2), 80–89 (2009)
14. Chen, X., Abraham, E., Sankaranarayanan, S.: Flow*: an analyzer for non-linear hybrid systems. In: Proceedings of the 25th International Conference on Computer Aided Verification, vol. 8044, pp. 258–263 (2013)
15. Cpałka, K.: A new method for design and reduction of neuro-fuzzy classification systems. IEEE Trans. Neural Netw. 20, 701–714 (2009)
16. Cpałka, K.: On evolutionary designing and learning of flexible neuro-fuzzy structures for nonlinear classification. In: Nonlinear Analysis Series A: Theory, Methods and Applications, vol. 71, pp. 1659–1672. Elsevier (2009)
17. Cpałka, K.: Design of Interpretable Fuzzy Systems. Springer (2017)
18. Cpałka, K., Rebrova, O., Nowicki, R., Rutkowski, L.: On design of flexible neuro-fuzzy systems for nonlinear modelling. Int. J. Gen. Syst. 42(6), 706–720 (2013)
19. Cpałka, K., Rutkowski, L.: Flexible Takagi-Sugeno fuzzy systems. In: Proceedings of the 2005 IEEE International Joint Conference on Neural Networks IJCNN '05, vol. 3, pp. 1764–1769 (2005)
20. Cyran, A.K., Kozielski, S., Peters, F.P., Stanczyk, U., Wakulicz-Deja, A.: Adaptable graphical user interfaces for player-based applications. Adv. Intell. Soft Comput. 59, 69–76 (2009)
21. Deb, K., Pratap, A., Agarwal, S., Meyarivan, T.: A fast and elitist multiobjective genetic algorithm: NSGA-II. IEEE Trans. Evol. Comput. 6, 182–197 (2002)
22. Duch, W., Korbicz, J., Rutkowski, L., Tadeusiewicz, R.: Biocybernetics and biomedical engineering EXIT, Warszawa (2013)
23. Duda, P., Hayashi, Y., Jaworski, M.: On the strong convergence of the orthogonal series-type kernel regression neural networks in a non-stationary environment. In: Artificial Intelligence and Soft Computing, vol. 7267, pp. 47–54. Springer (2012)
24. El-Samak, A.F., Ashour, W.: Optimization of traveling salesman problem using affinity propagation clustering and genetic algorithm. J. Artif. Intell. Soft Comput. Res. 5, 239–246 (2015)
25. Er, M.J., Duda, P.: On the weak convergence of the orthogonal series-type kernel regresion neural networks in a non-stationary environment. In: International Conference on Parallel Processing and Applied Mathematics. Lecture Notes in Computer Science, vol. 7203, pp. 90–98. Springer (2012)
26. Espinosa, J., Vandewalle, J.: Constructing fuzzy models with linguistic integrity from numerical data-AFRELI algorithm. IEEE Trans. Fuzzy Syst. 8, 591–600 (2000)
27. Farahbod, F., Eftekhari, M.: Comparsion of different T-norm operators in classification problems. Int. J. Fuzzy Logic Syst. 2(3), 33–41 (2012)
28. Fazendeiro, P., de Oliveira, J.V., Pedrycz, W.: A multiobjective design of a patient and anaesthetist-friendly neuromuscular blockade controller. IEEE Trans. Biomed. Eng. 54, 1667–1678 (2007)
29. Fraser, A., Burnell, D.: Computer Models in Genetics. McGraw-Hill, New York (1970)
30. Gabryel, M., Cpałka, K., Rutkowski, L.: Evolutionary strategies for learning of neuro-fuzzy systems. In: Proceedings of the I Workshop on Genetic Fuzzy Systems, Granada, vol. 119, p. 123 (2005)
31. Gacto, M.J., Alcalá, R., Herrera, F.: Integration of an index to preserve the semantic interpretability in the multi-objective evolutionary rule selection and tuning of linguistic fuzzy systems. IEEE Trans. Fuzzy Syst. 18(3), 515–531 (2010)
32. Gacto, M.J., Alcalá, R., Herrera, F.: Interpretability of linguistic fuzzy rule-based systems: an overview of interpretability measures. Inf. Sci. 181(20), 4340–4360 (2011)
33. Gorzalczany, M.B., Rudzinski, F.: Accuracy vs. interpretability of fuzzy rule-based classifiers: an evolutionary approach. In: Proceedings of the 2012 International Conference on Swarm and Evolutionary Computation SIDE'12, pp. 222–230 (2012)

34. Guillaume, S., Charnomordic, B.: Generating an interpretable family of fuzzy partitions from data. IEEE Trans. Fuzzy Syst. **12**(3), 324–335 (2004)
35. Ibrahim, S.S., Bamatraf, M.A.: Interpretation trained neural networks based on genetic algorithms. Int. J. Artif. Intell. Appl. (IJAIA) **4**(1), 13–22 (2013)
36. Icke, I., Rosenberg, A.: Multi-objective genetic programming for visual analytics. In: Silva, S., et al. (eds.) EuroGP 2011. LNCS, vol. 6621, pp. 322–334 (2011)
37. Ishibuchi, H., Nakashima, T., Murata, T.: Comparsion of the Michigan and Pittsburgh approaches to the design of fuzzy classification systems. Electron. Commun. Jpn. Part 3 **80**(12), 379–387 (1997)
38. Ishibuchi, H., Nakashima, T., Murata, T.: Performance evaluation of fuzzy classifier systems for multidimensional pattern classification problems. IEEE Trans. SMC B Cybern. **29**, 601–618 (1999)
39. Ishibuchi, H.: Rule weight specification in fuzzy rule-based classification systems. IEEE Trans. Fuzzy Syst. **13**(4), 428–436 (2005)
40. Ishibuchi, H., Nojima, Y.: Analysis of interpretability-accuracy tradeoff of fuzzy systems by multiobjective fuzzy genetics-based machine learning. Int. J. Approximate Reasoning **44**, 4–31 (2007)
41. Jaworski, M., Er, M.J., Pietruczuk, L.: On the application of the Parzen-type kernel regression neural network and order statistics for learning in a non-stationary environment. In: International Conference on Artificial Intelligence and Soft Computing. Lecture Notes in Artificial Intelligence, vol. 7267, pp. 90–98. Springer (2012)
42. Kacprzyk, J.: Studies in Computational Intelligence, vol. 143 (2008)
43. Kaczorek, T.: A modified state variable diagram method for determination of positive realizations of linear continous-time systems with delays. Int. J. Appl. Math. Comput. Sci. **22**(4), 897–905 (2012)
44. Kar, S., Das, S., Ghosh, P.K.: Applications of neuro-fuzzy systems: a brief review and future outline. Appl. Soft Comput. **15**, 243–259 (2014)
45. Kamyar, M.: Takagi-Sugeno fuzzy modeling for process control industrial automation. In: Robotics and Artificial Intelligence (EEE8005), School of Electrical, Electronic and Computer Engineering (2008)
46. Kenesei, T., Abonyi, J.: Interpretable support vector machines in regression and classification-application in process engineering. Hung. J. Ind. Chem. **35**, 101–108 (2007)
47. Klement, E.P., Mesiar, R., Pap, E.: Triangular Norms. Kluwer Academic Publishers (2000)
48. Leekwijck, W.V., Kerre, E.E.: Defuzzification: criteria and classification. Fuzzy Sets Syst. **108**(2), 159–178 (1999)
49. Leon, M., Xiong, N.: Adapting differential evolution algorithms for continuous optimization via greedy adjustment of control parameters. J. Artif. Intell. Soft Comput. Res. **6**(2), 103–118 (2016)
50. Liu, F., Quek, C., Ng, G.S.: A novel generic hebbian ordering-based fuzzy rule base reduction approach to Mamdani neuro-fuzzy system. Neural Comput. **19**, 1656–1680 (2007)
51. Loh, W.-Y.: Classification and regression trees. Wiley Interdisc. Rev.: Data Min. Knowl. Discovery **1**(1), 14–23 (2011)
52. Łapa, K., Cpałka, K., Wang, L.: New method for design of fuzzy systems for nonlinear modelling using different criteria of interpretability. Lect. Notes Comput. Sci. **8467**, 217–232 (2014)
53. Łapa, K., Szczypta, J., Venkatesan, R.: Aspects of structure and parameters selection of control systems using selected multi-population algorithms. Lect. Notes Comput. Sci. **9120**, 247–260 (2015)
54. Marquez, A.A, Marquez, F.A., Peregrin, A.: A multi-objective evolutionary algorithm with an interpretability improvement mechanism for linguistic fuzzy systems with adaptive defuzzification. IEEE Int. Conf. Fuzzy Syst. 1–7 (2010)
55. Mehran, K.: Takagi-Sugeno fuzzy modeling for process control. In: Industrial Automation, Robotics and Artificial Intelligence (EEE8005) (2008)

56. Mencar, C., Castellano, G., Fanelli, A.M.: Some fundamental interpretability issues in fuzzy modeling. In: Proceedings of the Joint 4th Conference of the European Society for Fuzzy Logic and Technology, pp. 100–105 (2005)
57. Mencar, C., Castellano, G., Fanelli, A.M.: On the role of interpretability in fuzzy data mining. Int. J. Uncertainty Fuzziness Knowl. Based Syst. 521–537 (2007)
58. Mencar, C., Castiello, C., Cannone, R., Fanelli, A.M.: Interpretability assessment of fuzzy knowledge bases: a cointension based approach. Int. J. Approximate Reasoning **52**(4), 501–518 (2011)
59. Miller, G.A.: The magical number seven, plus or minus two: some limits on our capacity for processing information. Psychol. Rev. **63**, 81–97 (1956)
60. Miyajima, H., Shigei, N., Miyajima, H.: Performance comparison of hybrid electromagnetism-like mechanism algorithms with descent method. J. Artif. Intell. Soft Comput. Res. **5**(4), 271–282 (2015)
61. Musa, A.A.H., Muawia, M.A.: Analysis of the DC motor speed control using state variable transition matrix. Int. J. Sci. Res. (IJSR) 2758–2763 (2012)
62. Nguyen, K.P., Fujita, G., Dieu, V.N.: Cuckoo search algorithm for optimal placement and sizing of static VAR compensator in large-scale power systems. J. Artif. Intell. Soft Comput. Res. **6**(2), 59–68 (2016)
63. Patan, K., Korbicz, J.: Nonlinear model predictive control of a boiler unit: a fault tolerant control study. Int. J. Appl. Math. Comput. Sci. **22**(1), 225–237 (2012)
64. Pietruczuk, L., Duda, P., Jaworski, M.: Adaptation of decision trees for handling concept drift. In: International Conference on Artificial Intelligence and Soft Computing. Lecture Notes in Artificial Intelligence, vol. 7894, pp. 459–473. Springer (2013)
65. Przybył, A., Cpałka, K.: A new method to construct of interpretable models of dynamic systems. Lect. Notes Artif. Intell. 697–705 (2012)
66. Pulkkinen, P., Koivisto, H.: A dynamically constrained multiobjective genetic fuzzy system for regression problems. IEEE Trans. Fuzzy Syst. **18**(1), 161–177 (2010)
67. Riid, A., Rustern, E.: Interpretability improvement of fuzzy systems: reducing the number of unique singletons in zeroth order Takagi-Sugeno systems. IEEE Int. Conf. Fuzzy Syst. 1–6 (2010)
68. Riid, A., Rustern, E.: Interpretability, interpolation and rule weights in linguistic fuzzy modeling. In: Petrosino, A., et al. (eds.) WILF 2011. LNAI, vol. 6857, pp. 91–98 (2011)
69. Riid, A., Rustern, E.: Adaptability, interpretability and rule weights in fuzzy rule-based systems. Inf. Sci. **257**(1), 301–312 (2014)
70. Rosfariedzah, R., Nagarajan, R., Rahim, M.: Fuzzy variable structure control with reduced-order observer for micro satellite stabilization in space. In: Proceedings of the International Conference on Man-Machine Systems (ICoMMS), pp. 11–13 (2009)
71. Rutkowski, L.: Flexible Neuro-Fuzzy Systems. Kluwer Academic Publishers (2004)
72. Rutkowski, L.: Computational Intelligence. Springer (2008)
73. Rutkowski, L., Cpałka, K.: A general approach to neuro-fuzzy systems. In: The 10th IEEE International Conference on Fuzzy Systems, 2001, Melbourne, pp. 1428–1431 (2001)
74. Rutkowski, L., Cpałka, K.: Compromise approach to neuro-fuzzy systems. In: 2nd Euro-International Symposium on Computation Intelligence, vol. 76, pp. 85–90, Kosice, Slovakia, 16–19 June 2002
75. Rutkowski, L., Cpałka, K.: A neuro-fuzzy controller with a compromise fuzzy reasoning. Control Cybern. **31**(2), 297–308 (2002)
76. Rutkowski, L., Cpałka, K.: Neuro-fuzzy systems derived from quasi-triangular norms. In: Proceedings of the IEEE International Conference on Fuzzy Systems, vol. 2, pp. 1031–1036, Budapest, 26–29 July 2004
77. Rutkowski, L., Cpałka, K.: Designing and learning of adjustable quasi-triangular norms with applications to neuro-fuzzy systems. IEEE Trans. Fuzzy Syst. **13**, 140–151 (2005)
78. Rutkowski, L., Cpałka, K.: Flexible neuro fuzzy systems. IEEE Trans. Neural Netw. **14**(2003), 554–574 (2013)

79. Rutkowski, L., Przybył, A., Cpałka, K.: Novel online speed profile generation for industrial machine tool based on flexible neuro-fuzzy approximation. IEEE Trans. Ind. Electron. **59**(2), 1238–1247 (2012)
80. Rutkowski, L., Przybył, A., Cpałka, K., Er, M.J.: Online speed profile generation for industrial machine tool based on neuro-fuzzy approach. Lect. Notes Artif. Intell. **114**, 645–650 (2010)
81. Sánchez, G., Jiménez, F., Sánchez, J.M., Alcaraz, J.M.: A multi-objective neuro-evolutionary algorithm to obtain interpretable fuzzy models. In: Current Topics in Artificial Intelligence. Lecture Notes in Computer Science, vol. 5988, pp. 51–60 (2010)
82. Scherer, R.: Neuro-fuzzy systems with relation matrix. Artif. Intell. Soft Comput. **6113**, 210–215 (2010)
83. Shukla, P.K., Tripathi, S.P.: A review on the interpretability-accuracy trade-off in evolutionary multi-objective fuzzy systems (EMOFS). Information **3**, 256–277 (2012)
84. Shukla, P.K., Tripathi, S.P.: Handling high dimensionality and interpretability-accuracy trade-off issues in evolutionary multiobjective fuzzy classifiers. Int. J. Sci. Eng. Res. **5**(6), 665–671 (2014)
85. Shukla, P.K., Tripathi, S.P.: A new approach for tuning interval type-2 fuzzy knowledge bases using genetic algorithms. J. Uncertainty Anal. Appl. **2**, 4 (2014)
86. Siminski, K.: Rule weights in a neuro-fuzzy system with a hierarchical domain partition. Int. J. Appl. Math. Comput. Sci. **20**(2), 337–347 (2010)
87. Singh, L., Kumar, S., Paul, S.: Automatic simultaneous architecture and parameter search in fuzzy neural network learning using novel variable length crossover differential evolution. In: IEEE International Conference on Fuzzy Systems, pp. 1795–1802 (2008)
88. Tadeusiewicz, R.: Place and role of intelligent systems in computer science. Comput. Methods Mater. Sci. **10**(4), 193–206 (2010)
89. Tan, Y., Shi, Y., Tan, K.C.: Fireworks algorithm for optimization. In: ICSI 2010, Part I. LNCS, vol. 6145, pp. 355–364 (2010)
90. Tan, C.: More than Accuracy: Interpretability. @MLDG 08/15/2013. https://chenhaot.com/pubs/mldg-interpretability.pdf (2013)
91. Tikk, D., Gedeon, T., Wong, K.: A feature ranking algorithm for fuzzy modeling problems. In: Interpretability Issues in Fuzzy Modeling, pp. 176–192. Springer (2003)
92. Tsanas, A., Xifara, A.: Accurate quantitative estimation of energy performance of residential buildings using statistical machine learning tools. Energy Build. **49**, 560–567 (2012)
93. Vanhoucke, V., Silipo, R.: Interpretability in multidimensional classification. In: Interpretability Issues in Fuzzy Modeling, pp. 193–217. Springer (2003)
94. Viharos, Z.J., Kis, K.B.: Survey on neuro-fuzzy systems and their applications in technical diagnostics. In: 13th IMEKO TC10 Workshop on Technical Diagnostics Advanced Measurement Tools in Technical Diagnostics for Systems' Reliability and Safety (2014)
95. Wang, H., Kwong, S., Jin, Y., Wei, W., Man, K.F.: Multi-objective hierarchical genetic algorithm for interpretable fuzzy rule-based knowledge extraction. Fuzzy Sets Syst. **149**(1), 149–186 (2005)
96. Yang, C.H., Moi, S.H., Lin, Y.D., Chuang, L.Y.: Genetic algorithm combined with a local search method for identifying susceptibility genes. J. Artif. Intell. Soft Comput. Res. **6**, 203–212 (2016)
97. Yeh, I.C.: Modeling slump flow of concrete using second-order regressions and artificial neural networks. Cement Concr. Compos. **29**(6), 474–480 (2007)
98. Yin, Z., O'Sullivan C, Brabazon A.: An analysis of the performance of genetic programming for realised volatility forecas. J. Artif. Intell. Soft Computing Res. **6**, 155–172 (2016)
99. Zalasiński, M.: New algorithm for on-line signature verification using characteristic global features. Adv. Intell. Syst. Comput. **432**, 137–146 (2016)
100. Zalasiński, M., Cpałka, K.: New algorithm for on-line signature verification using characteristic hybrid partitions. Adv. Intell. Syst. Comput. **432**, 147–157 (2016)
101. Zalasiński, M., Cpałka, K., Hayashi, Y.: A new approach to the dynamic signature verification aimed at minimizing the number of global features. Lect. Notes Comput. Sci. **9693**, 218–231 (2016)

102. Zalasiński, M., Cpałka, K., Rakus-Andersson, E.: An idea of the dynamic signature verification based on a hybrid approach. Lect. Notes Comput. Sci. **9693**, 232–246 (2016)
103. Żurada, J.M.: Introduction to Artificial Neural Systems. Jaico Publishing House (2005)

# On the Intuitionistic Fuzzy Sets of $n$-th Type

Krassimir T. Atanassov and Peter Vassilev

**Abstract**  A survey and new results, related to the intuitionistic fuzzy sets of $n$-th type are given. Some open problems are formulated.

## 1 Introduction

The idea for Intuitionistic Fuzzy Sets (IFSs, see [4, 5]) from $n$-th type (IFS-$n$T) was introduced by the first author in 1989 (see [2]) and illustrated for the case of second type in [3]. In [2], the geometrical interpretation of the IFS-2T is given. The results of this paper were extended sequentially in [3, 4, 17].

During last 2–3 years, some colleagues re-discovered the concept of IFS-2T and more general, IFS-$n$T, but using for them (**incorrectly**) the name Pythagorean fuzzy sets (see, e.g. [7–10, 12–14, 18–20, 26–31]). Really, the so-called Pythagorean fuzzy sets coincide exactly with IFS-2T and if we like to use the new name, the IFS-$n$T probably must be called Fermatian fuzzy sets. But, the truth is that these new names only generate a terminological chaos! Of course, this situation is not a new one. The IFSs were introduced in June 1983 in [1]. Using the same name, but in another sense, more than an year later, Takeuti and Titani published paper [22]. In 1993, changing cosmetically the form of IFSs, Gau and Buehrer introduced the concept of vague

---

K.T. Atanassov (✉) · P. Vassilev
Institute of Biophysics and Biomedical Engineering, Bulgarian Academy of Sciences, Acad. G. Bonchev Str., bl. 105, 1113 Sofia, Bulgaria
e-mail: krat@bas.bg; k.t.atanassov@gmail.com

P. Vassilev
e-mail: peter.vassilev@gmail.com

K.T. Atanassov
Intelligent Systems Laboratory Asen Zlatarov University, 8010 Bourgas, Bulgaria

© Springer International Publishing AG 2018
A.E. Gawęda et al. (eds.), *Advances in Data Analysis with Computational Intelligence Methods*, Studies in Computational Intelligence 738,
https://doi.org/10.1007/978-3-319-67946-4_10

sets [11]. For these sets, H. Bustince and P. Burillo proved in [6] that they coincide totally with IFSs.

With the aim to stop the use of different names for the IFS-$n$Ts and having in mind that this name exists already 28 years, below we give the basic theoretical results of IFS-$n$Ts and we hope that in future the colleagues will start using the original name of these sets. The sense of the name "intuitionistic" for the IFSs is discussed in details in [5] and all discussion from there is valid for the IFS-$n$Ts, too.

In the end of the paper, we formulate some problems, related to the IFS-$n$Ts.

## 2 A Second Type of IFSs

Following the definition of the concept of IFS, here we will introduce the concept of *IFS of second type* (IFS-2T) [3].

Let a set $E$ be fixed. An IFS-2T $A^*$ in $E$ is an object of the following form:

$$A^* = \{\langle x, \mu_A(x), \nu_A(x)\rangle | x \in E\}$$

where the functions $\mu_A : E \to [0, 1]$ and $\nu_A : E \to [0, 1]$ define respectively the degree of membership and the degree of non-membership of the elements $x \in E$, and for every $x \in E$:

$$0 \leq \mu_A(x)^2 + \nu_A(x)^2 \leq 1.$$

Every ordinary fuzzy set has the form:

$$\{\langle x, \mu_A(x), \sqrt{1 - \mu_A(x)^2}\rangle | x \in E\}.$$

If

$$\pi_A(x) = \sqrt{1 - \mu_A^2(x) - \nu_A^2(x)},$$

then $\pi_A(x)$ is the degree of non-determinacy of the element $x \in E$ to the set $A$. In case of ordinary fuzzy sets, $\pi_A(x) = 0$ for every $x \in E$.

For simplicity below we will write $A$ instead of $A^*$.

Obviously, for all real numbers $a, b \in [0, 1]$, if

$$0 \leq a + b \leq 1,$$

then

$$0 \leq a^2 + b^2 \leq 1.$$

Unlike the geometrical interpretation of the ordinary IFSs (see [4, 5]), the geometrical interpretation of the IFS–2Ts has the form shown in Fig. 1. The interpretation function is denoted by $g_A$, and $g_A : E \to F$.

**Fig. 1** Geometrical
interpretation of IFS-2Ts

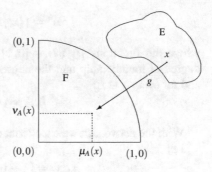

**Fig. 2** Geometrical
interpretation of the two
modal operators over IFS-2T

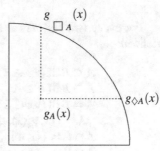

Here, the inequality

$$0 \leq a + b \leq 1$$

between coordinates $\langle a, b \rangle$ of the point $f_A(x) \in F$ changes to the inequality

$$0 \leq a^2 + b^2 \leq 1$$

between coordinates $\langle a, b \rangle$ of the point $g_A(x) \in F$.

Here we will define over the IFS-2Ts analogues of only first two modal operators over IFSs.

$$\Box A = \{\langle x, \mu_A(x), \sqrt{1 - \mu_A(x)^2}\rangle | x \in E\};$$
$$\Diamond A = \{\langle x, \sqrt{1 - \nu_A(x)^2}, \nu_A(x)\rangle | x \in E\}.$$

The geometrical interpretation of the two modal operators over IFS-2T is almost identical to its IFS version—the difference is only in the form of the figure $F$ (see Fig. 2).

## 3 IFS-$n$Ts

Let a set $E$ be fixed. Let $n > 0$ be a real number. An IFS-$n$T $A^*$ in $E$ is an object of the following form:

$$A^* = \{\langle x, \mu_A(x), \nu_A(x)\rangle | x \in E\} \tag{1}$$

where the functions $\mu_A : E \to [0,1]$ and $\nu_A : E \to [0,1]$ define respectively the degree of membership and the degree of non-membership of the elements $x \in E$, and for every $x \in E$:

$$0 \le \mu_A(x)^n + \nu_A(x)^n \le 1. \tag{2}$$

With the above aim we check that the new definition is correct.
Let

$$\pi_A(x) = \left(1 - ((\mu_A(x))^n + (\nu_A(x))^n)\right)^{\frac{1}{n}}. \tag{3}$$

For every two IFS-$n$Ts $A$ and $B$ the following relations and operations can be defined:

$A \subset B$ iff $(\forall x \in E)(\mu_A(x) \le \mu_B(x) \& \nu_A(x) \ge \nu_B(x))$;
$A \supset B$ iff $B \subset A$;
$A = B$ iff $(\forall x \in E)(\mu_A(x) = \mu_B(x) \& \nu_A(x) = \nu_B(x))$
$\neg A = \{\langle x, \nu_A(x), \mu_A(x)\rangle | x \in E\}$
$A \cap B = \{\langle x, \min(\mu_A(x), \mu_B(x)), \max(\nu_A(x), \nu_B(x))\rangle | x \in E\}$;
$A \cup B = \{\langle x, \max(\mu_A(x), \mu_B(x)), \min(\nu_A(x), \nu_B(x))\rangle | x \in E\}$;

We will introduce only the most important properties of these operations.

**Theorem 1** *Let $n > 0$ be a real number. For every three IFS-nTs $A, B$ and $C$:*

*(a)* $A \cup B = B \cup A$;
*(b)* $A \cap B = B \cap A$;
*(c)* $(A \cup B) \cup C = A \cup (B \cup C)$;
*(d)* $(A \cap B) \cap C = A \cap (B \cap C)$;
*(e)* $(A \cup B) \cap C = (A \cap C) \cup (B \cap C)$;
*(f)* $(A \cap B) \cup C = (A \cup C) \cap (B \cup C)$.

Let for every IFS-$n$T $A$, the IFS-topological operators have the forms

$$C(A) = \{\langle x, K, L\rangle | x \in E\}, \text{ where } K = \sup_{x \in E} \mu_A(x), L = \inf_{x \in E} \nu_A(x);$$

$$I(A) = \{\langle x, k, l\rangle | x \in E\}, \text{ where } k = \inf_{x \in E} \mu_A(x), l = \sup_{x \in E} \nu_A(x).$$

We call again these operators "closure" and "interior", respectively, and for them the following assertion holds:

**Theorem 2** *For each $n > 0$ and every two IFS-nTs $A$ and $B$:*

*(a)* $I(A) \subset A \subset C(A)$;
*(b)* $C(C(A)) = C(A)$;
*(c)* $C(I(A)) = I(A)$;

(d)  $I(C(A)) = C(A)$;
(e)  $I(I(A)) = I(A)$;
(f)  $C(A \cup B) = C(A) \cup C(B)$;
(g)  $C(A \cap B) \subset C(A) \cup C(B)$;
(i)  $I(A \cup B) \supset I(A) \cup I(B)$;
(j)  $I(A \cap B) = I(A) \cap I(B)$;
(l)  $I(A) = C(A)$.

Now, we can define for each real number $n > 0$:

$$\Box A = \{\langle x, \mu_A(x), (1 - ((\mu_A(x))^n)^{\frac{1}{n}} \rangle | x \in E\},$$

$$\Diamond A = \{\langle x, (1 - (\nu_A(x))^n)^{\frac{1}{n}}, \nu_A(x) \rangle | x \in E\}.$$

Obviously, for every IFS-$n$T $A$:

$$\Box A \subset A \subset \Diamond A.$$

These operators may be extended by analogy with the IFS-case (see, e.g. [5]) for every $\alpha, \beta \in [0, 1]$:

$$D_\alpha A = \{\langle x, ((\mu_A(x))^n + \alpha^n(\pi_A(x))^n)^{\frac{1}{n}}, ((\nu_A(x))^n + (1 - \alpha^n)(\pi_A(x))^n)^{\frac{1}{n}} \rangle | x \in E\},$$

$$F_{\alpha,\beta} A = \{\langle x, ((\mu_A(x))^n + \alpha^n(\pi_A(x))^n)^{\frac{1}{n}}, ((\nu_A(x))^n + \beta^n(\pi_A(x))^n)^{\frac{1}{n}} \rangle | x \in E\},$$

where $0 \le \alpha^n + \beta^n \le 1$,

$$G_{\alpha,\beta} A = \{\langle x, \alpha\mu_A(x), \beta\nu_A(x) \rangle | x \in E\},$$

$$H_{\alpha,\beta} A = \{\langle x, \alpha\mu_A(x), ((\nu_A(x))^n + \beta^n(\pi_A(x))^n)^{\frac{1}{n}} \rangle | x \in E\},$$

$$J_{\alpha,\beta} A = \{\langle x, ((\mu_A(x))^n + \alpha^n(\pi_A(x))^n)^{\frac{1}{n}}, \beta\nu_A(x) \rangle | x \in E\},$$

$$H^*_{\alpha,\beta} A = \{\langle x, \alpha\mu_A(x), ((\nu_A(x))^n + \beta^n(1 - \alpha^n(\mu_A(x))^n - (\nu_A(x))^n))^{\frac{1}{n}} \rangle | x \in E\},$$

$$J^*_{\alpha,\beta} A = \{\langle x, ((\mu_A(x))^n + \alpha^n(1 - (\mu_A(x))^n - \beta^n(\nu_A(x))^n))^{\frac{1}{n}}, \beta\nu_A(x) \rangle | x \in E\},$$

The basic properties of the standard IFSs are valid here, too. For example, the following assertions can be proved by the way, as for the standard IFS case.

**Theorem 3** *For each natural number $n > 0$, for each IFS-$n$T $A$, and for every two real numbers $\alpha, \beta \in [0, 1]$:*

$$\neg D_\alpha \neg(A) = D_{(1-\alpha^n)}(A),$$

$$\neg F_{\alpha,\beta} \neg(A) = F_{\beta,\alpha}(A), \text{ if } \alpha^n + \beta^n \le 1,$$

$$\neg G_{\alpha,\beta} \neg(A) = G_{\beta,\alpha}(A),$$

$$\neg H_{\alpha,\beta} \neg(A) = J_{\beta,\alpha}(A),$$

$$\neg J_{\alpha,\beta} \neg(A) = H_{\beta,\alpha}(A),$$

$$\neg H^*_{\alpha,\beta} \neg(A) = J^*_{\beta,\alpha}(A),$$

$$\neg J^*_{\alpha,\beta} \neg(A) = H^*_{\beta,\alpha}(A).$$

**Theorem 4** *For each natural number $n > 0$, for each IFS-nT $A$, and for every four real numbers $\alpha, \beta, \gamma, \delta \in [0, 1]$, so that $\alpha^n + \beta^n \le 1$ and $\gamma^n + \delta^n \le 1$:*

$$F_{\alpha,\beta}(F_{\gamma,\delta}(A)) = F_{\alpha^n+\gamma^n-\alpha^n\gamma^n-\alpha^n\delta^n,\beta^n+\delta^n-\beta^n\gamma^n-\beta^n\delta^n}(A),$$

$$G_{\alpha,\beta}(G_{\gamma,\delta}(A)) = G_{\alpha^n\gamma^n,\beta^n\delta^n}(A).$$

## 4  Uses of IFS-*n*T and Additional Results

IFS-2T find their use in image enhancement [15]. Another type of intuitionistic fuzzy sets also used in image enhancement are the intuitionistic fuzzy sets of root type [16], with (1), such that

$$\frac{\sqrt{\mu_A(x)}}{2} + \frac{\sqrt{\nu_A(x)}}{2} \le 1. \tag{4}$$

This definition does not conform to the general notion considered in [17], where the authors studied the properties of sets of IFS-*n*T, namely (1), satisfying (2), where $n \in (0, +\infty)$.

It is interesting to investigate if other types of IFS-*n*T may be successfully applied for image enhancement.

*Remark 1* Note that for $n \ge 1$ (2) may be also stated in an equivalent form:

$$\left(\mu_A(x)^n + \nu_A(x)^n\right)^{\frac{1}{n}} \le 1. \tag{5}$$

*Remark 2* (cf. [17]) If $0 < n < m < \infty$ it is fulfilled that an IFS-*n*T is also an IFS-*m*T.

It seems these results are not well known, since an article discussing a particular case of this investigation has appeared recently [21]. Further investigation was done by P. Vassilev in [23] for the extended modal operator analogous to $F_{\alpha,\beta}$ and $G^n_{\alpha,\beta}$ over the IFSs. More thorough investigation was done by P. Vassilev for the pointwise operator $F^n_{\alpha(x),\beta(x)}$ in his Ph.D. thesis [24]. Namely, the following result is established there:

**Theorem 5**  ([24, Theorem 2.47]) *Let A be an IFS-nT ($n \in (0, \infty)$) over E and B is an IFS over E. Then the pointwise operator*

$$F^n_B : IFS\text{-}nT(E) \to IFS\text{-}nT(E)$$

*is given by*

$$F^n_B(A) = \{\langle x, \hat{\mu}_A(x), \hat{v}_A(x)\rangle | x \in E\},$$

*where*

$$\hat{\mu}_A(x) = (\mu^n_A(x) + \mu_B(x)\pi^*_A(x))^{\frac{1}{n}}$$

$$\hat{v}_A(x) = (v^n_A(x) + v_B(x)\pi^*_A(x))^{\frac{1}{n}}$$

*and*

$$\pi^*_A(x) = 1 - \mu^n_A(x) - v^n_A(x)$$

*Remark 3* If $B$ is taken as an IFS-$n$T, the resulting operator will coincide with the one in the previous section.

Recognizing that (5) may be viewed as distance generated by Minkowski's norm $\varphi_n$ for $n \geq 1$ and by an appropriate subnorm for $n \in (0, 1)$ to the point $(0, 0)$, P. Vassilev introduced a unified metric approach to the notion IFS-$n$T by introducing the notion $d_\varphi$-IFS [24, 25]. It is noteworthy that while both notions describe the same triples as sets, one way that the "hesitancy function" may be defined for IFS-$n$T (a slightly different version of $\pi_A$ from (3)):

$$\pi^*_A(x) = 1 - \mu^n_A(x) - v^n_A(x) \tag{6}$$

does not, in general, coincide with the way it is defined for $d_\varphi$-IFS:

$$\pi_d(A)(x) = 1 - \varphi((\mu_A(x), v_A(x))). \tag{7}$$

which in the case of $\varphi_n$ norms coincides with:

$$\pi_{d_{\varphi_n}}(A)(x) = 1 - (\mu_A(x)^n + v_A(x)^n)^{\frac{1}{n}}$$

The exception to this is for $n = 1$, where (3), (6), (7) are identical, which in our view reinforces the idea that IFS are the most natural among the IFS-$n$Ts.

# 5 Analogues of Mappings of Complex Numbers

Following [5, p. 51], we can easily introduce by analogy the $n$-analogue (for $n \geq 1$) for complex numbers $a + \mathbf{i}b$ and $a - \mathbf{i}b$, with the constraints $a \in [0, 1]$, $b \in [-1, 1]$ and the condition

$$a^n + |b|^n \leq 1$$

Then a a transformation formula analogous to [5, (3.4)] is the following (for $n \geq 1$):

$$f(a, b) = \begin{cases} \left\langle \left(\frac{a^n}{2}\right)^{\frac{1}{n}}, \left(\frac{a^n}{2} + b^n\right)^{\frac{1}{n}} \right\rangle, & \text{for } b \geq 0 \\ \left\langle \left(\frac{a^n}{2} + |b|^n\right)^{\frac{1}{n}}, \left(\frac{a^n}{2}\right)^{\frac{1}{n}} \right\rangle, & \text{for } b \leq 0 \end{cases}$$

The fact that $f$ is a bijection is easy to check. We will start by showing that $f$ is an injection.

Let us be given $(a, b)$ and $(c, d)$ such that $|a - c| + ||b| - |d|| \neq 0$, $a, c \in [0, 1]$ $x_0, x_1 \in [0, 1]$, $b, d \in [-1, 1]$.

Then $f(a, b) \neq f(c, d)$. The case when $b$ and $d$ are of the same sign is obvious. Let us suppose, without loss of generality that $b \geq 0$ and $d \leq 0$.

Then $f(a, b) = f(c, d)$ is equivalent to:

$$\begin{cases} \frac{a^n}{2} = \frac{c^n}{2} + |d|^n \\ \frac{a^n}{2} + b^n = \frac{c^n}{2} \end{cases}$$

But the above is only possible when $a = b = c = d = 0$. Hence, $f$ is an injection.

It remains to prove that for any $(x, y) \in [0, 1] \times [0, 1]$ such that $x^n + y^n \leq 1$, there exists $(x_0, y_0) \in [0, 1] \times [-1, 1]$ with $x_0^n + |y_0|^n \leq 1$ and $f(x_0, y_0) = (x, y)$.

We will consider the three possible cases.

In Case 1: $x = y$, we have $x_0 = 2^{\frac{1}{n}} x$, $y_0 = 0$. Since, $2x^n \leq 1$, we have $x \leq \frac{1}{2^{\frac{1}{n}}}$, hence $x_0 \in [0, 1]$, $y_0 \in [-1, 1]$.

Let Case 2: $x > y$, be fulfilled. Then $y < \frac{1}{2^{\frac{1}{n}}}$. Hence, $x_0 = 2^{\frac{1}{n}} y < 1$ and we determine that $|y_0| = (x^n - y^n)^{\frac{1}{n}}$, i.e. $y_0 = -(x^n - y^n)^{\frac{1}{n}}$. It is easy to check that $x_0 \in [0, 1]$, $y_0 \in [-1, 0]$.

Analogously, let Case 3: $x < y$ be fulfilled. Then $x < \frac{1}{2^{\frac{1}{n}}}$. Hence, $x_0 = 2^{\frac{1}{n}} x < 1$ and we determine that $y_0 = (y^n - x^n)^{\frac{1}{n}}$. It is easy to check that $x_0 \in [0, 1]$, $y_0 \in [0, 1]$.

Thus we have shown that for any point $(x, y)$ there is a pre-image with $f$. Thus, $f$ is a bijection.

As a result of this, if we take two conjugate complex points $a + \mathbf{i}b$ and $a - \mathbf{i}b$, their repsective images with $f$ are in relation negation similarly to the the situation described in [5]. That is

$$\neg f(a, b) = f(a, -b),$$

i.e. the intuitionistic fuzzy pairs which after the transformation correspond to these two points are negations of one another.


# 6  Conclusion

In the future we plan to work on the development of the theory of IFS$n$T pointing our attention to:

**Open Problem 1**. What specific for particular IFS$n$T operators may be defined?
**Open Problem 2**. Can the defined above operators be modified in the sense of [5]?
**Open Problem 3**. What other negation operators may be defined over IFS$n$T?
**Open Problem 4**. What other implications may be defined over IFS$n$T?

**Acknowledgements**  The authors are thankful for the support provided by the Bulgarian National Science Fund under Grant Ref. No. DFNI-I-02-5 "InterCriteria Analysis: A New Approach to Decision Making".


# References

1. Atanassov K.: Intuitionistic fuzzy sets. VII ITKR's Session, Sofia, June 1983 (Deposed in Central Sci.—Techn. Library of Bulg. Acad. of Sci., 1697/84) (in Bulg.). Reprinted in: Int. J. Bioautomation, Vol. 20(S1), 2016, S1-S6 (in English)
2. Atanassov K.: Geometrical interpretations of the elements of the intuitionistic fuzzy objects. Preprint IM-MFAIS, 1–89. Sofia (1989). Reprinted in: Int. J. Bioautomation. **20**(S1), S27–S42 (2016)
3. Atanassov, K.: A second type of intuitionistic fuzzy sets. BUSEFAL **56**, 66–70 (1993)
4. Atanassov, K.: Intuitionistic fuzzy sets. Springer, Heidelberg (1999)
5. Atanassov, K.: On Intuitionistic Fuzzy Sets Theory. Springer, Heidelberg (2012)
6. Bustince, H., Burillo, P.: Vague sets are intuitionistic fuzzy sets. Fuzzy. Sets. Syst. **79**(3), 403–405 (1996)
7. Dick, S., Yager, R., Yazdanbakhsh, O.: On Pythagorean and complex fuzzy set operations. IEEE Trans. Fuzzy. Syst. **24**(5), 1009–1021 (2016)
8. Garg, H.: A novel correlation coefficients between Pythagorean fuzzy sets and its applications to decision-making processes. Int. J. Intell. Syst. **31**(12), 1234–1252 (2016)
9. Garg, H.: A new generalized Pythagorean fuzzy information aggregation using Einstein operations and its application to decision making. Int. J. Intell. Syst. **31**(9), 886–920 (2016)
10. Garg, H.: A novel accuracy function under interval-valued pythagorean fuzzy environment for solving multicriteria decision making problem. J. Intell. Fuzzy. Syst. **31**(1), 529–540 (2016)
11. Gau, W.L., Buehrer, D.J.: Vague sets. IEEE. Trans. Syst. Man. Cybern. **23**, 610–614 (1993)
12. Gou, X., Xu, Z., Ren, P.: The properties of continuous Pythagorean fuzzy information. Int. J. Intell. Syst. **31**(5), 401–424 (2016)
13. Liu, J., Zeng, S., Pan, T.: Pythagorean fuzzy dependent ordered weighted averaging operator and its application to multiple attribute decision making. Gummi. Fasern. Kunststoffe. **69**(14), 2036–2042 (2016)

14. Ma, Z., Xu, Z.: Symmetric pythagorean fuzzy weighted geometric/averaging operators and their application in multicriteria decision-making problems. Int. J. Intell. Syst. **31**(12), 1198–1219 (2016)

15. Palaniappan, N., Srinivasan, R.: Applications of intuitionistic fuzzy sets of root type in image processing. In: North American Fuzzy Information Processing Society (NAFIPS). Annual Conference (2009)

16. Palaniapan, N., Srinivasan, R., Parvathi, R.: Some operations on intuitionistic fuzzy sets of root type. Notes. Intuit. Fuzzy Sets. **12**(3), 20–29 (2006)

17. Vassilev, P., Parvathi, R., Atanassov, K.: Note on intuitionistic fuzzy sets of $p$-th type. Issues. Intuit. Fuzzy Sets. Gener. Nets. **6**, 43–50 (2008)

18. Peng, X., Yang, Y.: Fundamental properties of interval-valued pythagorean fuzzy aggregation operators. Int. J. Intell. Syst. **31**(5), 444–487 (2016)

19. Peng, X., Yang, Y.: Pythagorean fuzzy Choquet integral based MABAC method for multiple attribute group decision making. Int. J. Intell. Syst. **31**(10), 989–1020 (2016)

20. Ren, P., Xu, Z., Gou, X.: Pythagorean fuzzy TODIM approach to multi-criteria decision making. J. Appl. Soft. Comput. **42**, 246–259 (2016)

21. Srinivasan, R., Begum, S.S.: Some properties on intuitionistic fuzzy sets of third type. Ann. Fuzzy Math. Inform. **10**(5), 799–804 (2015)

22. Takeuti, G., Titani, S.: Intuitionistic fuzzy logic and intuitionistic fuzzy set theory. J. Symb. Log. **49**(3), 851–866 (1984)

23. Vassilev, P.: The generalized modal operator $F^p_{\alpha,\beta}$ over $p$-intuitionistic fuzzy sets. Notes. Intuit. Fuzzy. Sets. **15**(4), 19–24 (2009)

24. Vassilev, P.: Intuitionistic fuzzy sets with membership and non-membership functions in metric relation, Ph.D. thesis defended on 18.03.2013, Institute of Biophysics and Biomedical Engineering, Bulgarian Academy of Sciences (in Bulgarian)

25. Vassilev, P.: Intuitionistic fuzzy sets generated by Archimedean metrics and ultrametrics. In: Recent Contributions in Intelligent Systems, Studies in Computational Intelligence 657, pp. 339–378 Springer, Cham (2017)

26. Yager, R.R.: Pythagorean membership grades in multi-criteria decision making. IEEE Trans. Fuzzy Syst. **22**, 958–965 (2014)

27. Yager, R.R.: Properties and applications of Pythagorean fuzzy sets. Stud. Fuzziness. Soft. Comput. **332**, 119–136 (2016)

28. Zeng, S., Chen, J., Li, X.: A hybrid method for Pythagorean fuzzy multiple-criteria decision making. Int. J. Inf. Technol. Decis. Mak. **15**(2), 403–422 (2016)

29. Zhang, C., Li, D., Ren, R.: Pythagorean fuzzy multigranulation rough set over two universes and its applications in merger and acquisition. Int. J. Intell. Syst. **31**(9), 921–943 (2016)

30. Zhang, X.: A Novel approach based on similarity measure for Pythagorean fuzzy multiple criteria group decision making. Int. J. Intell. Syst. **31**(6), 593–611 (2016)

31. Zhang, X.: Multicriteria Pthagorean fuzzy decision analysis: a hierarchical QUALIFLEX approach with the closeness index-based ranking methods. Inf. Sci. **330**, 104–124 (2016)

# Part IV
# Intelligent Technologies in Decision Making, Optimization and Control

# MCTS/UCT in Solving Real-Life Problems

Jacek Mańdziuk

**Abstract** Monte Carlo Tree Search (MCTS) supported by the Upper Confidence Bounds Applied to Trees (UCT) method, i.e. MCTS/UCT, since its onset in 2006, has been one of the state-of-the-art techniques in game-playing domain. In particular, the recent breakthroughing success of this method (combined with deep neural networks trained with the reinforcement learning algorithm) in the game of Go, made its leading position even stronger than before. In this paper we summarize our studies in application of MCTS/UCT to domains other than games, with particular emphasis on hard real-life problems which possess a large degree of uncertainty due to existence of certain stochastic factors in their definition. The two example problems of this nature considered in this work are Capacitated Vehicle Routing Problem with Traffic Jams and Risk-Aware Project Scheduling Problem. Our results show that MCTS/UCT is a viable method in these two domains, efficiently dealing with uncertainty by means of on-line adaptation of the core MCTS simulations to the current situation (actual realization of the stochastic components).

**Keywords** Monte Carlo Tree Search · Upper Confidence Bounds Applied to Trees · Dynamic Vehicle Routing Problem · Traffic jams · Project scheduling

## 1 Introduction

Monte Carlo Tree Search (MCTS) [2] is a simulation-based method of searching a problem space represented in a tree-based form. A typical example are classical board games (e.g. chess, checkers, Othello, Go, etc.) in which possible game continuations from the current game state can be represented in the form of the so-called *game tree*. The method is particularly well-suited to games for which a meaningful

J. Mańdziuk (✉)
Faculty of Mathematics and Information Science,
Warsaw University of Technology, Warsaw, Poland
e-mail: j.mandziuk@mini.pw.edu.pl

© Springer International Publishing AG 2018     277
A.E. Gawęda et al. (eds.), *Advances in Data Analysis with Computational Intelligence Methods*, Studies in Computational Intelligence 738,
https://doi.org/10.1007/978-3-319-67946-4_11

and compact evaluation function is not known (e.g. Go [6, 21], Havannah [27] or Arimaa [26]).

The most popular implementation of the MCTS method was proposed in 2006 by Kocsis and Szepesvari [10] under the name Upper Confidence Bounds Applied to Trees (UCT) and soon after became one of the state-of-the-art approaches in game-playing domain. In this paper we summarize our recent results related to application of MCTS/UCT to domains other than games, showing plausibility and strong potential of the method in solving hard and varying in time real-life optimization problems.

The remainder of the paper is structured as follows. In the next section the MCTS/UCT approach is presented and its application to games is briefly discussed. In Sect. 3 the Capacitated Vehicle Routing Problem with Traffic Jams is introduced along with proposed UCT-based solution method, and its comparative results with two most popular swarm optimization algorithms, specifically developed and tuned for solving this problem. In Sect. 4 the Resource Constrained Projet Scheduling Problem (frequently considered in scheduling problems domain) is briefly recalled and its non-deterministic version—Risk-Aware Project Scheduling Problem—proposed, accompanied by the UCT-based approach and its results versus the outcomes of the heuristic solver, typically applied in this area. The main qualitative observations related to *general applicability of the UCT method* and its *synergetic combinations with domain knowledge heuristics* conclude the paper is Sect. 5.

## 2 Monte Carlo Tree Search

The MCTS algorithm is an iterative simulation-based method of searching the problem space, in the case potential solutions of the problem can be represented as paths in a specifically designed problem tree. In the problem tree, nodes represent possible problem states (or partial solutions) and edges correspond to possible actions. Application of an action in a given state leads to a new state, located one level below in the problem tree (a child node). In the current state of the problem, application of MCTS relies on performing massive simulations, from the node representing that state (the *root node* of a tree). Certainly for complex problems the respective tree based representation is too big to fit in memory and only a part of it is kept and maintained online by the method. The problem tree is gradually extended usually by adding one new node in each of the performed simulations. More precisely, each simulation consists of the four following phases, depicted in Fig. 1.

**Selection**: starting from the root node, traverse the tree down until a leaf node is reached. In each node, choose the child node according to some *in-tree node selection policy*;
**Expansion**: if the leaf node is not terminal, choose a continuation which falls out of the tree and allocate a new child node. This new node is added to the tree and serves as the starting point for the next phase;

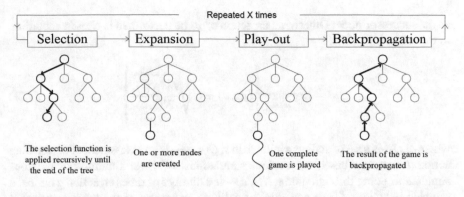

**Fig. 1** Four phases of the MCTS algorithm. The figure replicates an illustration presented in [3] for games domain. In other optimization problem domains, instead of playing a game and propagating its outcome, the system explores a full path to the solution state and back-propagates the corresponding goal value

**Simulation**: starting from a state associated to the newly expanded node, perform a full game simulation (i.e. to the terminal state) choosing the subsequent states according to some *out-of-the-tree selection policy*;

**Backpropagation**: once the simulation reaches a terminal state (i.e. the one which corresponds to some solution of the problem being solved) read out the solution value (score) and propagate it along the solution path, all the way back to the root node. Update the average solution score and increment the number od visits of each in-tree node including the newly-added one.

When the time allotted for the simulations is exhausted, an action leading to the state with the highest average score among the root child states is selected to be performed as a part of *real* (i.e. not simulated) solution. In the *out-of-the-tree selection* procedure the next node to be visited is usually selected uniformly among all possible choices (as the *Monte Carlo* part of the method's name suggests), the *in-tree selection policy* may vary, depending on particular problem being solved. One of the most popular choices is the Upper Confidence Bounds Applied to Trees (UCT) [10] selection method, described in the next section.

## 2.1 Upper Confidence Bounds Applied to Trees

The main purpose of the *in-tree selection policy* is to choose the nodes in a way that maintains a balance between *exploration* of the less frequently simulated actions (nodes) and *exploitation* of the already promising ones. In order to serve this purpose the action selection process follows the following procedure:

If, in the currently simulated node *s*, there exist some actions which have not been yet visited, one of them is uniformly selected, and the corresponding state is assigned as the successor node. Otherwise, an action $a^*$ to be performed is chosen according to the following rule:

$$a^* = arg \max_{a \in A(s)} \left\{ Q(s,a) + C \sqrt{\frac{ln\,[N(s)]}{N(s,a)}} \right\} \qquad (1)$$

where $A(s)$—is a set of actions available in $s$; $Q(s, a)$—is an assessment of performing action $a$ in state $s$ based on previous simulations; $N(s)$—is a number of previous simulations going through state $s$; $N(s, a)$—is a number of times an action $a$ has been sampled in state $s$; $C$—is a coefficient defining an impact of the latter component (*exploration*) in (1).

It can be theoretically proven that the simulation-based assessment of $Q(s,a)$, for each node $s$ and each action $a$ in $s$, converges to its true (real) value when the number of simulations tends to infinity.

## 2.2 MCTS/UCT in Games — a Short Overview

MCTS/UCT is renowned for being the state-of-the-art algorithm for searching a game tree in a variety of games. It is particularly useful in complex games with high branching factors such as Go or Hex [1], for which a compact evaluation function is not known. Since the introduction of MCTS/UCT to a domain of General Game Playing (GGP) [8, 28] in 2007, it has also become a backbone of almost all the strongest players [25]. GGP deals with creating autonomous agents capable of playing many games with a high level of competence. The term was proposed by Stanford Logic Group in 2005, together with the introduction of the GGP Competition as the official world championships. Our player, called MINI-Player [23, 24], has been our annual entry to the competition in the years 2012, 2013 and 2014.

The most remarkable success of MCTS/UCT in games was related to the game of Go, where majority of the strongest programs, e.g. MoGo [7], CrazyStone [5] or the latest accomplishment AlphaGo [21] use variants of MCTS. In contrast to all variations of the min-max alpha-beta search, the MCTS is an *aheuristic*, knowledge-free [12, 13] method, which means that it does not require any game-specific knowledge except for the rules of move generation and definition of the goal states with their assigned payoffs. Consequently, in principle, the MCTS/UCT method is applicable to a wide variety of search and decision-making problems. Two examples of such application domains are presented in the following sections.

# 3 Capacitated Vehicle Routing Problem with Traffic Jams

Capacitated Vehicle Routing Problem (CVRP) is a widely-known NP-hard optimization problem, whose goal is to define a set of routes of a minimum cumulative length (cost), given a certain number of homogeneous trucks (with some pre-defined capacity) and a certain set of clients (each of them defined by a 2D location and requested demand of goods to be delivered). The trucks start and end their routes in a depot (having a certain 2D location). Each client must be served by exactly one truck and in one shot, i.e. multiple visits to one client are not allowed. Roughly speaking, CVRP combines the multiple-tour formulation of the Traveling Salesman Problem with the Bin Packing Problem. For its formal definition and a review of Operational Research and Computational Intelligence approaches please refer, for example, to [17].

In our previous paper [14], the baseline problem formulation was further extended and complicated by adding stochastically defined events—traffic jams—occurring on the atomic parts of the routes (edges) and resulting in temporal increase of the cost of traversal (of such a jammed edge). The effective problem formulation is abbreviated as CVRPwTJ, i.e. CVRP with Traffic Jams.

Highly dynamic nature of CVRPwTJ (stemming from frequently changing traffic conditions), requires the methods used to solving it to be able to swiftly and almost instantly adapt to frequent, on-line changes of the cost function values.

## 3.1 MCTS/UCT Approach to CVRPwTJ

In the approach proposed in [14], the MCTS/UCT is applied to a specifically defined set of UCT trees, each devoted to a particular truck and representing possible continuations of the currently committed part of the route of that truck. In the first step an initial solution is build, for the static version of the problem (there are no traffic jams imposed yet) using the modified version [19] of the Savings algorithm [4].

Suppose the initial solution is composed of $k$ routes, i.e. $k$ trucks are employed. In such a case the initial set of UCT trees is composed of $k$ degenerated trees—each in the form of a path with the first and the last elements being a depot and the internal nodes representing clients to be served in a given order defined by the initial solution.

At each time step the internal UCT simulations are performed simultaneously for all $k$ trees, which are gradually extended by adding one leaf node at each simulation (cf. Sect. 2.1). While the general simulation scheme follows the classical UCT pattern, there are several differences reflecting the specificity of the CVRPwTJ [14]. In particular, since shorter solutions are preferred over the longer ones, the UCT formula (1) is modified to the following version (2), which favors lower $Q(s, a)$ outcomes:

$$a^* = \arg \max_{a \in A(s)} \left\{ C \sqrt{\frac{ln\,[N(s)]}{N(s, a)}} - Q(s, a) \right\} \qquad (2)$$

The other pertinent difference is that the next compound step (simultaneous movement of all $k$ trucks) is a result of a combined knowledge obtained from all $k$ trees
(not *one* tree as in the typical UCT implementation). To this end, once in each tree the
most promising action is selected, then these $k$ selected actions are sorted in descend-
ing order based on their UCT values (i.e. values of $C\sqrt{\frac{ln[N(s)]}{N(s,a)}} - Q(s,a)$ in (2)) and
afterwards executed in this order. This way, execution of higher-ranked actions may
disable some of lower-ranked ones. In such a case the next-best candidates in the
respective UCT trees are selected as replacements (for these disabled actions). Please
consult [14] for the details.

After an assumed number of internal simulations are executed, the real (actual)
movement of $k$ trucks is simultaneously made (concurrently in all $k$ trees) following
the smallest $Q(s,a)$ values among the child nodes.

## 3.2   Possible Actions in the UCT Trees

As previously stated nodes in the $j$th UCT tree represent possible variants of the
remaining part of the $j$th route, i.e. the order of service of clients remaining for the
$j$th truck. The edges coming out from a given node in the $j$th tree represent potential
actions to be applied to the $j$th route or to a combination of this route and some other
route. Three types of actions, differing by the level of complexity, were proposed,
denoted by: *level-0*, *level-1* and *level-2*, which modify 0, 1 and 2 existing routes,
respectively. Each action has some legality conditions which must be fulfilled in
order for this action to be available (see [14] for further explanation).

*Level-0* and *level-1* actions are listed in Table 1. All of them are self-explanatory.
This selection of actions was complemented by the set of four more complex, *level-
2* actions (denoted $A9$–$A12$), which operate on (any) two routes. Since, all pairs of
routes are considered, there can be many realizations of each of these actions in
one time step, depending on the number of route pairs fulfilling legality conditions.
While these four actions differ by implementation details (described in [14]), their
underpinning idea is to exchange customers between two routes so as to *locally* minimize the total travel cost. As a special case of this exchange mechanism a merge
operation is considered as action $A12$.

Generally speaking all 13 actions are rooted in the following rationale: if the considered candidate edge is not jammed then traverse it, otherwise make an attempt
to enhance the planned route (by avoiding a traffic jam) by means of local changes
in the planned orders of visited clients. In theory, one might proceed with defining even more complex actions, e.g., the ones involving three or more routes, but
such approach immediately becomes infeasible due to computational complexity
explosion.

**Table 1** Actions of the types *level-0* and *level-1*

| Ac. | L. | Action description |
|---|---|---|
| A0 | 0 | Continue the planned (non-jammed) route |
| A1 | 0 | Continue the planned (jammed) route |
| A2 | 1 | Move the current client at the end of a route (just before returning to the depot) |
| A3 | 1 | Move the current client $X$ into locally optimal place in a route, i.e. between clients $B$ and $C$ so as to minimize $\|BX\| + \|XC\| - \|BC\|$ |
| A4 | 1 | Insert the first found client to whom there is no TJ before the current client (as the first one) |
| A5 | 1 | Reverse the route (except for the depot which remains the closing element) |
| A6 | 1 | Insert the client to whom the edge from the current state is the cheapest as the first one |
| | | Due to greedy nature of this action, the score $Q$ in (2) is multiplied by a penalty (discouraging) factor $> 1$ |
| A7 | 1 | Insert the client to whom the edge from the current state is the second cheapest as the first one |
| | | Due to greedy nature of this action, the score $Q$ in (2) is multiplied by a penalty (discouraging) factor $> 1$ |
| A8 | 1 | The current route is finished (by immediately moving to a depot) |
| | | A new route is commenced from the depot with all customers left inherited from the finished route |

**Ac.** denotes the code of an action, **L.** is level-type

## 3.3 Results

The above-described UCT-based approach was experimentally verified on a set of widely-known static benchmarks downloaded from the CVRP webpage [18]. In each case the initial conditions (i.e. the number of available trucks, their capacity, clients requests' sizes, and the coordinates of a depot and customers) are included in the benchmark set definition. These (static) CVRP instances were transformed into dynamic versions (CVRPwTJ) by imposing traffic jams with uniform probability distributions. More precisely, at (the beginning of) each time step of the solving method, on each edge $e_{ij}$ a traffic jam was defined with a certain probability $P$. In such a case, the regular cost $c_{ij}$ of traversing this edge was multiplied by the traffic intensity $I(e_{ij})$ (sampled from a certain probability distribution) for a randomly selected number of steps $L(e_{ij})$. In order to prevent intensities of traffic jams from exponential growth, if a traffic jam was selected for an already jammed edge, then its intensity remained unchanged and only its length ($L(e_{ij})$) was increased by a newly-sampled value. Traffic jams' steering parameters tested in the experiments belonged to the following ranges:

$$P \in \{0.02, 0.05, 0.15\}, \qquad I = U_{INT}[10, 20], \qquad L = U_{INT}[2, 5], \qquad (3)$$

where $U_{INT}[a, b]$ denotes random uniform selection of any integer $x$, such that $a \leq x \leq b$. Based on the initial calibration tests, the value of $C$ in (2) was set to 1.8 multiplied by the length of the initial solution found for the static instance. The size of benchmark sets ranged from 19 to 150 and the number of available trucks varied from 2 to 14.

The proposed simulation-based approach was compared on a common ground with selected population-based methods. The selection of comparative methods included Ant Colony Optimization (ACO) [14, 15], Tabu Search (TS) [16], Genetic Algorithms (GA) [16], and Particle Swarm Optimization (PSO) [15]. Since the focus of this paper is on making qualitative conclusions related to the universality of the MCTS/UCT in solving dynamic optimization problems the exact numerical results are not presented - they can be accessed in the above-cited papers [14–16].

On a general note, the UCT method applied to CVRPwTJ outperformed the competitive methods by a clear margin, except for the GA version specifically tailored and optimized for this task, which nevertheless turned out to be slightly inferior to proposed UCT forest. The main advantage of the UCT was visible in the case of $P = 0.15$, i.e. the most dynamic situation with frequent traffic jams generated in subsequent time steps. In this case all differences in results were statistically significant (in favor of UCT).

Generally speaking, UCT is a much more repeatable (stable) method, with significantly lower standard deviation of results. Furthermore, it is easier to parameterize than GA, i.e. its main competitor. UCT yielded results of similar quality over a wide range of $C$ selections, between 0.9 and 1.8 (note that theoretically advisable value of that parameter equals $\sqrt{2}$).

## 4 Risk-Aware Project Scheduling Problem

The other problem considered in this paper to illustrate the usage of the UCT algorithm is the Resource-Constrained Project Scheduling Problem (RCPSP) and its extension Risk Aware Project Scheduling Problems (RAPSP). RCPSP is a popular NP-complete [22] optimization problem, especially interesting due to its almost direct applicability to real-life scenarios. While the problem closely fits to various static project scheduling cases, it is—on the other hand—too simplistic to cover a wide range of possible situations that may potentially occur during project execution.

For this reason, a new class of related problems, namely RAPSP, was proposed in our previous papers [29, 30], so as to address unpredictable, non-deterministic aspects of real-life project scheduling and execution. In addition to standard formulation of RCPSP (whose goal is to find a legal schedule that minimizes the makespan of the project), the new model built on top of RCPSP introduces several new concepts:

1. **non-deterministic activity durations** taking into account unavoidable mistakes in estimations;
2. **non-renewable resources**—as in multi-mode RCPSP;
3. **risks**—unpredictable external events that may influence the project characteristics in various ways;
4. **risk responses**—special, optional activities (not required for completion of the project) whose effects influence project parameters (analogically to risks).

In particular the last facet of RAPSP, i.e. optional risk responses, which can be *proactively* or *post-factum* applied to mitigate (or eliminate) potential risks, makes the problem interesting for at least two reasons. First of all, the inclusion of risk management process makes the problem fit even closer to actual real-life scenarios. Second of all, a dynamic characteristic of the RAPSP formulation makes it an especially well-suited testbed for various AI- and CI-based approaches.

As mentioned above, RCPSP is a relatively popular NP-complete optimization problem, with various practical realizations. For the sake of brevity of the paper, we'll skip its formal definition here (which can be found, for instance, in [22]) and describe the problem informally. A single-mode deterministic RCPSP instance spans a number of activities as well as capacitated renewable resources. All activities are required to be completed in order for the whole project to be finished. Each activity has a certain time span and requires certain resources, i.e. cannot be started unless sufficient amounts of resources of each required type are available. Furthermore, activities may have predecessors (activities that must be performed beforehand). Typically, activities are not preemptive (once started they cannot be split into several smaller activities).

Below we briefly sketch the newly-added RAPSP concepts that extend the above RCPSP formulation.

**Non-deterministic activities** Unlike in RCPSP, RAPSP activities' durations are not constant, but rather are random variables sampled from a pre-defined probability distributions. In our implementation, the actual values of variables (their realizations) become known at start of the corresponding activity.

**Risks** Risks represent unpredictable (typically external) events which may possibly influence project's execution, and as such are non-deterministic in nature. Despite differences stemming from their practical meaning, they share a common description pattern, which includes realization conditions, occurrence probabilities, and effects. Three types of risks (temporal decrease of renewable resource amount, disappearance of non-renewable resource, and underestimation of certain project's activities) were employed in our experiments model. Please consult our previous papers [30, 32] for their exact definitions and parameterizations.

Observe that in practice effects of a currently active risks may influence some activities in progress, making them, for instance, illegal (e.g., the amount of available renewable resource may drop below the required level). There are several ways to deal with this kind of situation, for example, an activity may be canceled or split into two. For the sake of simplicity, in the current system implementation, we assume that risks do not affect activities in progress.

**Risk responses** Risk responses represent various actions that may be taken to manage or handle project risks. They may be performed both reactively (*post factum*) and proactively, and their effects will take place whether or not any risks have actually occurred. In this sense, risk responses are independent of risks themselves. Each risk response consists in increase of a certain renewable/non-renewable resource amount using (other) non-renewable resources (e.g. financial budget). Observe that risk responses are activities which are not (in principle) required to be performed for the project to be successfully finished. Just like any activity, risk responses may have positive duration and may require resources. After completion of a risk response, its effect will materialize and influence the project's realization.

**Heuristic solver** Due to NP-completeness of RCPSP and RAPSP, application of any brute-force method is infeasible except for small instances of a problem. Instead, a typical approach is based on using some heuristics that guide the scheduling (search) process. A particular heuristic solver (HS) used in our studies employs a prioritization rule and Parallel Schedule Generation Scheme (PSGS) to generate a schedule. In each time step, all legal activities are commenced according to the order defined by the selected prioritization rule. Only when all eligible activities have been started does the algorithm advances to the next time step.

Employing the above approach in solving RAPSP requires certain modifications so as to accommodate risks and risk responses. In short, simulating typical human behavior, HS realizes the project by devising a baseline schedule, then following it as long as possible, and regenerating it in case of major deviations. Therefore, decision points (time steps in which a new baseline schedule is devised) are defined in any of the three following situations: either the effects of a new risk or risk response have just materialized, or there is a certain delay in the current baseline schedule (greater than 2 time units/steps), or for the first time in the project's execution a new risk response has become eligible. In all other cases (time units) a valid currently adopted baseline schedule should be followed.

A new baseline schedule, whenever necessary, is defined by the HS by creating a number of randomly chosen legal combinations of risk responses that can be started immediately and a randomly sampled priority rule (from a set of such rules, see below). For each such combination a new schedule is simulated starting with execution of the selected risk responses. Afterwards, the RAPSP project is converted to a simplified deterministic version, in which all activities durations are set as the expected values of their distributions. Next, for this deterministic project a schedule is generated according to PSGS and selected priority rule. This schedule combined with the respective set of risk responses, becomes a candidate for a baseline schedule. Among certain number of such candidate schedules, the one with the shortest makespan is finally chosen, as the baseline schedule for the RAPSP project under consideration and lasts till the next decision point.

Our current implementation of HS relies on a set of five simple priority rules:

1. Duration—preference for activities with greatest duration;

2. LateFinish (LF)—choosing activities with earlier LF first;
3. LateStart (LS)—choosing activities with earlier LS first;
4. Slack (SL)—preferring activities with low Slack values;
5. DurationWithSuccessors—considering summed duration of the activity and its direct successors;
6. SuccessorsCount—based on the total number of direct and indirect successors of the activity.

LF, LS and SL are calculated using a classical technique called Critical Path Analysis [9].

## 4.1 MCTS/UCT Approach to RAPS

This section describes our UCT-based approach to solving RAPSP instances. First, a straightforward application of UCT to the above-described model is proposed, which is then enhanced by adding heuristic domain knowledge to guide the UCT simulations.

**Basic UCT** Basic UCT (BasicUCT) method is a straightforward application of the simulation-based UCT approach to solving RAPSP problem. In short, each BasicUCT simulation covers full realization of the project from its current state until its completion (a success) or detection it can no longer be completed (a failure). Three types of actions are eligible in each node of the UCT tree: (1) starting a new (legal) activity, (2) starting a new (legal) risk response, and (3) noop—i.e. waiting till next time unit. Observe, that only the third action (noop) advances the project in time. Consequently, it is possible to start multiple activities and/or risk responses in the same time unit.

Due to highly dynamic nature of RAPSP stemming from various possible realizations of a significant number of random variables, the number of possible project states can be expected to be explosive for real-life situations. On the other hand, one can safely assume that minute differences in the current project realization or historical information about the project's development can be safely ignored in the current decision-making process. The above reasoning leads to the idea of clustering possible states in the UCT tree, i.e. associate UCT statistics not with the exact project states but with their "generic" (simplified) representations. The detailed description of this state-simplification process can be found in [30, 32]. State-simplification procedure imposes adequate modification of the UCT formula (1):

$$
\begin{aligned}
Q(s, a) &:= Q(s^*, a) \\
N(s, a) &:= N(s^*, a) \\
N(s) &:= \sum_{a \in A(s)} N(s^*, a)
\end{aligned}
\tag{4}
$$

where $s^*$ denotes a simplified project state representation.

Finally, specificity of the RAPSP imposes some modifications in the Monte-Carlo rollouts policy, as fully random state selection can be easily proven not sensible as, for instance, it never makes sense to wait till the next time unit (the end of the current unit) when there are no activities, risk responses or effects in progress. In order to address this property of RAPSP a fairly simple rule-of-thumb policy was developed:

1. if there are any legal activities, then start a randomly sampled one, with probability 0.9;
2. otherwise, if there are any legal risk responses, then start a randomly sampled one with probability 0.5;
3. otherwise move to next time unit.

The above probability values were optimized based on some initial tests.

**Proactive UCT** The other realization of the UCT method in project scheduling is Proactive UCT (ProUCT) method [29, 30], which incorporates domain knowledge (the heuristic solver) into BasicUCT simulations. In more detail, while performing simulations there are only two kinds of available actions: either starting a legal risk response or letting the HS algorithm create a baseline schedule (as described above) which is then followed for a number of time units. As soon as a decision point is reached or a predefined maximum number of time units passes, the control of the system is transferred back to the UCT part. In other words, each UCT action encompasses and governs the process of HS application for several time units. Please note, that since risk responses are handled by the UCT component of ProUCT, they are excluded from the HS operational scheme.

Furthermore, these two components are combined in a truly synergetic way, i.e. the task duration statistics gathered during the UCT simulations are additionally passed to the HS module in the form of the expected activity durations to improve HS accuracy. Finally, since there are only two kinds of possible actions (starting a legal risk response or invoking the HS algorithm to create a new baseline schedule), whenever at least one eligible risk response is available, a random one is started with probability 0.8. Otherwise, the HS module is called.

## 4.2   Results

In this section both UCT approaches (BasicUCT and ProUCT) are briefly compared with the application of a plain heuristic RAPSP solver HS (which, as stated above, relies on exactly the same heuristical principles as the ProUCT implementation).

**Problem Instances** Test cases of the RAPSP were generated by modifying RCPSP instances provided by the PSPLIB Library [11, 20]—a standard and widely-known reference site in this domain. Transformation process developed to that extent consisted of two phases. Firstly, all activity duration values were replaced with random variables with known distributions—thus converting RCPSP into Stochastic RCPSP (with the use of Beta distribution—the *de facto* standard in project management area). Secondly, three types of risks and corresponding risk responses were added (the exact parameters, e.g. lengths, realization probabilities, and budget constraints are presented in detail in [32]). Observe that even though the transformation procedure is deterministic, the resulting RAPSP instances are not, i.e. multiple realizations of the same project may have different durations even with the same strategy due to different realizations of random variables describing project risks.

In order to thoroughly compare tested methods on projects with various characteristics, 3 transformation modes were introduced in the source paper [32], namely:

**Temporary Effects with Separate Budgets (TSep)** In this mode non-renewable resource risks and risk responses would have temporary effects, lasting 10–30 and 40 time units, respectively. Consequently, no combination of risks and risk responses could render the project unsuccessful by making it impossible to finish all required activities. Each risk response category would have a separate *budget* (a non-renewable resource).

**Temporary Effects with Shared Budget (TSh)** TSh differed from TSep only in that a common *budget* was introduced for all 3 types of risk responses. Projects of TSh type allowed for more flexibility in deciding about risk responses since more legal risk responses combinations were available, thus making the task even more complex and more dynamic than in the previous case.

**Permanent Effects with Separate Budgets (PSep)** In this mode risk response budgets were again separate, but non-renewable resource risk and risk-response effects were permanent. Consequently, certain combinations of risks could, in principle, lead to a project failure. This threat could be multiplied by poor risk response budget management.

For the sake of space savings the third type of transformation, which significantly differs from the former two by the possibility of the project's failure and therefore requires application of slightly different success measures, will be omitted here.

**Testing Procedure** Due to highly non-deterministic nature of the task, the solvers were tested on several thousand problem instances, in total, so as to obtain meaningful results. Each tested problem instance was solved by each and every algorithm. Furthermore, for each instance, the same seed was used for a random number generator for all solvers, i.e. should all the solvers make the same decisions for a given problem instance, the yielded results would all be the same (the same risks would materialize at the same times).

Projects with 30, 60 and 120 activities were considered in the experiment, with the problem instance quantities respectively equal to 480, 480 and 100, for each project size and each of the two transformation modes (2120 test runs in total). Two statistics were calculated: the average relative project duration and the win rate. In the first (length-based) comparison, for each project instance the best performing method (or more than one method in the case of ties) was assigned a result of 100%, and the remaining ones had values proportionally higher. The latter statistic (win rate) simply equaled the number of experiments in which a given solver accomplished

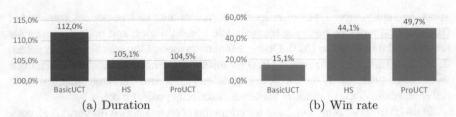

(a) Duration                                   (b) Win rate

**Fig. 2** TSep: relative project durations (left) and solvers win rates (right). **a** Duration **b** Win rate

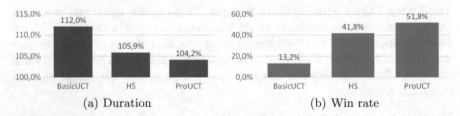

(a) Duration                                   (b) Win rate

**Fig. 3** TSh: relative project durations (left) and solvers win rates (right). **a** Duration **b** Win rate

the best result (the relative project duration of 100%) divided by the total number of experiments. Win rates might not sum up to 100% as multiple solvers could achieve the same result for any given problem instance. Solvers' internal parameters were set up based on some number of preliminary tests devoted to their calibration.

**Results** Results for the projects obtained using the TSep transformation are presented in Fig. 2a and b, respectively. It can be seen from the figures that BasicUCT fared visibly worse than the two other methods, and that ProUCT accomplished slightly (though statistically significantly) better outcomes than HS.

Similar qualitative results were obtained in the case of TSh experiment, which involved one shared non-renewable resource treated as a risk-response budget. This new feature added another layer of complexity to the problem, and consequently more sophisticated risk-management strategies were possible and also more risk response related decisions were available in practice. Figure 3a and b show that also in this case ProUCT achieved a (statistically significant) victory over competitive methods. This can be attributed to the proactive nature of the ProUCT algorithm and its clearly being better-suited for dealing with risk responses via the use of UCT algorithm than BasicUCT. Statistical significance of the differences in the results obtained by any two solvers was verified using Wilcoxon signed-rank test [31].

## 5 Conclusions

This study and related works indicate that the applicability of MCTS/UCT method extends beyond games domain where it has already established itself the state-of-the-art approach. While the method is, in principle, *aheuristic* it turns out that taking advantage of domain knowledge has a twofold advantage over vanilla UCT. First of all, efficient application of domain knowledge allows for shrinking the UCT tree, by either grouping some of the states or by restricting the edges in the tree to truly relevant actions only. Both these possibilities were demonstrated in this paper in the case of CVRPwTJ and RAPSP—two NP-hard scheduling tasks. For both of them a problem-tuned application of the MCTS/UCT algorithm proved to be a stronger (or at least competitive) approach than heuristic-based solution techniques used hitherto. The results suggest that in the case of complex and highly dynamic problems the synergistic *UCT + heuristic* approach outperforms application of each of the component methods in isolation.

# References

1. Arneson, B., Hayward, R.B., Henderson, P.: Monte Carlo tree search in hex. IEEE Trans. Comput. Intell. AI Games **2**(4), 251–258 (2010)
2. Browne, C.B., Powley, E., Whitehouse, D., Lucas, S.M., Cowling, P.I., Rohlfshagen, P., Tavener, S., Perez, D., Samothrakis, S., Colton, S.: A survey of Monte Carlo tree search methods. IEEE Trans. Comput. Intell. AI in Games **4**(1), 1–43 (2012)
3. Chaslot, G., Winands, M.H., Szita, I., Van den Herik, H.J.: Cross-entropy for Monte-Carlo tree search. ICGA J. **31**(3), 145–156 (2008)
4. Clarke, G., Wright, J.: Scheduling of vehicles from a central depot to a number of delivery points. Operat. Res. **12**(4), 568–581 (1964)
5. Coulom, R.: Efficient selectivity and backup operators in Monte-Carlo tree search. In: Computers and Games, pp. 72–83. Springer (2007)
6. Gelly, S., Kocsis, L., Schoenauer, M., Sebag, M., Silver, D., Szepesvári, C., Teytaud, O.: The grand challenge of computer go: Monte Carlo tree search and extensions. Commun. ACM **55**(3), 106–113 (2012)
7. Gelly, S., Silver, D.: Achieving master level play in 9 × 9 computer go. AAAI. **8**, 1537–1540 (2008)
8. Genesereth, M.R., Love, N., Pell, B.: General game playing: overview of the AAAI competition. AI Mag. **26**(2), 62–72 (2005)
9. Kelley, J.E., Walker, M.R.: Critical-path planning and scheduling. IRE-AIEE-ACM '59 (Eastern), ACM (1959) 160–173
10. Kocsis, L., Szepesvári, C.: Bandit based Monte-Carlo planning. In: Proceedings of the 17th European conference on Machine Learning. ECML'06, pp. 282–293. Springer, Berlin, Heidelberg (2006)
11. Kolisch, R., Sprecher, A.: PSPLIB—a project scheduling library. Eur. J. Oper. Res. **96**, 205–216 (1996)
12. Mańdziuk, J.: Computational intelligence in mind games. In: Duch, W., Mańdziuk, J. (eds.) Challenges for Computational Intelligence. Studies in Computational Intelligence, vol. 63, pp. 407–442. Springer, Berlin, Heidelberg (2007)
13. Mańdziuk, J.: Knowledge-Free and Learning-Based Methods in Intelligent Game Playing. Volume 276 of Studies in Computational Intelligence. Springer, Berlin, Heidelberg (2010)
14. Mańdziuk, J., Świechowski, M.: Simulation-based approach to vehicle routing problem with traffic jams. In: 4th IEEE Symposium on Computational Intelligence for Human-like Intelligence (CIHLI16), pp. 1–8. Athens, Greece, IEEE (2016)
15. Mańdziuk, J., Świechowski, M.: Swarm intelligence in solving stochastic capacitated vehicle routing problem. In: International Conference on Artificial Intelligence and Soft Computing (ICAISC), Zakopane, Poland, LNAI, vol. 10246, pp. 543–552. Springer (2017)

16. Mańdziuk, J., Świechowski, M.: UCT in capacitated vehicle routing problem with traffic jams. Inf. Sci. vol. 406–407, pp. 42–56. Elsevier (2017)
17. Mańdziuk, J., Żychowski, A.: A memetic approach to vehicle routing problem with dynamic requests. Appl. Soft Comput. **48**, 522–534 (2016)
18. NEO. Networking and Emerging Optmization: (2013). http://neo.lcc.uma.es/vrp/vrp-instances/capacitated-vrp-instances/
19. Pichpibul, T., Kawtummachai, R.: An improved Clarke and Wright savings algorithm for the capacitated vehicle routing problem. Sci. Asia, 307–318 (2012)
20. PSPLIB: Project Scheduling Problem Library—PSPLIB. http://www.om-db.wi.tum.de/psplib/main.html
21. Silver, D., Huang, A., Maddison, C.J., Guez, A., Sifre, L., Van Den Driessche, G., Schrittwieser, J., Antonoglou, I., Panneershelvam, V., Lanctot, M., et al.: Mastering the game of go with deep neural networks and tree search. Nature **529**(7587), 484–489 (2016)
22. Słowiński, R., Wę glarz, J.: Advances in Project Scheduling. Studies in Production and Engineering Economics. Elsevier Science Limited (1989)
23. Świechowski, M., Mańdziuk, J.: Self-adaptation of playing strategies in general game playing. IEEE Trans. Comput. Intell. AI Games **6**(4), 367–381 (2014)
24. Świechowski, M., Mańdziuk, J., Ong, Y.S.: Specialization of a UCT-based general game playing program to single-player games. IEEE Trans. Comput. Intell. AI Games **8**(3), 218–228 (2016)
25. Świechowski, M., Park, H., Mańdziuk, J., Kim, K.: Recent advances in general game playing. Sci. World J. (2015). http://dx.doi.org/10.1155/2015/986262
26. Syed, O., Syed, A.: Arimaa—a new game designed to be difficult for computers. ICGA **26**, 138–139 (2003)
27. Teytaud, F., Teytaud, O.: Creating an upper-confidence-tree program for havannah. In: Advances in Computer Games, pp. 65–74. Springer (2010)
28. Walędzik, K., Mańdziuk, J.: An automatically-generated evaluation function in general game playing. IEEE Trans. Comput. Intell. AI Games **6**(3), 258–270 (2014)
29. Walędzik, K., Mańdziuk, J., Zadrożny, S.: Proactive and reactive risk-aware project scheduling. In: 2nd IEEE Symposium on Computational Intelligence for Human-like Intelligence (CIHLI14), pp. 94–101. Orlando, FL, USA, IEEE (2014)
30. Walędzik, K., Mańdziuk, J., Zadrożny, S.: Risk-aware project scheduling for projects with varied risk levels. In: 3rd IEEE Symposium on Computational Intelligence for Human-like Intelligence (CIHLI15), pp. 1642–1649. Cape Town, South Africa, IEEE (2015)
31. Wilcoxon, F.: Individual comparisons by ranking methods. Biom. Bull. **1**(6), 80–83 (1945)
32. Walędzik, K., Mańdziuk, J., Applying Hybrid Monte Carlo Tree Search Methods to Risk-Aware Project Scheduling Problem, Inf. Sci, (2017). (in press). http://dx.doi.org/10.1016/j.ins.2017.08.049

# Interactive Cone Contraction for Evolutionary Mutliple Objective Optimization

Miłosz Kadziński, Michał K. Tomczyk and Roman Słowiński

**Abstract** We present a new interactive evolutionary algorithm for Multiple Objective Optimization (MOO) which combines the NSGA-II method with a cone contraction method. It requires the Decision Maker (DM) to provide preference information in form of a reference point and pairwise comparisons of solutions from a current population. This information is represented with a compatible Achievement Scalarizing Function (ASF) which is used to guide the evolutionary search towards the most preferred region of the Pareto front. The performance of the proposed algorithm is illustrated on a set of benchmark problems. The experimental results confirm its ability to converge quickly to the DM's most preferred region. Its competitive advantage over the state-of-the-art method, called NEMO-0, is increasing when the DM provides a richer preference information composed of a greater number of pairwise comparisons of solutions.

**Keywords** Multiple objective optimization · Achievement scalarizing function · Preference cone · Cone contraction · Pairwise comparisons

M. Kadziński · M.K. Tomczyk · R. Słowiński (✉)
Institute of Computing Science, Poznań University of Technology,
60-965 Poznań, Poland
e-mail: roman.slowinski@cs.put.poznan.pl

M. Kadziński
e-mail: milosz.kadzinski@cs.put.poznan.pl

M.K. Tomczyk
e-mail: michal.tomczyk@cs.put.poznan.pl

R. Słowiński
Systems Research Institute Polish Academy of Sciences, 01-447 Warsaw, Poland

© Springer International Publishing AG 2018                                       293
A.E. Gawęda et al. (eds.), *Advances in Data Analysis with Computational Intelligence Methods*, Studies in Computational Intelligence 738,
https://doi.org/10.1007/978-3-319-67946-4_12

# 1  Introduction

Multiple Objective Optimization (MOO) is concerned with problems involving several objectives to be optimized simultaneously, subject to a set of constraints [3]. It has been applied in many domains, including engineering design, economics, management, transportation, and production. In all these fields, decision making needs to account for multiple conflicting viewpoints, which implies that there is no objectively best solution. Consequently, MOO methods aim at identifying a set of non-dominated solutions, which form a Pareto front in the objective space.

In the recent years, Evolutionary Multiple Objective Optimization (EMO) methods are prevailing in decision contexts where an entire Pareto front needs to be approximated [1]. These algorithms mimic the process of natural evolution by progressively modifying a set of solutions through mutation, recombination, and selection [9]. Due to a smart simulation of the principles of reproduction and survival of the fittest, EMO methods have proven their suitability for finding a well-distributed set of non-dominated solutions, being a good approximation of the Pareto front, in many real-world optimization problems [3].

Nonetheless, when using EMO for approximating an entire Pareto front, one needs to be aware of some concerns. Firstly, evaluating a large set of very diverse solutions contained in the approximation of the Pareto front with the aim of identifying the most preferred solution may be demanding for the Decision Maker (DM) in terms of the required cognitive effort. Secondly, generation of an entire representation of the Pareto front may be resource intensive. Thirdly, when the number of objectives increases, a selection pressure imposed by the traditional EMO methods may be insufficient to ensure a satisfactory quality of approximation and distribution.

To address these problems, one has proposed to guide the evolutionary optimization by integrating interactively some preference information provided by the DM. In this way, the search may be focused on the DM's most preferred region of the Pareto front, the pace of convergence increases, and the pressure is strengthened enough to effectively deal with many-objective problems.

The existing interactive evolutionary algorithms account for preference information provided in different forms and employ various preference models. When it comes to the former, the prevailing trend in MOO consists in asking the DM to compare some pairs of solutions from a current population (see, e.g., [4] and [13]). On the practical side, this allows to avoid by the DM a direct reference to some technical parameters. From the methodological viewpoint, the inference of a preference model reproducing such natural holistic preferences necessitates looking for the rational basis through which the desired pairwise comparisons were made [10].

As far as models used to represent DM's preferences are concerned, the majority of interactive evolutionary algorithms incorporate value functions. In particular, in [7] and [2] one employed, respectively, polynomial or non-linear complex functions. Conversely, [13] and [4] proposed to use an additive value model. Whichever the form of the function, the scores it provides allow comprehensive evaluation of the solutions in a given population. Such results can be used to guide the evolutionary search.

In this paper, we propose a new interactive approach to MOO, which combines an evolutionary algorithm, called NSGA-II [8], with an interactive cone contraction method originally conceived to deal with a limited number of solutions [10]. We require the DM to define a reference point and provide – at regular intervals – pairwise comparisons of some solutions. These pairwise comparisons are represented with an Achievement Scalarizing Function (ASF) [15]. The directions of the iso-quants of all compatible ASFs form a preference cone in the objective space, which is gradually contracted when more pairwise comparisons are provided. The solutions which are situated within the cone correspond better to the DM's preferences. To promote such solutions in the optimization run, we modify NSGA-II so that it accounts for distances of the solutions from the reference point according to a compatible ASF which is the most discriminant with respect to the solutions compared pairwise by the DM. The phases of preference elicitation and evolutionary search alternate until the population is well-converged.

The use of the proposed algorithm is illustrated by examples revealing its ability to focus the search on the DM's most preferred region. We also discuss results of the experiments concerning the quality of population constructed by the method, as well as its convergence speed for different benchmark problems and various simulated decision making policies.

## 2 Concepts: Definitions and Notation

**Multiple Objective Optimization**. We consider MOO problem in which a set of solutions $A = \{a_1, a_2, \ldots\}$ is evaluated in view of $m$ conflicting objectives, $F = \{f_1, f_2, \ldots, f_m\}$. Each solution $a_i \in A$ is associated with an evaluation vector denoted by $f(a_i) = [f_1(a_i), f_2(a_i), \ldots, f_m(a_i)]$. Without loss of generality, we assume that for all objectives less is preferred to more. Thus, a general formulation of a MOO problem is:

$$\text{Minimize } \{f_1(a_i), f_2(a_i), \ldots, f_m(a_i)\}$$
$$\text{subject to } a_i \in S, \tag{1}$$

where $S$ is a non-empty feasible region.

**Non-dominated Solutions**. Solution $a_i \in A$ is *non-dominated* if and only if there is no other $a_k \in A$ such that $a_k$ is at least as good as $a_i$ with respect to all objectives, and strictly better for at least one objective.

**Preference Model.** We require the DM to provide some desired values on all objectives which contribute to the definition of a *reference point* $\bar{z} = \{\bar{z}_1, \dots, \bar{z}_m\}$. Most often, reference points correspond to objective values that the DM would like to achieve. To represent preferences of the DM and comprehensively judge the quality of solutions, we use an *achievement scalarizing function*. It provides a distance of solution $a_i \in A$ from reference point $\bar{z}$. ASF is defined as follows [15]:

$$s(a_i, \lambda, f) = max_j\{\lambda_j(f_j(a_i) - \bar{z}_j)\} + \rho \sum_{j=1}^{m}(f_j(a_i) - \bar{z}_j), \qquad (2)$$

where $\lambda = [\lambda_1, \dots, \lambda_m]$ is a weighting vector, $\lambda_j \geq 0$, $j = 1, \dots, m$, and $\rho > 0$ is an augmentation multiplier. Clearly, the less $s(a_i, \lambda, f)$, the more $a_i$ is preferred to the DM.

**Preference Information.** During the optimization run, we expect the DM to answer pairwise elicitation questions in the form "which one do you prefer, $a_i$ or $a_k$?" for $a_i, a_k \in A$ [6]. The pairs to be compared are either selected by the DM or drawn randomly. Using ASF, we assume that it is suitable for representing the DM's preferences. Thus, answering $a_i$ by the DM (denoted by $a_i \succ_{DM} a_k$) imposes $s(a_i, \lambda, f) < s(a_k, \lambda, f)$, and indication of $a_k$ as more preferred implies an inverse inequality. Hence, $a_i \succ_{DM} a_k \Rightarrow \exists_j \forall_l \ \lambda_l(f_l(a_i) - \bar{z}_l) + \gamma \leq \lambda_j(f_j(a_k) - \bar{z}_j)$, where $\gamma = \rho \sum_{j=1}^{m} \lambda_j(f_j(a_i) - f_j(a_k))$. In this regard, the set of constraints $E(DM)$ given below translates all pairwise comparisons provided by the DM to a compatible ASF:

$$\left.\begin{array}{l} \text{for all } a_i \succ_{DM} a_k : \\ [\ \left(\lambda_1(f_1(a_i) - \bar{z}_1) + \gamma + \varepsilon \leq \lambda_j(f_j(a_k) - \bar{z}_j)\right) \wedge \\ \left(\lambda_2(f_2(a_i) - \bar{z}_2) + \gamma + \varepsilon \leq \lambda_j(f_j(a_k) - \bar{z}_j)\right) \wedge \\ \qquad\qquad \dots \\ \left(\lambda_m(f_m(a_i) - \bar{z}_m) + \gamma + \varepsilon \leq \lambda_j(f_j(a_k) - \bar{z}_j)\right)\ ], \\ \qquad\qquad\qquad \text{for some } j = 1, \dots, m \end{array}\right\} E(DM)$$

where $\varepsilon$ is an arbitrarily small positive value.

If $\varepsilon^* = max\ \varepsilon$, s.t. $E(DM)$, is greater than 0, the set of compatible ASFs $s(DM)$ is non-empty. Otherwise, the provided preference information is inconsistent with the assumed preference model, which means that there is no ASF that would reproduce all pairwise comparisons provided by the DM.

**Representative Achievement Scalarizing Function.** If $s(DM) \neq \emptyset$, there is usually more than one compatible ASF. In this paper, to evaluate a set of solutions we will use a single compatible ASF $s^R \in s(DM)$, which is obtained by maximizing $\varepsilon$, subject to $E(DM)$. It aims at discriminating comprehensive values of solutions compared pairwise by the DM. However, since $E(DM)$ is non-linear, to identify the most discriminant ASF, we will use Monte Carlo simulation [14]. For this purpose, we sample a large set of weights $\lambda$ from a uniform distribution, and select the one for which $s(a_k, \lambda, f) - s(a_i, \lambda, f)$ for any $a_i \succ_{DM} a_k$ is maximal. Clearly, $s^R$ imposes a complete order on set $A$. In particular, $a_i$ is preferred to $a_k$ according to $s^R$ ($a_i \succ_{s^R} a_k$) iff $s^R(a_i, \lambda^R, f) < s^R(a_k, \lambda^R, f)$.

**Dealing with Incompatibility of Preference Information**. If there is no ASF compatible with the DM's preferences, some constraints underlying the inconsistency need to be removed from $E(DM)$. For this purpose, the procedure removes the constraints representing the oldest pairwise comparisons. Once the set of constraints becomes feasible, the method reintroduces the removed constraints starting from these corresponding to the newest pairwise comparisons until feasibility is maintained [5].

# 3 Interactive Cone Contraction for Evolutionary Multiple Objective Optimization

The EMO methods aim to approximate an entire Pareto front. In this paper, we refer to Non-dominated Sorting Genetic Algorithm II (NSGA-II) [8], which starts the search with the initialization of random population $P_0$ composed of $N$ solutions. In iteration $t$, $N$ offspring solutions $Q_t$ are created using the usual genetic operators applied to the parents $P_t$. Then, both sets of solutions are combined to obtain population $R_t = P_t \cup Q_t$ of size $2N$. The new population ($P_{t+1}$) is constructed by:

- incorporating the best Pareto fronts ($\mathcal{F}_1, \mathcal{F}_2, ..., \mathcal{F}_l$) from $R_t$ that entirely fit in $P_{t+1}$ (i.e., $\sum_{k=1}^{l} |\mathcal{F}_k| \leq N$);
- filling the remaining slots ($N - \sum_{k=1}^{l} |\mathcal{F}_k|$) in $P_{t+1}$ with the best solutions from $\mathcal{F}_{l+1}$ according to the crowding distance operator.

The process is iterated until a stopping criterion is met (usually, the algorithm is run for a fixed number of generations).

The proposed algorithm combines NSGA-II with an interactive cone contraction method [10]. The major difference consists in asking the DM to provide reference point $\bar{z}$ and to answer a single pairwise comparison question at regular elicitation intervals ($EI$). Analogously to NSGA-II, the method incorporates fast nondominated sorting algorithm as a primary criterion when constructing a new population. When it comes to the secondary criterion, instead of promoting solutions with the greatest crowding distance, we favor these whose distance from $\bar{z}$ is the least according to a representative ASF, $s^R$. Algorithm 1 describes the use of the proposed method for the $t$-th generation.

**Algorithm 1** A single iteration of the interactive evolutionary cone contraction method for constructing the $t$-th generation (adapted from [12]).

---

$R_t = P_t \cup Q_t$
**if** Time to ask the DM **then**
    Elicit DM's pairwise comparison
    Determine the most discriminant achievement scalarizing function $s^R$
**end if**
$\mathcal{F} = \text{fast-non-dominated-sort}(R_t)$
$P_{t+1} = \emptyset$ and $i = 1$
**while** $|P_{t+1}| + |\mathcal{F}_i| \leq N$ **do**
    $P_{t+1} = P_{t+1} \cup \mathcal{F}_i$
    $i = i + 1$
**end while**
$\text{Sort}(\mathcal{F}_i, \succ_{s^R})$
$P_{t+1} = P_{t+1} \cup \mathcal{F}_i[1 : (N - |P_{t+1}|)]$
$Q_{t+1} = \text{make-new-pop}(P_{t+1})$

$t = t + 1$

---

## 4    Experimental Results

In this section, we study the performance of the proposed interactive evolutionary algorithm on a set of benchmark MOO problems involving from 2 to 5 objectives. Our method (let us denote it by ECC-MRW – *evolutionary cone contraction – the most representative weights*) is compared against NSGA-II and NEMO-0 [4]. The latter method is similar to ECC-MRW with the proviso that it incorporates a general additive value function as an internal preference model.

In order to model the interaction, we simulate an artificial DM applying some predefined preference model $U_{DM}$ for indicating more preferred option in each pair of solutions selected by the algorithm. In particular, we use either linear or Chebycheff function defined as follows:

$$U_{DM}^{LIN}(a_i) = \sum_{j=1}^{m} w_j f_j(a_i), \tag{3}$$

$$\text{and } U_{DM}^{CHEB}(a_i) = max_{j=1,\dots,m}\{w_j f_j(a_i)\}, \tag{4}$$

where $w_j, j = 1, \dots, m$, is a weight of the $j$-th objective. Note that since all objectives are of cost-type, these functions are to be minimized.

The solutions presented to the DM are non-dominated and, whenever possible, they have the least distance from $\bar{z}$ for some compatible ASF. In out tests, we assumed $\bar{z}_j = 0$, for $j = 1, \dots, m$, thus, focusing more on investigating the impact of pairwise comparisons.

Regarding generation of offspring population, to fill the mating pool we perform tournament selection with size of 5. We generate offspring by simulated binary crossover with probability of 1.0 and distribution index of 10 for NSGA-II. For

NEMO-0 and ECC-MRW the distribution index is set to 1.0 in order to ensure the convergence of the population. Furthermore, we apply polynomial mutation with probability of $1/v$ (where $v$ is the number of decision variables) and distribution index of 10 for all the algorithms.

## 4.1 Illustrative Examples

In this subsection, we use 2-objective benchmark problems ZDT1 and DTLZ2 for an initial graphical presentation of the convergence and accuracy of the proposed method. We assume that elicitation interval $EI$ is equal to 8, population size is set to $N = 50$, whereas two accounted artificial DM's Chebycheff value functions are parameterized with the following weights $(w_1, w_2)$ for: $U_{DM,1}^{CHEB} - (0.5, 0.5)$ and $U_{DM,2}^{CHEB} - (0.3, 0.7)$.

Figure 1 shows the typical results of ECC-MRW, NEMO-0, and NSGA-II after 40, 80, and 200 generations for ZDT1 with a convex Pareto front and a DM whose value system is simulated with $U_{DM,1}^{CHEB}$. NSGA-II attempts to approximate the entire Pareto front and gets there only after 200 generations. Conversely, ECC-MRW and NEMO-0 focus the search only on the region which is relevant from the DM's perspective. For example, after 80 generations both interactive algorithms converged to the DM's most preferred region, however, ECC-MRW reached a desired part of the

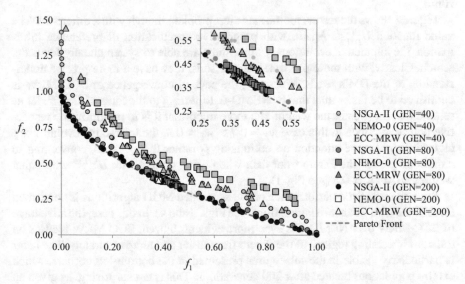

**Fig. 1** Exemplary results of ECC-MRW, NEMO-0, and NSGA-II on ZDT1 after 40, 80, and 200 generations for $U_1^{CHEB} ((w_1, w_2) = (0.5, 0.5))$

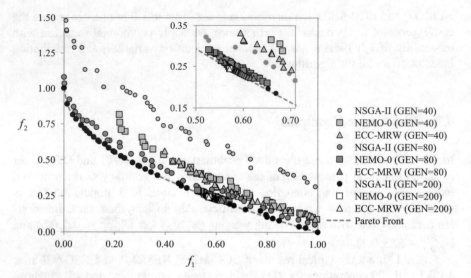

**Fig. 2** Exemplary results of ECC-MRW, NEMO-0, and NSGA-II on ZDT1 after 40, 80, and 200 generations for $U_2^{CHEB}$ $((w_1, w_2) = (0.3, 0.7))$

Pareto front faster. This proves that incorporation of pairwise comparisons into the evolutionary search allows speeding up the convergence of the optimization algorithm.

Figure 2 shows the results for the same test problem though with a different DM's value function $U_{DM,2}^{CHEB}$. Again, with progressive specification of preference information, the interactive evolutionary algorithms are able to systematically focus the search. Clearly, with more pairwise comparisons, they have a more precise understanding of the DM's needs. Moreover, the pace of convergence of ECC-MRW is confirmed to be faster than that of NEMO-0. Referring to the form of $U_{DM,2}^{CHEB}$, let us remind that the greater the weight, the more important it is to minimize the respective objective. Since in this case $w_2 = 0.7 > w_1 = 0.3$, the interactive evolutionary algorithms put more attention on optimizing $f_2$ rather than $f_1$, thus, converging to a different part of the Pareto front than when being guided with $U_{DM,1}^{CHEB}$ with equal weights $w_1 = w_2 = 0.5$ (see Fig. 1).

Such characteristic performance of the accounted MOO algorithms is confirmed for DTLZ2 with a concave Pareto front (see Figs. 3 and 4). In this case, the advantage of ECC-MRW over NEMO-0 is even more evident. Indeed, ECC-MRW focuses on the most interesting regions of the Pareto front faster and more accurately. The latter is particularly visible in the sub-figures presented in the bottom-left corners, which exhibit populations created after 200 generations. This is not surprising, as given an artificial DM with the Chebycheff value function, an ASF-based preference model is able to better capture the DM's preferences.

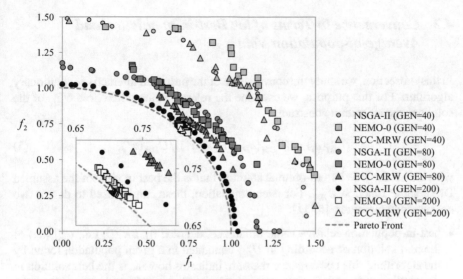

**Fig. 3** Exemplary results of ECC-MRW, NEMO-0, and NSGA-II on DTLZ2 after 40, 80, and 200 generations for $U_1^{CHEB}$ $((w_1, w_2) = (0.5, 0.5))$

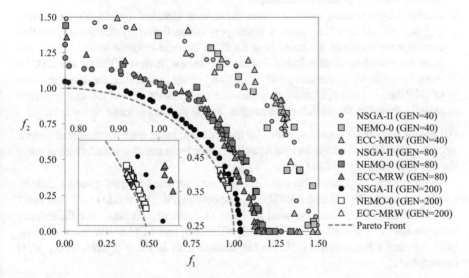

**Fig. 4** Exemplary results of ECC-MRW, NEMO-0, and NSGA-II on DTLZ2 after 40, 80, and 200 generations for $U_2^{CHEB}$ $((w_1, w_2) = (0.3, 0.7))$

## 4.2 Convergence in Terms of the Best-in-population and Average-of-population Values

In this subsection, we study the convergence of the proposed interactive evolutionary algorithm. For this purpose, we examine the relative value differences $U_{DM}^{rel}$ of the solutions the algorithm constructs:

$$U_{DM}^{rel}(a_i) = [U_{DM}(a_i) - U_{DM}(a^w)]/U_{DM}(a^w), \qquad (5)$$

where $a^w$ is the true Pareto-optimal solution that is the best in view of the assumed DM's value function $U_{DM}$. For each population, these can be used to derive two convergence measures [5, 11]:

- best-in-population relative value difference, denoted by *BRVD*, i.e., $U_{DM}^{rel}(a^*)$ of the best solution $a^*$ according to $U_{DM}$ contained in a given population found by the algorithm; this convergence measure indicates how far is the best solution in the population from the Pareto-optimal solution that is truly the best in view of the assumed DM's value function $U_{DM}$; the same value function is used to designate the best solution in the population;
- average-of-population relative value difference, denoted by *ARVD*, i.e., mean $U_{DM}^{rel}(a_i)$ for all solutions $a_i$ in a given population found by the algorithm; this convergence measure indicates how far is an average solution in the population from the Pareto-optimal solution that is truly the best in view of the assumed DM's value function $U_{DM}$; the same value function is used to calculate the average value of solutions in the population; this measure says whether the algorithm is appropriately focusing the search on the region of the greatest interest to the DM.

All results have been averaged over 100 independent runs, each for different weight vectors of the DM's assumed value function drawn from a uniform distribution using the Hit-And-Run algorithm [14].

Figures 5 and 6 present the convergence plots for the best and average convergence measures (BRVD and ARVD) for, respectively, ZDT2 and DTLZ2 (the exact simulation parameters are provided in Table 1). Note that to make the differences between BRVD and ARVD for NSGA-II, NEMO-0, and ECC-MRW more evident, we have used a logarithmic scale for the vertical axes in these figures. These plots demonstrate:

- a competitive advantage that the interactive evolutionary algorithms gain over NSGA-II when they accumulate a sufficient number of pairwise comparisons to get a good understanding of the DM's preferences;
- a difference between best-in-population and average-of-population convergence measures; for example, this difference tends to be small for interactive algorithms and very large for NSGA-II;
- a generation number at which the performance of algorithms stabilizes due to either converging to the most preferred region (in case of ECC-MRW and NEMO-0) or getting stuck as a result of insufficient selection pressure (in case of NSGA-

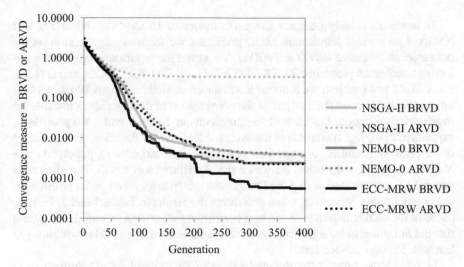

**Fig. 5** Best and average convergence measures (BRVD and ARVD) in successive generations of ECC-MRW, NEMO-0, and NSGA-II applied to ZDT2 with $m = 2$ objectives and the DM's Chebycheff value function

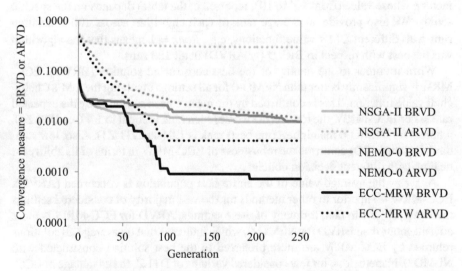

**Fig. 6** Best and average convergence measures (BRVD and ARVD) in successive generations of ECC-MRW, NEMO-0, and NSGA-II applied to DTLZ2 with $m = 3$ objectives and the DM's Chebycheff value function

II); note that such a number can be interpreted in terms of accuracy in identifying the DM's most preferred solution, or the Pareto front; in fact, low values of convergence measures BRVD and ARVD for ECC-MRW confirm that the proposed algorithm works well.

To comprehensively compare the performance of ECC-MRW, NEMO-0, and NSGA-II on various benchmark MOO problems, we focused our attention on the convergence measures *BRVD* and *ARVD*. We tested the algorithms against the following benchmark problems: ZDT1, ZDT2, DTLZ1, DTLZ2, DTLZ3, and DTLZ4. For DTLZ1 test problem we reduced the distance-related bias from 100.0 to 1.0 in order to focus more on the analysis of the convergence of the algorithms towards the most preferred region. For each of the problems, in Tables 1 and 2 we provide an experimental setting. It consists of the assumed DM's value function, characteristics of the problem (numbers of objectives $m$ and decision variables $v$), population size $N$, and number of generations for which the algorithms was run $G$. The elicitation interval $EI$ was adjusted so that the DM provided 50 pairwise comparisons throughout $G$ generations. Moreover, when presenting the results in Tables 1 and 2, for each problem we distinguish in bold the best performing algorithm as well as the one(s) that did not prove to be significantly worse than it according to a Mann-Whitney-U test with 5% significance level.

In Table 1, we present the minimal values of BRVD and ARVD throughout $G$ generations for three algorithms and eleven different settings. For clarity of presentation, the results presented in the main part of the table have been multiplied by factor $p$ whose value (from $10^3$ to $10^5$, reported in the table) depends on the specific setting. We also provide an average rank of each algorithm across 100 simulation runs with different DM's value functions, e.g., $\overline{Rank} = 1$ means that the algorithm was the best with respect to BRVD (or ARVD) in all 100 runs.

When it comes to the quality of the best constructed solution (BRVD), ECC-MRW is significantly better than NEMO-0 for all settings involving the DM's Chebycheff value function. This is confirmed by the mean values as well as by the expected ranks. For ECC-MRW, the lowest (the worse) such rank is equal to 1.4 (see DTLZ4), whereas for NEMO-0 the highest (the best) rank is 1.8 (see DTLZ1). Also, low standard deviation (SD) confirms the robustness of ECC-MRW in terms of its ability for dealing with different decision policies.

As far as the average value of the entire best population is concerned (ARVD), ECC-MRW is superior to other methods for the vast majority of considered settings involving $U_{DM}^{CHEB}$. In fact, for many of these settings, ARVD for ECC-MRW is more advantageous than BRVD for NEMO-0, which indicates that on average all solutions returned by ECC-MRW are more preferred to the best solution constructed with NEMO-0. Nonetheless, for few considered variants of DTLZ, the advantage of ECC-MRW over NEMO-0 is not statistically significant.

Conversely, for a unique test problem ZDT1 that incorporated the DM's linear value function $U_{DM}^{LIN}$, NEMO-0 proved to be more advantageous than ECC-MRW. Finally, whichever the considered problem, NSGA-II is much worse than both interactive evolutionary algorithms. It is not surprising since the latter ones construct only solutions that are relevant from the DM's point of view.

To evaluate the performance of the algorithms throughout all $G$ generations, we also consider the average values of BRVD and ARVD over $G$ generations. These are presented in Table 2, using the same convention as in Table 1. The conclusions that

**Table 1** Minimal (best) values of BRVD (first row) and ARVD (second row) throughout $G$ generations, and the expected rank $\overline{Rank}$ (results averaged over 100 runs; SD = standard deviation)

| Problem | Parameters | | | NSGA-II Mean | SD | Rank | NEMO-0 Mean | SD | Rank | ECC-MRW Mean | SD | Rank |
|---|---|---|---|---|---|---|---|---|---|---|---|---|
| ZDT1 | $m = 2$ | $v = 30$ | EI = 6 | 39.42 | 15.60 | 2.6 | **2.22** | **3.82** | **1.1** | 58.20 | 106.64 | 2.3 |
| $U_{DM}^{LIN}$ | $N = 50$ | $G = 300$ | $p = 10^5$ | 1580.16 | 1051.17 | 3.0 | **4.28** | **7.76** | **1.1** | 107.33 | 163.01 | 1.9 |
| ZDT1 | $m = 2$ | $v = 30$ | EI = 6 | 35.36 | 3.05 | 3.0 | 0.55 | 2.62 | 2.0 | **0.00** | **0.00** | **1.0** |
| $U_{DM}^{CHEB}$ | $N = 50$ | $G = 300$ | $p = 10^5$ | 269.94 | 70.97 | 3.0 | 1.08 | 2.91 | 1.6 | **0.90** | **1.76** | **1.4** |
| ZDT2 | $m = 2$ | $v = 30$ | EI = 8 | 3.52 | 2.90 | 2.8 | 2.37 | 8.35 | 2.1 | **0.55** | **2.84** | **1.2** |
| $U_{DM}^{CHEB}$ | $N = 50$ | $G = 400$ | $p = 10^3$ | 327.61 | 131.02 | 3.0 | 3.72 | 11.10 | 1.8 | **2.25** | **5.99** | **1.2** |
| DTLZ1 | $m = 3$ | $v = 5$ | EI = 8 | 10.59 | 6.71 | 2.9 | 5.03 | 8.54 | 2.0 | **0.69** | **1.46** | **1.1** |
| | $N = 80$ | $G = 400$ | $p = 10^5$ | 145.13 | 31.90 | 3.0 | 7.45 | 11.98 | 1.8 | **2.52** | **5.79** | **1.2** |
| $U_{DM}^{CHEB}$ | $m = 5$ | $v = 5$ | EI = 12 | 34.73 | 18.83 | 3.0 | 7.50 | 10.42 | 1.8 | **2.37** | **2.74** | **1.3** |
| | $N = 200$ | $G = 600$ | $p = 10^3$ | 397.99 | 84.63 | 3.0 | 9.16 | 11.85 | **1.5** | **6.33** | **7.12** | **1.5** |
| DTLZ2 | $m = 3$ | $v = 10$ | EI = 5 | 9.73 | 7.80 | 2.8 | 9.88 | 25.03 | 2.0 | **0.73** | **1.99** | **1.2** |
| | $N = 80$ | $G = 250$ | $p = 10^3$ | 296.92 | 90.03 | 3.0 | 11.42 | 27.37 | **1.5** | **4.15** | **7.47** | **1.5** |
| $U_{DM}^{CHEB}$ | $m = 5$ | $v = 10$ | EI = 7 | 26.74 | 17.26 | 2.8 | 19.14 | 34.26 | 2.0 | **2.13** | **3.73** | **1.2** |
| | $N = 200$ | $G = 350$ | $p = 10^3$ | 339.21 | 61.35 | 3.0 | 20.77 | 34.79 | 1.6 | **10.92** | **16.84** | **1.4** |
| DTLZ3 | $m = 3$ | $v = 10$ | EI = 10 | 50.91 | 35.08 | 2.9 | 26.25 | 31.31 | 2.1 | **3.61** | **3.85** | **1.0** |
| | $N = 80$ | $G = 500$ | $p = 10^3$ | 475.36 | 105.47 | 3.0 | 52.63 | 44.31 | 2.0 | **8.30** | **8.31** | **1.0** |
| $U_{DM}^{CHEB}$ | $m = 5$ | $v = 10$ | EI = 14 | 171.23 | 140.37 | 3.0 | 15.25 | 26.39 | 1.8 | **5.74** | **9.11** | **1.2** |
| | $N = 200$ | $G = 700$ | $p = 10^3$ | 2150.83 | 269.62 | 3.0 | 39.04 | 42.97 | 1.8 | **16.10** | **20.23** | **1.3** |
| DTLZ4 | $m = 3$ | $v = 10$ | EI = 10 | 6.08 | 5.97 | 2.4 | 16.84 | 42.70 | 2.2 | **5.01** | **20.69** | **1.4** |
| | $N = 80$ | $G = 500$ | $p = 10^3$ | 276.24 | 92.33 | 3.0 | 17.19 | 42.63 | 1.6 | **9.48** | **24.97** | **1.4** |
| $U_{DM}^{CHEB}$ | $m = 5$ | $v = 10$ | EI = 14 | 7.02 | 6.60 | 2.7 | 7.95 | 11.49 | 2.2 | **2.93** | **6.04** | **1.2** |
| | $N = 200$ | $G = 700$ | $p = 10^3$ | 230.47 | 45.93 | 3.0 | **14.47** | **23.32** | **1.4** | 10.72 | 19.23 | 1.6 |

**Table 2** Average values of BRVD (first row) and ARVD (second row) throughout $G$ generations and the expected rank $\overline{Rank}$ (results averaged over 100 runs; SD = standard deviation)

| Problem | Parameters | | | NSGA-II Mean | SD | Rank | NEMO-0 Mean | SD | Rank | ECC-MRW Mean | SD | Rank |
|---|---|---|---|---|---|---|---|---|---|---|---|---|
| ZDT1 | m = 2 | v = 30 | EI = 6 | 6.63 | 2.97 | 2.9 | **5.09** | **2.26** | **1.6** | **5.08** | **2.50** | **1.5** |
| $U_{DM}^{LIN}$ | N = 50 | G = 300 | p = $10^3$ | 23.26 | 6.06 | 3.0 | **7.72** | **3.09** | **1.6** | **7.68** | **3.51** | **1.4** |
| ZDT1 | m = 2 | v = 30 | EI = 6 | 7.72 | 3.59 | 2.9 | 5.92 | 3.11 | 1.9 | **5.07** | **2.76** | **1.2** |
| $U_{DM}^{CHEB}$ | N = 50 | G = 300 | p = $10^3$ | 40.16 | 11.73 | 3.0 | 13.16 | 6.60 | 1.9 | **11.28** | **5.94** | **1.2** |
| ZDT2 | m = 2 | v = 30 | EI = 8 | 15.68 | 9.98 | 2.5 | 14.75 | 9.62 | 2.0 | **13.15** | **8.29** | **1.6** |
| $U_{DM}^{CHEB}$ | N = 50 | G = 400 | p = $10^2$ | 46.17 | 17.33 | 3.0 | 19.01 | 11.29 | 1.7 | **17.19** | **9.46** | **1.3** |
| DTLZ1 | m = 3 | v = 5 | EI = 8 | 2.00 | 1.17 | 2.7 | 1.67 | 1.29 | 2.1 | **0.85** | **0.49** | **1.2** |
| $U_{DM}^{CHEB}$ | N = 80 | G = 400 | p = $10^2$ | 23.05 | 4.04 | 3.0 | 8.14 | 2.50 | 2.0 | **4.75** | **1.17** | **1.1** |
| | m = 5 | v = 5 | EI = 12 | 4.32 | 3.04 | 2.9 | 1.16 | 1.17 | 1.5 | **0.74** | **0.37** | **1.6** |
| $U_{DM}^{CHEB}$ | N = 200 | G = 600 | p = $10^2$ | 69.08 | 16.24 | 3.0 | 5.04 | 2.32 | 1.9 | **2.81** | **0.95** | **1.1** |
| DTLZ2 | m = 3 | v = 10 | EI = 5 | 1.39 | 0.97 | 2.6 | 1.89 | 3.43 | 1.7 | **0.84** | **0.98** | **1.6** |
| $U_{DM}^{CHEB}$ | N = 80 | G = 250 | p = $10^2$ | 31.74 | 8.74 | 3.0 | 5.18 | 3.90 | 1.7 | **3.55** | **1.59** | **1.3** |
| | m = 5 | v = 10 | EI = 7 | 4.25 | 2.66 | 2.7 | **2.44** | **3.43** | **1.7** | **1.22** | **0.89** | **1.6** |
| $U_{DM}^{CHEB}$ | N = 200 | G = 200 | p = $10^2$ | 53.03 | 10.32 | 3.0 | 4.74 | 3.70 | 1.6 | **3.43** | **1.43** | **1.4** |
| DTLZ3 | m = 3 | v = 10 | EI = 10 | 13.20 | 9.22 | 2.6 | 13.12 | 9.75 | 2.3 | **7.72** | **5.65** | **1.1** |
| $U_{DM}^{CHEB}$ | N = 80 | G = 500 | p = $10^2$ | 105.00 | 21.32 | 2.9 | 74.83 | 27.78 | 2.1 | **40.09** | **14.48** | **1.1** |
| | m = 5 | v = 10 | EI = 14 | 22.35 | 15.45 | 2.9 | 6.25 | 5.25 | 1.7 | **3.88** | **2.27** | **1.4** |
| $U_{DM}^{CHEB}$ | N = 200 | G = 700 | p = $10^2$ | 307.38 | 69.46 | 3.0 | 35.98 | 17.18 | 1.9 | **18.87** | **7.79** | **1.1** |
| DTLZ4 | m = 3 | v = 10 | EI = 10 | **0.94** | **0.73** | **1.7** | 2.35 | 4.05 | 2.1 | 1.81 | 2.37 | 2.2 |
| $U_{DM}^{CHEB}$ | N = 80 | G = 500 | p = $10^2$ | 28.40 | 9.49 | 3.0 | **4.59** | **4.23** | **1.5** | 3.93 | 2.45 | 1.5 |
| | m = 5 | v = 10 | EI = 14 | **1.06** | **1.06** | **1.8** | **1.63** | **2.62** | **1.7** | 1.64 | 1.17 | 2.5 |
| $U_{DM}^{CHEB}$ | N = 200 | G = 700 | p = $10^2$ | 27.32 | 9.66 | 3.0 | **3.79** | **2.97** | **1.4** | 3.82 | 1.65 | 1.6 |

can be derived from their analysis are similar to those for the minimal (best found) values. The major differences are the following:

- for a problem involving the DM's linear value function, ECC-MRW is competitive to NEMO-0; this suggests that using ASF as an internal preference model within the algorithm might get to the more preferred region faster, even when pairwise comparison questions are answered with $U_{DM}^{LIN}$;
- when the DM's Chebycheff value function is used, the difference between NEMO-0 and ECC-MRW is smaller; this indicates that in the initial generations, the algorithms attain similar results, while ECC-MRW gains a competitive advantage only when it cumulates more preference information;
- NSGA-II is more advantageous in terms of BRVD, compared to the results reported in Table 1; this derives from the fact that even if NSGA-II is not able to discover solutions which are very relevant from the DM's point of view, it finds some reasonably good solution quickly.

# 5  Conclusions and Future Research

In this paper, we presented an interactive evolutionary algorithm, called ECC-MRW, for dealing with multiple objective optimization problems. The proposed method requires the DM to provide a reference point and successively compare pairs of solutions from a current population. Such preference information is represented by a compatible instance of an achievement scalarizing function. We use the scores of solutions derived from an application of the most discriminant compatible function to modify the selection procedure originally used in NSGA-II. This allows the algorithm to guide the evolutionary search towards the region of the greatest interest to the DM.

Our empirical results show that the proposed method works well. First, we illustrated its ability to focus the search by presenting populations obtained at different stages of the optimization run. We proved that the algorithm was able to converge to different parts of the Pareto front depending on various weights put in the value function of a hypothetical DM. Secondly, we presented results concerning the speed of convergence to the DM's most preferred region and accuracy in finding the user-preferred solutions. Our method proved to vastly outperform NSGA-II for all considered test functions. Its competitive advantage over the state-of-the-art method, called NEMO-0, was particularly visible when the DM provided a richer preference information, i.e., relatively more pairwise comparisons of solutions. Then, for many benchmark problems, an average quality of the solutions constructed with ECC-MRW was more advantageous than a quality of the best solution obtained with NEMO-0.

We envisage the following future research directions:

- designing interactive evolutionary algorithms that would take into all compatible achievement scalarizing functions instead of solely the most discriminant one, and incorporating the outcomes of robustness analysis involving all these functions into the selection procedure;

- studying the performance of algorithms for different elicitation intervals, starting generations for preference elicitation [11], and interaction patterns [7];
- proposing adaptive strategies for preference elicitation that would ask the DM for comparing pairs of solutions only when it is needed, thus, decreasing the cognitive effort required from the DM;
- developing methods for graphical presentation of the populations constructed by the algorithms during the entire optimization run rather than only these obtained after arbitrarily selected few generations.

**Acknowledgements** Miłosz Kadziński and Michał Tomczyk acknowledge financial support from the Polish National Science Center (grant no. DEC-2013/11/D/ST6/03056).

# References

1. Abraham, A., Jain, L.C., Goldberg, R.: Evolutionary Multiobjective Optimization: Theoretical Advances and Applications (Advanced Information and Knowledge Processing). Springer, New York (2005)
2. Battiti, R., Passerini, A.: Brain-computer evolutionary multiobjective optimization: a genetic algorithm adapting to the decision maker. IEEE Trans. Evol. Comput. **14**(5), 671–687 (2010)
3. Branke, J., Deb, K., Miettinen, K., Słowiński, R. (eds.): Multiobjective optimization: interactive and evolutionary approaches. LNCS, vol. 5252. Springer, Berlin (2008)
4. Branke, J., Greco, S., Słowiński, R., Zielniewicz, P.: Learning value functions in interactive evolutionary multiobjective optimization. IEEE Trans. Evolut. Comput. **19**(1), 88–102 (2015)
5. Branke, J., Corrente, S., Greco, S., Słowiński, R., Zielniewicz, P.: Using Choquet intergral as preference model in interactive evolutionary multiobjective optimization. Eur. J. Oper. Res. **250**(3), 884–901 (2016)
6. Ciomek, K., Kadziński, M., Tervonen, T.: Heuristics for prioritizing pair-wise elicitation questions with additive multi-attribute value models. Omega **71**, 27–45 (2017)
7. Deb, K., Sinha, A., Korhonen, P., Wallenius, J.: An interactive evolutionary multiobjective optimization method based on progressively approximated value functions. IEEE Trans. Evolut. Comput. **14**(5), 723–730 (2010)
8. Deb, K., Agrawal, S., Pratap, A., Meyarivan, T.: A fast and elitist multi-objective genetic algorithm: NSGA-II. IEEE Trans. Evolut. Comput. **6**(2), 182–197 (2002)
9. Fonseca, C., Fleming, P.: Genetic algorithms for multiobjective optimization: Formulation, discussion, and generalization. In: Proceedings of the Fifth International Conference on Genetic Algorithms, pp. 416–423 (1993)
10. Kadziński, M., Słowiński, R.: Interactive robust cone contraction method for multiple objective optimization problems. Int. J. Inf. Technol. Decis. Making **11**(2), 327–357 (2012)
11. Kadziński, M., Tervonen, T., Tomczyk, M., Dekker, R.: Evaluation of multi-objective optimization approaches for solving green supply chain design problem. Omega **68**, 168–184 (2017)
12. Kadziński, M., Tomczyk, M.: Interactive Evolutionary Multiple Objective Optimization for Group Decision. Group Decision and Negotiation **26**(4), 693–728 (2017)
13. Phelps, S., Köksalan, M.: An interactive evolutionary metaheuristic for multi-objective combinatorial optimization. Manag. Sci. **49**(12), 1726–1738 (2003)

14. Tervonen, T., van Valkenhoef, G., Basturk, N., Postmus, D.: Hit-And-Run enables efficient weight generation for simulation-based multiple criteria decision analysis. Eur. J. Oper. Res. **224**(3), 552–559 (2013)
15. Wierzbicki, A.P.: On the completeness and constructiveness of parametric characterizations to vector optimization problems. OR Spektrum **8**, 73–87 (1986)

# A Review of Fuzzy and Mathematic Methods for Dynamic Parameter Adaptation in the Firefly Algorithm

Oscar Castillo, Carlos Soto and Fevrier Valdez

**Abstract** The firefly algorithm is a bioinspired metaheuristic based on the firefly's behavior. This paper presents a review on previous works on parameters analysis and dynamical parameter adjustment, using different mathematical approches and fuzzy logic.

**Keywords** Firefly algorithm · Parameter adaptation · Optimization problems · Mathematical functions · Fuzzy adaptation

## 1 Introduction

The metaheuristic algorithms for search and optimization divide their work into two tasks. The first consist in making an exploration or localization of promising areas, where the best solutions may be, and the second task, the exploitation, which consist on concentrating in the areas where the best solutions were found to continue with the search.

The firefly algorithm (FA) has been proved to be very efficient in solving multimodal, nonlinear, global optimization problems, classification problems, and image processing. The fireflies are beetles from the Lampyridae family; which the main characteristic are their wings. Exist over 2000 species of fireflies; their brightness comes from the special luminal organs under the abdomen. The flashing light of the fireflies shines in a specific way for each species. Each shining way is an optical sign that helps the fireflies to find their couple. The light can also work as a defense mechanism. This characteristic inspired Xin-She Yang in 2008 to design the firefly algorithm [27].

O. Castillo (✉) · C. Soto · F. Valdez
Tijuana Institute of Technology, Tijuana, BC, Mexico
e-mail: ocastillo@tectijuana.mx

© Springer International Publishing AG 2018                                                311
A.E. Gawęda et al. (eds.), *Advances in Data Analysis with Computational Intelligence Methods*, Studies in Computational Intelligence 738,
https://doi.org/10.1007/978-3-319-67946-4_13

This paper is organized as follow. Section 2 describes the firefly algorithm. Section 3 is about parameter tuning and dynamic adjustment. Section 4 presents a summary of previous works with FA and where a dynamical adjustment of the FA parameters is carried out. Section 5 presents some final comments.

## 2 Firefly Algorithm

The FA uses three idealized rules:

1. Fireflies are unisex so that one firefly will be attracted to other fireflies regardless of their sex.
2. The attractiveness is proportional to the brightness, and decreases when the distance increases between two fireflies. Thus for any two flashing fireflies, the less bright one will move towards the brighter one. If there is no brighter one than a particular firefly, then it will move randomly.
3. The brightness of a firefly is determined by the landscape of the objective function.

As it is shown in Fig. 1, the algorithm begins defining the number of fireflies to use and the number of iterations. The parameters $\alpha$, $\beta$ and $\gamma$, controlling the exploration and exploitation, must be initialized, and the optimization function must be defined. The next step in the algorithm is the main cycle. The firefly algorithm is based on the attraction between the fireflies depending its light intensity, in the algorithm one firefly is selected and then its light intensity is compared with all the other fireflies. If the light intensity of a firefly A is less bright than the firefly B, then the firefly A is move towards the firefly B. If not, the firefly has only a random movement. This process continues until all the iterations are over or a stop criterion is reached.

### 2.1 Firefly's Movement

The firefly moves according to the following equation.

$$x_{i+1} = x_i + \beta_0 e^{-\gamma r_{ij}^2}(x_i - x_j) + \alpha \varepsilon_1 \tag{1}$$

The FA movement consists of three terms to determine the next position of the firefly. The actual position of the firefly is represented by the vector $x_i$, the second term manage the exploitation where $\beta_0$ is the initial attraction of the firefly, $\gamma$ is the constriction factor, $r$ is the Euclidean distance between the position of the firefly $x_i$ and the firefly $x_j$. The last term manage the exploration where $\alpha$ is the parameter that controls how much randomness is allow the firefly to have in its movement and $\varepsilon_1$

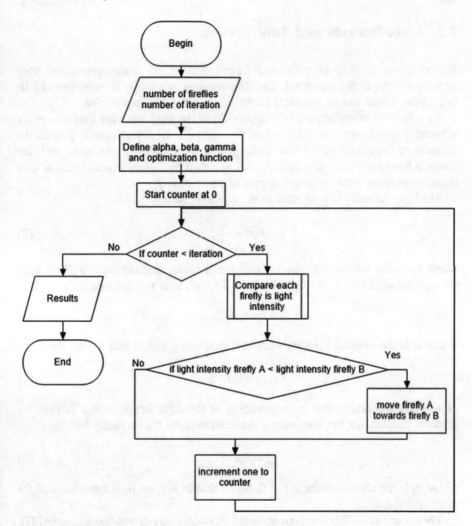

**Fig. 1** FA flowchart

is a vector that contains random numbers drawn from a Gaussian distribution or uniform distribution at time $t$. If $\beta_0 = 0$, it becomes a simple random walk, or $\beta_0$ is the attractiveness at the distance $r = 0$. On the other hand, if $\gamma = 0$, FA reduces to a variant of particle swarm optimization.

## 2.2  Light Intensity and Attractiveness

The variation of light intensity and formulation of the attractiveness are very important in the firefly algorithm. The attractiveness of a firefly is determined by its brightness, which can be associated with the encoded fitness function.

As a firefly's attractiveness is proportional to the light intensity that is seen by adjacent fireflies, we can now define the variation of attractiveness $\beta$ with the distance $r_{ij}$ between firefly $i$ and firefly $j$. The light intensity decreases with the distance from its source, and light is also absorbed by the enviroment, this is why the attractiveness varies with the degree of absorption $\gamma$.

The light intensity is presented in the equation below,

$$I(r) = \frac{I_s}{r^2}, \tag{2}$$

where $I_s$ is the intensity at the source. For a given medium with a fixed light absorption coefficient $\gamma$, the light intensity $I$ varies with the distance $r$.

$$I = I_0 e^{-\gamma r}, \tag{3}$$

where $I_0$ is the original light intensity and combining Eqs. 2 and 3, we have.

$$I(r) = I_0 e^{-\gamma r^2}, \tag{4}$$

as a firefly's attractiveness is proportional to the light intensity that is seen by adjacent fireflies, we can now define the attractiveness $\beta$ of a firefly by

$$\beta = \beta_0 e^{-\gamma r^2}, \tag{5}$$

where $\beta_0$ is the attractiveness at r = 0. Some studies suggest $\beta_0 = 1$ can be used for most applications.

The attractiveness function $\beta(r)$ can be any monotonically decreasing functions [27].

## 2.3  Restriction Coefficient

The $\gamma$ parameter is the absorption coefficient for the light and controls the variation of the attractiveness (and light intensity), and its values determine the speed of the convergence of the FA. In theory, $\gamma \in |0, \infty]$. But if $\gamma$ is very large, then the light intensity decreases too quickly and would result in stagnation, thus the second term (4) becomes negligible. On the other hand if $\gamma$ is very small then the exponential factor: $e^{-\gamma r_{ij}^2} \to 1$ and would suffer from premature convergence [27]. But this rule is not always true as was report in [9] were the experiments show that with $\gamma = 0.008$ the highest success rate was achieved.

In [3] is mentioned that for most applications its value varies from 0.01 to 100, but they used $\gamma = 1$ for their simulations. Other experiments set the light absorption coefficient as $0.00006 \leq \gamma \leq 0.4$ and they recommend the fixed value of 0.02 of their method [28].

However, we can set $\gamma = \sqrt{L}$, where L is the scale. Or we can set $\gamma = O(1)$ if the scaling variations are not significant, then.

For the population size $n$ we can use $15 \leq n \leq 100$, or $25 \leq n \leq 40$ [24]. But there are some cases when a small number of fireflies is more efficient as in [28] although increasing the population helps to improve the search because more fireflies are distributed throughout the search space.

## 2.4  Distance

The variable $r$ in the second term on (1) is the distance between any two fireflies $i$ and $j$ at vector $x_i$ and vector $x_j$, is the Cartesian distance,

$$r_{ij} = \sqrt{\sum_{k=1}^{d} \left( x_{i,k} - x_{j,k} \right)^2}, \tag{6}$$

where $x_{i,k}$ is the $k$th component of the spatial coordinate $x_i$ of the $i$th firefly. Although the distance $r$ is not limited to the Euclidean distance. If necessary another type of distance can be use in the n-dimensional hyperspace depending on the type of problem or our interest. Any measure that can effectively characterize the quantities of interest in the optimization problem can be used as the distance $r$ [1]. For example [18] uses distance $r$ as the difference between the scores of two fireflies (solutions). The Hamming distance is used in [9] to represent the structural differences between candidate (firefly or solution) genotypes.

## 2.5  Randomization

The component of uncertainty in (1) is the third term in the equation. This component adds randomization to the algorithm that helps to explore various regions of the search space and has diversity of solutions [24]. With the adequate control of exploration, the algorithm can jump out of any local optimum and the global optimum can be reached.

Some experiments suggest $\alpha \rightarrow 1$ at the start of the iterations and finishing with a $\alpha \rightarrow 0$. The formula presented below is representing this idea.

$$\alpha_t = \alpha_0 \delta^t, 0 < \delta < 1 \qquad (7)$$

where $\alpha_0$ is the initial randomness scaling factor, and $\delta$ is a cooling factor, and it can be use as $\delta = 0.95$ to 0.97. If $\alpha_0$ is associated with the scalings of design variables FA will be more efficient. Let $L$ be the average scale of the problem of interest, we can set $\alpha_0 = 0.01L$ initially. The factor 0.01 comes from the fact that random walks requires a number of steps to reach the target while balancing the local exploitation without jumping too far in a few steps, for most problems 0.001 to 0.01 can be used [24].

Another formula for the control of randomness was presented in [27], and the implementation is as follows:

$$\Delta = 1 - \left(\frac{10^{-4}}{0.9}\right)^{\frac{1}{MAX\_GEN}},$$

$$\alpha^{(t+1)} = (1 - \Delta) * \alpha^t, \qquad (8)$$

where $\Delta$ determines the step size of the random walk, MAX_GEN is the maximum number of iterations and t is the generation counter. The parameter $\alpha^{(t+1)}$ descends with the increasing of the generation counter.

In [3] they replaced the $\alpha$ by $\alpha * S_j$, where the scaling parameters $S_j$ in the d dimensions are determined by the actual scales of the problem of interest, and is calculated by:

$$S_j = u_j l_j \qquad (9)$$

where $j = 1,2,\dots,d$, $u_j$ and $l_j$ are the lower and upper bound. Also in this paper it is reported that value of the parameter $\alpha$ less than 0.01 do not affect the optimization results.

With the goal of giving a better exploration behavior to the FA the following formula is proposed:

$$\alpha = \alpha_\infty + (\alpha_0 - \alpha_\infty)e^{-t}, \qquad (10)$$

where $t \in [0, Max\_Iteration]$ is the time for simulations and Max_Iteration is the maximum number of iterations. $\alpha_0$ is the initial randomization parameter while $\alpha_\infty$ is the final value [25].

Referring to the vector of random numbers $\varepsilon_1$ it can be drawn from a Gaussian distribution or Uniform distribution [27]. However, the appropriate distribution depends on the problem to be solved, more precisely, on a fitness landscape that maps each position in the search space into fitness value. When the fitness landscape is flat, uniform distribution is more preferable for the stochastic search process, whilst in rougher fitness landscapes Gaussian distribution should be more appropriate. Various probability distributions were used to study their impact on the algorithm, they used Uniform, Gaussian, Lévi flights, chaotic maps, and the

Random sampling in turbulent fractal cloud. Here's concluded that in some cases the selection of the more appropriate randomized method can even significantly improve the results of the original FA (Gaussian). The best results were observed by the Random Sampling in Turbulent Fractal Cloud and Kent chaotic map [8].

The original FA doesn't consider the current global best "*gbest*", and adding an extra term can make an improvement [27].

$$\lambda \epsilon_2 (gbest - x_i) \tag{11}$$

In [24] an analysis on the number of iterations needed to achieve a certain level of accuracy is showed, here we see that the number of iterations it isn't affected much by dimensionality and for higher dimensional problems the number of iterations does not increase significantly. The analysis above mentioned was on the algorithm worst-case scenario.

# 3 Parameter Control

Parameter control is an open question problem, a quest for the right technique that can show mathematically how the performance of the algorithm is affected by the parameters and use this information to make the right adjustment for improvement. The works that have been done until this moment propose very simple cases, strict conditions and sometimes unrealistic assumptions, but there are no theoretical results, so the problem remains unsolved [24].

## 3.1 Parameter Tuning

The convergence rate of the algorithms is related to the eigenvalues that control the parameters and the randomness when we represent as a vector equation once establishing an algorithm as a set of interacting Markov chains. The difficulty of finding this eigenvalue makes the tuning of parameters a very hard problem. The aim of parameter tuning is to find the best parameter setting so that an algorithm can perform well for a wider range of problems, at the end, this optimization problem requires a higher level of optimization methods [24]. One approach to develop a successful framework for self-tuning algorithms was in [23] having good results. Here the FA algorithm is used to tune itself.

## 3.2 Parameter Control

After tuning the parameters of an algorithm very often, they remain fixed during iterations, on the other hand for dynamic parameter control they vary during the iterations, searching for the global optimal. This is also an optimization problem unresolved. In the next lines, we are going to point out the considered factors to construct a fuzzy controller in others studies.

The use of fuzzy logic has been widely use for controlling the parameters of metaheuristic to get an improve in the performance. Knowing the difficult task of choosing the correct values of the parameters to have a good balance between exploration and exploitation, [15] use ACO for doing this work focusing on the alpha parameter who has a big effect on the diversity and can control the convergence, for the fuzzy control inputs they use the error and change of error with respect to an average branching factor reference level. And an improvement was observed but when they try to attack optimization problems with benchmark functions their proposed strategy fail due to the lack of heuristic information.

Another proposal that improves the performance of the algorithm in this case PSO is [16] where three approaches using fuzzy control was presented, here two parameters, the cognitive factor, and social factor, are change dynamically during execution via a control using fuzzy logic; the inputs consider were iteration, diversity and error. The results show two of the fuzzy controllers helps to improve the algorithm. In another study, only using three tests functions, the inputs considered are the current best performance evaluation and the current inertia weight; the output variable is the inertia weight [5].

The importance of knowing how much influence a parameter has in the algorithm is crucial for implementing an efficient fuzzy control system [17]. In [4] only one input (generations) is use for the fuzzy control system and one parameter its values is dynamically decreasing.

## 4 FA Applications

FA is potentially more powerful than other existing algorithms such as PSO. And it's being used in many areas, for example: benchmark reliability-redundancy optimization problems [19], benchmark engineering problems [22], path planning [21], dynamic environment optimization problems [1, 6, 14], neural network training [13], image and data clustering [10, 29], generate optimized fuzzy models [12]. Some study leaves the door open for more research where varying an added parameter can make an improvement on the convergence of the algorithm [26].

## 4.1   Firefly Algorithm Parameter Adjustment

The strategy for setting the optimal values of the adjustable parameters on [18] is trial-and-error, the results show an Hybridizing Firefly Algorithm improvement in the fitness of the optimal solution was obtained with a significant reduction of the execution time.

Using the combination of different techniques can obtain good results: Learning Automata for adapting a parameter, Genetic Algorithm for sharing information between the populations [7].

For a job scheduling application the FA's parameters were set doing experiments and statistical analysis, in this case (the values may be specific to this problem) the best values were:

100 fireflies, 25 generations, $\alpha = 0.5$, $\beta = 1$ and $\gamma = 0.1$. They found the most influencing factor was $\alpha$, followed by $\beta$, a number of fireflies, generations and finally $\gamma$ [11].

The virtue of FA is a natural attraction and how it works for a firefly and his close neighbors.

## 4.2   Fuzzy Control for Parameter Adjustment

An important factor to consider in the design of a fuzzy parameter controller is the iteration, diversity, and error which are used in [20]. But also choosing the right combination of parameters is important, in this case, the results obtained do not improve the original algorithm, this may be because they only focus on controlling the parameters that are responsible for the convergence but forgot to reduce the parameter responsible for the randomness.

Different analysis leaves to different implementations, [2] considers the parameters $\alpha$ (control the exploration) and $\gamma$ (coefficient of light restriction) to be ones to control. As inputs to the fuzzy control system they use the variable Count for referring to the generations and Delta that is defined in 12:

$$
\begin{aligned}
Delta(count) = & fitnessofthebestsolution(count) \\
& - fitnessofbestsolution(tillcount)
\end{aligned}
\tag{12}
$$

With the proposed method they reported an increased performance in FA for solving traveling salesman problems.

# 5 Conclusions

There is no correct technique for tuning or dynamically controlling the parameters of an algorithm, so following a framework is very helpful. It is worth pointing out the need for investigating how to improve the frameworks that already exists or create new ones.

For parameter tuning, experiments and stadistical studies is often use and proven to improve the performance so it shoul be the first technique to try, but the disadvantage is setting will only work for this specific problem. In the case of parameter control a combination of fuzzy logic or another metaheurictic or computational intelligent technique as neural network or learning automata improves the performance.

# References

1. Abshouri, A.A., et al.: New Firefly Algorithm based on Multi swarm & Learning Automata in Dynamic Environments
2. Bidar, M., Rashidy Kanan, H.: Modified firefly algorithm using fuzzy tuned parameters. In: 2013 13th Iranian Conference on Fuzzy Systems (IFSC), pp. 1–4. IEEE (2013)
3. Brajevic, I., Tuba, M.: Cuckoo Search and Firefly Algorithm: Theory and Applications. Presented at the (2014)
4. Castillo, O., et al. (eds.): Recent Advances on Hybrid Approaches for Designing Intelligent Systems. Springer International Publishing, Cham (2014)
5. Eberhart, R.C.: Fuzzy adaptive particle swarm optimization. In: Proceedings of the 2001 Congress on Evolutionary Computation (IEEE Cat. No. 01TH8546), pp. 101–106. IEEE (2001)
6. Farahani, S.M., et al.: A multiswarm based firefly algorithm in dynamic environments 3, 68–72 (2011)
7. Farahani, S.M., et al.: Some hybrid models to improve firefly algorithm performance. 8(12), 97–117 (2012)
8. Fister, I., et al.: Cuckoo Search and Firefly Algorithm: Theory and Applications. Presented at the (2014)
9. Husselmann, A.V, Hawick, K.A.: Cuckoo Search and Firefly Algorithm: Theory and Applications. Presented at the (2014)
10. Jitpakdee, P., et al.: Fuzzy-Based Firefly Algotithm for Data Clustering (2013)
11. Khadwilard, A., et al.: Application of firefly algorithm and its parameter setting for job shop scheduling. J. Ind. Technol. 8 (2012)
12. Kumar, S., et al.: Fuzzy model identification: a firefly optimization approach. Int. J. Comput. Appl. 58(6), 1–8
13. Nandy, S., et al.: Analysis of a Nature Inspired Firefly Algorithm based Back-propagation Neural Network Training (2012). arXiv:1206.5
14. Nasiri, B., Meybodi, M.R.: Speciation based firefly algorithm for optimization in dynamic environments (2012). http://www.ceser.in/ceserp/index.php/ijai/article/view/2359
15. Neyoy, H., et al.: Fuzzy Logic Augmentation of Nature-Inspired Optimization Metaheuristics: Theory and Applications. Presented at the (2015)
16. Olivas, F., et al.: Fuzzy Logic Augmentation of Nature-Inspired Optimization Metaheuristics. Springer International Publishing, Cham (2015)

17. Pérez, J., et al.: Fuzzy Logic Augmentation of Nature-Inspired Optimization Metaheuristics: Theory and Applications. Presented at the (2015)
18. Salomie, I., et al.: Cuckoo Search and Firefly Algorithm: Theory and Applications. Presented at the (2014)
19. dos Santos Coelho, L., et al.: A chaotic firefly algorithm applied to reliability-redundancy optimization. In: 2011 IEEE Congress of Evolutionary Computation (CEC), pp. 517–521. IEEE (2011)
20. Solano-Aragón, C., Castillo, O.: Fuzzy Logic Augmentation of Nature-Inspired Optimization Metaheuristics: Theory and Applications. Presented at the (2015)
21. Wang, G., et al.: A Modified firefly algorithm for UCAV path planning. Int. J. Hybrid Inf. Technol. 5(3), 123–144
22. Yang, X.S.: Firefly algorithm, stochastic test functions and design optimisation. Int. J. Bio-Inspired Comput. 2(2), 78 (2010)
23. Yang, X.-S., et al.: A framework for self-tuning optimization algorithm. Neural Comput. Appl. 23(7–8), 2051–2057 (2013)
24. Yang, X.-S. (ed.): Cuckoo Search and Firefly Algorithm. Springer International Publishing, Cham (2014)
25. Yang, X.-S.: Engineering Optimization: An Introduction with Metaheuristic Applications (2010)
26. Yang, X.-S.: Firefly algorithm, levy flights and global optimization 10 (2010)
27. Yang, X.-S.: Nature-Inspired Metaheuristic Algorithms (2008)
28. Yousif, A., et al.: Cuckoo Search and Firefly Algorithm: Theory and Applications. Presented at the (2014)
29. Image Clustering using Fuzzy-based Firefly Algorithm | Parisut Jitpakdee—Academia.edu. https://www.academia.edu/5870258/Image_Clustering_using_Fuzzy-based_Firefly_Algorithm

# Part V
# Applications of Intelligent Technologies

Part V
Applications of Intelligent Technologies

# Computational Intelligence Methods in Personalized Pharmacotherapy

Adam E. Gawęda and Michael E. Brier

**Abstract** Effective pharmacologic therapy of chronic diseases remains a challenge to physicians. Individual dose-response characteristics of patients may vary significantly across patient populations. In addition, due to the chronic nature of the process, they may change over time within individual patients as well. Current state of the art protocols for dose adjustment of pharmacologic agents rely heavily on data from drug approval process and physician's expertise. However, they do not directly incorporate the wealth of knowledge hidden in patient data collected in the course of the treatment. In this chapter, we review the application of two Computational Intelligence methods, Artificial Neural Networks and Fuzzy Set Theory, to personalized pharmacologic treatment of a chronic condition using patient data stored in Electronic Medical Records. As the application example, we use anemia management in patients with renal failure. To demonstrate the potential of Computational Intelligence methods in improving the disease management, we discuss three human studies in which the discussed methods proved to be an effective decision support aid to the physician.

## 1 Introduction

Pharmacologic treatment of chronic conditions frequently resembles a trial and error process within a feedback loop. An initial dose of a pharmacologic agent is first selected based on a standard reference. The patient is then monitored for therapeutic response and adverse events. Subsequently, the dose is adjusted following the observed patient response. If the response is not sufficient, the dose may be increased. If one or more adverse events are observed, the dose is decreased or

A.E. Gawęda (✉) · M.E. Brier
Department of Medicine, University of Louisville, Louisville, KY, USA
e-mail: adam.gaweda@louisville.edu

M.E. Brier
e-mail: michael.brier@louisville.edu

© Springer International Publishing AG 2018
A.E. Gawęda et al. (eds.), *Advances in Data Analysis with Computational
Intelligence Methods*, Studies in Computational Intelligence 738,
https://doi.org/10.1007/978-3-319-67946-4_14

even withheld. This trial and error process continues until an optimal balance is achieved between the desired response and adverse events.

For many pharmacologic agents, the relationship between the dose and response is non-linear and time-varying. To facilitate the administration of such agents, practitioners have traditionally used standard protocols derived from official drug approval data collected. Traditionally, such data have been analyzed using well established population statistics looking at an average dose-response relationship within a population of studied subjects. To achieve optimal drug response in an individual patient, a medical practitioner must combine the knowledge encoded in such "population-based" protocols with a considerable amount of experience and expertise. Achieving a desired response on an individual basis is further complicated by other concurrent medications and comorbidities, specific for each patient.

In this chapter, we use anemia of End Stage Renal Disease (ESRD) as an example of a chronic condition. Anemia is a very common comorbidity in ESRD and is the result of insufficient production of erythropoietin, a hormone responsible for stimulating red cell production in the bone marrow. Clinically, anemia is defined as insufficient amount of healthy red blood cells or hemoglobin. Hemoglobin is the main part of red blood cells and is responsible for oxygen delivery to tissues to promote energy metabolism. For a long time, repeated blood transfusions had been the standard of care for anemia treatment in ESRD. Nowadays, Erythropoiesis Stimulating Agents (ESA) are the primary form of treatment [1, 2]. However, effective ESA dosing is challenging due to significant variability in hemoglobin response across patient populations. Furthermore, large ESA doses without observable hemoglobin response have been associated with adverse cardiovascular events [3, 4]. Because of this, the US Food and Drug Administration (FDA) stipulates ESA treatment individualization to decrease the risk of blood transfusions. Furthermore, changes in reimbursement policy by Medicare, which provides coverage for a large majority of ESRD patients, have led to an evolution of ESA dosing patterns that results in lower average hemoglobin levels and more transfusions [5].

Ever since the introduction of ESA's, nephrologists have been administering this agent according to standardized protocols combining information from the regulatory guidelines with their own experience and expertise. In January 2010, the FDA called for clinical studies to establish better dosing approaches for ESA's, including computer-directed algorithms [6]. In this chapter we specifically focus on the application of two Computational Intelligence methods, Artificial Neural Networks (ANN) and Fuzzy Set Theory (FST) to perform dose-response modeling and individualized dosing of ESA's in anemia management in ESRD patients.

The chapter is organized as follows. In Sect. 2, we review reported examples in which ANN's and Fuzzy Models have been applied to data-driven ESA dose-response modeling in ESRD patients. We look at the application of Fuzzy Models for the same purpose. In Sect. 3, we review the results of several clinical studies in which these techniques have been employed in conjunction with modern control theoretic tools, to facilitate individualized ESA dosing in ESRD patients. Section 4 summarizes the review.

# 2 Computational Intelligence Approaches to Drug Response Modeling

## 2.1 Artificial Neural Networks

Until the late twentieth century, traditional approaches to pharmacologic modeling heavily relied on physiology-based models. Due to the complex multidimensional nature of the human body, these physiologic models contained large numbers of interrelated parameters which posed an extreme challenge when estimating such models from clinical data. In certain situations, where the predictive capabilities of a pharmacologic model are of primary interest, nonlinear data-driven black-box modeling techniques, such as Artificial Neural Networks are a valuable alternative as the modeling tool of choice in pharmacology.

One of the first reported applications of ANN's in pharmacology dates back to 1995 [7, 8]. These early examples include the prediction of peak and trough concentrations of gentamicin [7] and the delayed renal allograft function as a guide to initiate immunosuppression therapy in kidney transplant recipients [8]. In both cases, the ANN's were found to be superior to the state-of-the-art approach at the time, the nonlinear mixed effect modeling (NONMEM). A similar finding was reported by Chow et al. [9] who used the ANN to model the serum concentration of tobramycin in pediatric patients. Camps-Valls et al. [10] investigated the application of three types of ANN's: Multilayer Perceptron (MLP), Finite Impulse Response (FIR) network, and Elman network to prediction of cyclosporine concentration in kidney transplant recipients. They found all three ANN models to be robust and accurate for this purpose.

The first applications of ANN to ESA dose-response modeling were reported in 2003 [11, 12]. Martin-Guerrero et al. [12] applied MLP, FIR, and Elman networks to perform longitudinal modeling of hemoglobin response to subcutaneous ESA. The models were trained to predict hemoglobin concentration one month ahead from an input vector containing patient characteristics (age, weight), current hemoglobin and iron stores concentration, as well as previously received dose of ESA and iron. The networks were trained on data from 110 patients. Model selection was performed using cross-validation. The authors found that all three analyzed ANN types achieved similar performance measures in terms of mean absolute (MAE) and root mean square (RMSE) prediction error. Due to the short time sequence of the input information, the authors concluded that the recurrent ANN's (Elman, FIR) did not offer a significant advantage over the MLP, as demonstrated by the validation results. In [11], we reported the results of using two different ANN types: MLP and Radial Basis Function (RBF) network, to the task of predicting average hematocrit response to ESA dose. Hematocrit is the measure of red blood cell volume in relation to the total volume of the blood and has been used interchangeably with hemoglobin as the clinical biomarker of anemia. The ANN models were trained on data from 209 patients undergoing hemodialysis treatments at the University of Louisville Kidney Disease Program. Patient-specific Leave-One-Out cross-validation was performed to train the models. The models predicted one month ahead average hematocrit from an input vector containing monthly average hematocrit levels, their standard deviations, as well as average ESA doses over past two months. No patient specific covariates were included in the input vector. The optimal MLP network contained 10 hidden neurons with hyperbolic tangent activation functions (Fig. 1). The ANN's were benchmarked against a linear autoregressive (ARX) model. Statistical analysis of the performance measures (RMSE and normalized RMSE) revealed superiority of both ANN models over the ARX model, with the MLP slightly outperforming the RBF network.

In a largest study of this kind to date, Barbieri et al. [13] developed an ANN model for 3 month ahead prediction of hemoglobin concentration in response to ESA and iron dose. They used data from a cohort of 1558 ESRD patients and similarly to other researchers, found that the ANN models delivered a highly accurate predictive performance.

Yet another approach to ANN-based ESA dose-response modeling in anemia patients was proposed by Gabutti et al. [14]. They evaluated two types of ANN models: MLP and Generalized Feedforward Network (GFN) as a tool to identify most important clinical and biological covariates determining patient's responsiveness to ESA and to predict monthly ESA dose requirement in individual

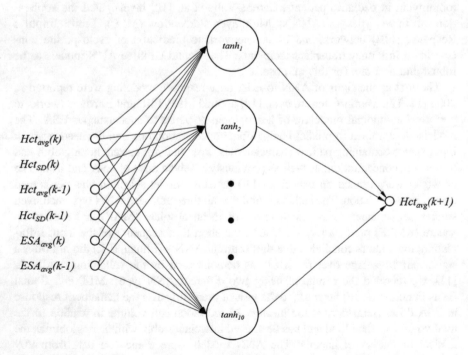

**Fig. 1** Architecture of the Multilayer Perceptron network for one step ahead prediction of hematocrit in response to ESA [11]. Legend: $Hct_{avg}$—average hematocrit, $Hct_{SD}$—standard deviation of hematocrit, $ESA_{avg}$—average ESA dose, k—time step (month)

patients. The ANN models were trained on a classification problem of predicting hemoglobin <11.0 g/dL using data of 432 patients from 29 dialysis facilities. The performance measures used were sensitivity and specificity. In the context of ESA dose selection, hemoglobin < 11.0 g/dL was used as a trigger to increase the ESA dose. In the task of ESA dose requirement prediction, the ANN model was found to be superior to a linear regression (sensitivity 78% versus 40% at 50% specificity). Furthermore, the ANN model for predicting dose adjustments proved superior to a nephrologist following European best practice guidelines detecting 48 versus 25% patients requiring ESA dose increase.

## 2.2 Fuzzy Set Models

First reported use of fuzzy set theory in pharmacology dates back to 1997 [15]. In this application, fuzzy rule-based model was created to predict serum concentration of lithium. In addition to dosing data and serum creatinine levels, the model incorporated as input information patient specific characteristics, such as age and weight. Interestingly, these covariates were found not to be instrumental in predicting the lithium concentration. Based on the reported performance metrics, the authors concluded that fuzzy set models are a feasible alternative to standard analytical methods used in pharmacology.

In [16] fuzzy sets were applied as an alternative to probabilistic methods to create a physiologically based pharmacokinetic model of diazepam disposition. In this approach, fuzzy sets were used as a means to represent uncertainty in physiologic model parameters. The authors postulated that fuzzy set theory can successfully represent the vagueness and imprecision associated with minimal drug discovery data.

In the context of modeling ESA response in ESRD patients, we applied fuzzy sets to represent imprecision involved in the classification of the dose-response profile in an individual patient [17]. We developed a model to predict hemoglobin response to a change in ESA dose associated with different dose-response profiles. We compared three methods of dose-response classification: one group (population approach), crisp classification, and fuzzy classification. Comparison of the mean square prediction error of hemoglobin revealed that fuzzy set classification of the dose-response profile significantly improves the predictive capacity of the model over the population approach and the crisp classification.

## 3 Computational Intelligence Approach to Drug Dosing

In this section we review three clinical studies in human subjects, in which we successfully demonstrated the use of Artificial Neural Networks and Fuzzy Set Models to optimize ESA dosing in patients with anemia due to End Stage Renal

Disease. We specifically focus on two control methods: Neural Predictive Control and Fuzzy Multiple Model Predictive Control.

## 3.1 Neural Predictive Control

In [18] we proposed an approach to personalized ESA dosing based on the concept of Model Predictive Control (MPC). MPC is a modern control technique based on the interaction between a process model and an optimization algorithm to minimize an objective function defining the control goal (Fig. 2). In our application, the model was represented as a Multi-Layer Perceptron ANN predicting hemoglobin one month ahead based on the previous two monthly hemoglobin levels and eight weekly ESA doses. The MPC was developed from clinical data of 186 ESRD patients at the University of Louisville. In silico testing was first performed in a cohort of 60 virtual patients to compare the MPC approach to a standard anemia management protocol (AMP). The in silico testing proved that our proposed ANN-based MPC (Neural Predictive Control, NPC) approach to ESA dosing achieved the target hemoglobin level (11.5 g/dL) more precisely and more consistently than a standard AMP. Following the in silico evaluation, we performed a human study in a cohort of 9 subjects who received ESA based on MPC recommendation for a period of 6 months. While the difference in achieved hemoglobin between the AMP and MPC was not as impressive as in the in silico trial, the NPC approach achieved better hemoglobin stability.

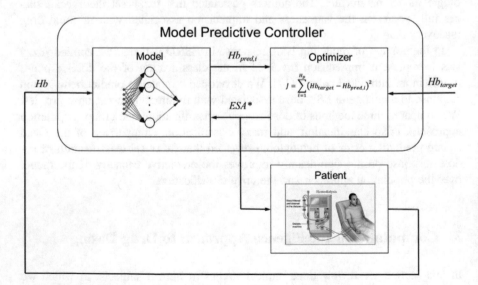

**Fig. 2** Block diagram of Neural Model Predictive Control applied in [18] and [19]. Legend: Hb—hemoglobin, Hb$_{pred,i}$—hemoglobin predicted at time step i, Hb$_{target}$—target hemoglobin level, ESA*—optimal ESA dose, J—objective function, H$_p$—prediction horizon

Following this study, we designed and performed a full scale randomized controlled clinical trial of an NPC algorithm [19] for ESA dosing in ESRD patients. This trial was performed in a cohort of 60 hemodialysis patients receiving treatment at the University of Louisville. One half of the subjects were randomly assigned to a control arm (ESA dose determined by a standard AMP), the other half were assigned to a treatment arm (ESA dose determined by NPC). The subjects were followed for 8 months. The main performance metric used in the study was the proportion of hemoglobin levels between 11 and 12 g/dL. Hemoglobin variability was defined as an absolute difference between the measured hemoglobin and the median of the hemoglobin target range (11.5 g/dL). In this study, we again demonstrated that the NPC approach resulted in a more stable hemoglobin control when compared to a standard AMP.

## 3.2 Fuzzy Multiple Model Predictive Control

The NPC algorithms presented in [18, 19] used a fixed ANN model. To facilitate truly personalized ESA dosing, we developed an MPC-based approach based on the concept of multiple controllers, where each controller was optimized to a specific ESA dose-response profile [20]. Each MPC corresponded to one of five dose-response classes: extreme hyper-responder, hyper-responder, moderate hyper-responder, intermediate responder, and hypo-responder. At each dosing interval, the individual MPC generated a dose recommendation to achieve a target hemoglobin specific for its dose-response class. The dose-response class was matched to an individual patient based on an average weekly ESA dose received over 4 weeks before dose adjustment using fuzzy sets as described previously in [17]. The block diagram of the overall algorithm is shown in Fig. 3. Data required by the algorithm were directly abstracted from Electronic Medical Record database.

To validate the algorithm, we performed a single-center randomized controlled trial and compared it against the existing standard of care AMP at the University of Louisville. We enrolled 62 hemodialysis patients and followed them for 12 months. The primary performance metric used in the study was the proportion of hemoglobin measurements between 10 and 12 g/dL. Over the course of the study, subjects assigned to have ESA dose guided by the MMPC algorithm achieved 10.6% more hemoglobin levels within target range, compared to a standard AMP. Furthermore, subjects receiving MMPC-guided ESA doses, achieved significantly lower percentage of hemoglobin levels less than 10 g/dL, which translated into two-fold decrease in blood transfusion rate.

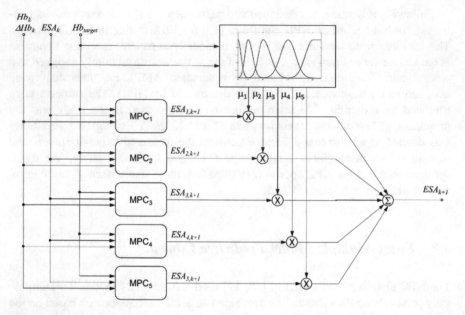

**Fig. 3** Block diagram of Fuzzy Multiple Model Predictive Control applied in [20]. Legend: $Hb_k$—hemoglobin at time k, $\Delta Hb_k$—hemoglobin rate at time k, $ESA_k$—ESA dose at time k, $MPC_{1,\ldots,5}$—Model Predictive Controller for dose-response profile 1 through 5, $ESA\text{-}_{1,\ldots,5,\ k+1}$—optimal ESA dose at time k for dose-response profile 1 through 5, $\mu_{1,\ldots,5}$—patient's membership degree in dose-response profile 1 through 5, $ESA_{k+1}$—optimal ESA dose at time k+1

## 4 Conclusions

This chapter provided a brief review of two mainstream Computational Intelligence methodologies, Artificial Neural Networks and Fuzzy Set Theory applied to the problem of personalized pharmacotherapy using anemia management in End Stage Renal Disease patients as a demonstration platform. Described real world application examples, backed by the results from rigorous human studies, prove that Computational Intelligence techniques are a viable alternative to the existing standard of care currently used by physicians and pharmacists.

## References

1. Adamson, J.W., Eschbach, J.W.: Treatment of the anemia of chronic renal failure with recombinant human erythropoietin. Ann. Rev. Med. **41**, 349–360 (1990)
2. Eschbach, J.W., et al.: Recombinant human erythropoietin in anemic patients with end-stage renal disease. Results of a phase III multicenter clinical trial. Ann. Intern. Med. **111**(12), 992–1000 (1989)
3. Pfeffer, M.A., et al.: Baseline characteristics in the Trial to Reduce Cardiovascular Events With Aranesp Therapy (TREAT). Am. J. Kidney Dis. **54**(1), 59–69 (2009)

4. Singh, A.K., et al.: Correction of anemia with epoetin alfa in chronic kidney disease. N. Engl. J. Med. **355**(20), 2085–2098 (2006)
5. Collins, A.J., et al.: Effect of facility-level hemoglobin concentration on dialysis patient risk of transfusion. Am. J. Kidney Dis. **63**(6), 997–1006 (2014)
6. Unger, E.F., et al.: Erythropoiesis-stimulating agents–time for a reevaluation. N. Engl. J. Med. **362**(3), 189–192 (2010)
7. Brier, M.E., Zurada, J.M., Aronoff, G.R.: Neural network predicted peak and trough gentamicin concentrations. Pharm. Res. **12**(3), 406–412 (1995)
8. Brier, M.E., Aronoff, G.R.: Application of artificial neural networks to clinical pharmacology. Int. J. Clin. Pharmacol. Ther. **34**(11), 510–514 (1996)
9. Chow, H.H., et al.: Application of neural networks to population pharmacokinetic data analysis. J. Pharm. Sci. **86**(7), 840–845 (1997)
10. Camps-Valls, G., et al.: Prediction of cyclosporine dosage in patients after kidney transplantation using neural networks. IEEE Trans. Biomed. Eng. **50**(4), 442–448 (2003)
11. Gaweda, A.E., et al.: Pharmacodynamic population analysis in chronic renal failure using artificial neural networks–a comparative study. Neural Netw. **16**(5–6), 841–845 (2003)
12. Martin Guerrero, J.D., et al.: Use of neural networks for dosage individualisation of erythropoietin in patients with secondary anemia to chronic renal failure. Comput. Biol. Med. **33**(4), 361–373 (2003)
13. Barbieri, C., et al.: Performance of a predictive model for long-term hemoglobin response to Darbepoetin and iron administration in a large cohort of hemodialysis patients. PLoS ONE **11** (3), e0148938 (2016)
14. Gabutti, L., et al.: Would artificial neural networks implemented in clinical wards help nephrologists in predicting epoetin responsiveness? BMC Nephrol. **7**, 13 (2006)
15. Sproule, B.A., et al.: Fuzzy logic pharmacokinetic modeling: application to lithium concentration prediction. Clin. Pharmacol. Ther. **62**(1), 29–40 (1997)
16. Seng, K.Y., Nestorov, I., Vicini, P.: Physiologically based pharmacokinetic modeling of drug disposition in rat and human: a fuzzy arithmetic approach. Pharm. Res. **25**(8), 1771–1781 (2008)
17. Gaweda, A.E., Jacobs, A.A., Brier, M.E.: Application of fuzzy logic to predicting erythropoietic response in hemodialysis patients. Int. J. Artif. Organs **31**(12), 1035–1042 (2008)
18. Gaweda, A.E., et al.: Model predictive control of erythropoietin administration in the anemia of ESRD. Am. J. Kidney Dis. **51**(1), 71–79 (2008)
19. Brier, M.E., et al.: Randomized trial of model predictive control for improved anemia management. Clin. J. Am. Soc. Nephrol. **5**(5), 814–820 (2010)
20. Gaweda, A.E., et al.: Individualized anemia management reduces hemoglobin variability in hemodialysis patients. J. Am. Soc. Nephrol. **25**(1), 159–166 (2014)

# Embodying Intelligence in Autonomous and Robotic Systems with the Use of Cognitive Psychology and Motivation Theories

Kowalczuk Zdzisław and Czubenko Michał

**Abstract** The article discusses, on a certain level of abstraction and generalization, a coherent anthropological approach to the issue of controlling autonomous robots or agents. A contemporary idea can be based on appropriate modeling of the human mind using the available psychological knowledge. One of the main reasons for developing such projects is the lack of available and effective top-down approaches resulting from the known research on autonomous robotics. On the other hand, there is no system that models human psychology sufficiently well for the purpose of constructing autonomous systems. Nevertheless, to combat this lack, several ideas have been proposed for embodying human intelligence. We review recent progress in our understanding of the mechanisms of cognitive computations underlying decision-making and discuss some of the pertinent challenges identified and implemented in several systemic solutions founded on cognitive ideas (like LIDA, CLARION, SOAR, MANIC, DUAL, OpenCog). In particular, we highlight the idea of an Intelligent System of Decision-making (ISD) based on the achievements of cognitive psychology (using the aspect of 'information path'), motivation theory (where the needs and emotions serve as the main drives, or motivations, in the mechanism of governing autonomous systems), and several other detailed theories, which concern memory, categorization, perception, and decision-making. In the ISD system, in particular, an xEmotion subsystem covers the psychological theories on emotions, including the appraisal, evolutionary and somatic theories.

**Keywords** Cognitive architecture · Cognitive development · Decision-making · Human-computer interaction · Perception · Intelligent agents

K. Zdzisław (✉) · C. Michał
Faculty of Electronics, Telecommunications and Informatics, Gdańsk University
of Technology, Narutowicza 11/12, 80-233 Gdańsk, Poland
e-mail: kova@pg.edu.pl

C. Michał
e-mail: michal.czubenko@pg.edu.pl

© Springer International Publishing AG 2018
A.E. Gawęda et al. (eds.), *Advances in Data Analysis with Computational Intelligence Methods*, Studies in Computational Intelligence 738,
https://doi.org/10.1007/978-3-319-67946-4_15

# 1 Introduction

Creating a system functioning in a human-like way, has long been a principal subject of artificial intelligence and robotics. As can be seen from the many known results of robotics, a significant number of artificial creatures and humanoids have been constructed [32], and some of them even try to communicate in natural language [55]. Moreover, considering the inner aspect, the well-known artificial neural networks (of a convolutional type) have been conceived and applied for different system control and recognition purposes [9, 23]. All such minor steps are being made towards creating an artificial humanoid, synthetic organism, or android robot, designed to look and act like a human.

Artificial Intelligence is being developed in a continued effort to solve engineering problems, such as reasoning, problem solving, knowledge representation, machine learning, natural language processing, machine perception, and others. Eventually, solving these problems should lead to an invented humanoid system similar to a human being, to a certain extent. A few principal types of approaches to artificial intelligence are worth mentioning here:

- cybernetic—which postulates to follow an imitation of some aspects of real, physical, or biological systems in a virtual world (using neural networks, evolution algorithms, swarm algorithms, etc.) [53],
- statistic—which seeks to build rigorous, usually sophisticated, mathematical tools necessary for statistical modeling of processes [49],
- symbolic (top-down, synthetic, 'neats', clean)—which uses high-level logic (simplistic, black-box) mathematical modeling, knowledge-based processing, and machine learning [47],
- sub-symbolic (bottom-up, analytic, 'scruffies', ad hoc, embodied)—which involves the use of small (white-box, physical, neuronal) models to first create a low-level, and next, by the ad hoc rules, higher-level solutions [8].

The variety of known AI branches strive for (usually partially) modeling of the human mind, and none of them fulfills this objective fully. Modeling the human mind can be performed by applying the symbolic (top-down) approach and the sub-symbolic (bottom-up) method. These two approaches are complementary, and both are related to the cybernetic method. Certainly, the statistical tools developed in a mathematical way are of great use. Probably, solely an intelligent combination of many methods will be able to satisfactorily reflect the effects of the human brain.

**Embodied Intelligence** (EI) represents the sub-symbolic approach. It is an extension of the genuine cybernetic projects from the 50s, which tried to reproduce simple phenomena of 'intelligence' identified at a low level of cognition [3, 7, 8, 17]. We can recall here the early cybernetic projects, like the construction of *homeostat*, a device which retains stable despite external disturbances, or *tortois*, a robot which follows an assumed intensity of light [53]. Quite promising results can be obtained by following *baby steps*, that is, by simulating a certain basic functionality using simple elements (note that the tortois had only two neurons, for instance). On the basis of such affirmative experience, a new branch of behavior-based robotics has emerged [4].

Most issues, such as finding an optimal trajectory or recognition of environmental objects, require rather complex operation, whereas inference and reasoning are relatively simple (from the biological and computer science points of view). It is Moravec's paradox that applies to this problem [46]:

Encoded in the large, highly evolved sensory and motor portions of the human brain is a billion years of experience about the nature of the world and how to survive in it. The deliberate process we call reasoning is, I believe, the thinnest veneer of human thought, effective only because it is supported by this much older and much more powerful, though usually unconscious, sensorimotor knowledge. We are all prodigious olympians in perceptual and motor areas, so good that we make the difficult look easy. Abstract thought, though, is a new trick, perhaps less than a hundred thousand years old. We have not yet mastered it. It is not all that intrinsically difficult; it just seems so when we do it.

It seems natural that different achievements from the fields of embodied intelligence, behavior-based robotics, and top-down approaches in AI, are indispensible in modeling the effect of the human mind. However, to reach an intelligent interaction of an artificial agent with the environment it is also important to clearly define what 'embodied intelligence' means [60].

In this paper the concept of 'embodied intelligence' will be understood in a slightly different way than the 'classical' notion. Recall that mathematical modeling providing a description of a hypothetical fragment of an existing reality, reflects the behavior of a real system in a particular environment. Such an environment generates different distal signals determining the so-called experimental setting. At each stage of the process of modeling of physical phenomena, the results of the next simplified mathematical model are thoroughly referenced to the previously conducted experiments. This is in line with the bottom-up approach (analytic, physical, white-box). On the other hand in natural sciences, psychology, philosophy, and cybernetics, the top-down approach (synthetic, mathematical, black-box) is most frequently in use. Ignorance of the aforementioned principles may easily lead to confusion and inadequate interpretations.

## 1.1 ... Intelligence

One of the first definitions of intelligence has been proposed by Spearman in [59]:

...all branches of intellectual activity have in common one fundamental function, whereas the remaining or specific elements of the activity seem in every case to be wholly different from that in all the others.

It appears, however, too vague for the aim of determining the intelligence for robot purposes. Though clear, other definitions like: "*The ability to deal with cognitive complexity*" or "*Goal-directed adaptive behavior*" [20, 61] also seem to be overly general. Nevertheless, due to such definitions, you can at least imagine what is the essence of human-like intelligence:

**Definition 1** Intelligence is the ability of active processing of cognitive information in order to adapt to the changing environment and to gain own, specific purposes or common goals.

In an extremely simple case, an intelligent agent, by being completely focused on searching for a source of energy necessary to survive, can function completely selfishly. Clearly, the latter brings to the mind the aforementioned tortois and cybernetic theories.

## 1.2  Embodied ...

Embodiment in the human case means that the entire perception of the real world completely relies on its physical components and senses. Embodiment is also associated with the philosophy of mind, and, in particular, with the whole mind-body problem as formulated by Descartes [2].

Certainly, intelligence could not be developed without embodiment [60]. It is also clear that any virtual or robotic agent ought to be designed for, and located in, a certain environment to have a chance to implement a two-way interaction. Then one can talk about *engineered intelligence*, having the *environmental embodiment* (or foundation) defined as:

> Mechanism under the control of an intelligence core that contains sensors and actuators connected with this core via communication channels.

Such embodiment of a robot or agent can be easily extended with various kinds of tools (like glasses, spectacles, drives, or even a mobile or car), which augment both the agent's perception and possibilities of reaction.

## 2  Decision Systems

The idea is to build a system that—in line with the increasing capabilities of computers and their power—would be able to take autonomous decisions, according to current circumstances. Certainly, there exist, and are being developed, increasingly sophisticated decision-support systems, such as: expert systems [1, 5], and systems based on Bayesian networks [16, 65] or neural networks [57, 66]. Such systems usually support human decision making (for diagnostic purposes, for instance). In most cases they are strictly tailored to pre-defined conditions. In general, however, there are two known paths for decision-making:

- classical, which finds the most optimal decision for a well-defined problem,
- cognitive, aiming at finding a solution to real problems defined or recognized only partially.

Thus the classical decision theory treats about taking decisions in a strictly optimal sense for mathematically well-modeled tasks and well-defined problems. Whereas the cognitive theory shows how to take proper decisions for difficult real-world problems, which are usually uncertain and not well defined [19].

An early elaboration on human decision-making processes was delivered in 1910 by Dewey [15]. According to him, there are five stages in the decision making process: Defining the problem, Indication of its character, Finding possible solutions, their Evaluation, and Selection of the appropriate solution. A similar and a bit more universal division, referred to as *GOFER*, presented in 1991 [41] suggests the following phases:

1. Goals—searching for selecting the objectives,
2. Options—considering a wider spectrum of alternative actions concerning the goals currently considered,
3. Facts—gathering additional knowledge about actions (options) and goals,
4. Effects—evaluating (usually hypothetically) the results of the chosen options,
5. Rating—final evaluation of the decisions, and selecting the best one.

In addition, there are many other interesting approaches to the analysis of complete decision processes [6, 44, 54]. Not far from, in its simplest form, the decision making process can always be described in solely three phases [58]:

1. definition of the problem,
2. finding possible solutions,
3. selection of the optimal solution.

In order to achieve the effect of autonomous decision-making suitable for a current situation, the system should not only take the opportunity of learning (knowledge extension), understanding and recognizing (known) objects, but also it should have some motivations which compel it to take action.

There are a great number of decision-making systems based on human motivation factors. Human is the highest of all species in terms of adaptation to the changing environment, thus the human system of motivation appears to be most adequate as a template of behavior. Ethical foundations for such systems can be derived from the existing variety of the available models of psychology and human intelligence. These achievements have also notably contributed to artificial intelligence. Among them one can distinguish the following types of conceptual solutions:

- behavioral [4, 14],
- BDI (Beliefs-Desires-Intentions) [13, 21, 25, 52],
- emotional [33, 42] (sometimes they are assigned to BDI),
- driven by needs [22, 43, 45, 50, 56],
- cognitive (**LIDA**, **CLARION**, **SOAR**, MANIC, DUAL, OpenCog, ...).

To give you a taste of the existing spectrum of complex systems, we will discuss below three (in bold) of the above-listed representatives of cognitive systems.

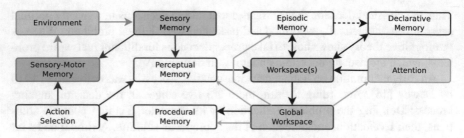

**Fig. 1** Cognitive architecture of Lida: the grey lines represent interaction with the environment, blue lines show low-level processing, orange lines indicate learning process, and dotted lines portray consolidation of the memory

## 2.1  LIDA

Learning Intelligent Distribution Agent, LIDA, originally developed by Stan Franklin [18], is a cognitive system which intends to model biological cognition [18, 40]. It implements an architecture of sub-sumption [8] and other aspects of the sub-symbolic branch of AI. This is one of the most advanced projects aiming at modeling the results of psychological and neuro-psychological theories, in particular, embodied knowledge, symbolic systems of perception, different types of memory, and the different ways of learning mechanisms, overt attention and motivation in the form of emotion (Fig. 1).

LIDA is executed using cognitive cycles (repeated in each executive run), each of which consists of the subprocesses of perception, selection of appropriate response (relative to the perceived environmental facts), and implementation of the selected reaction. Advanced cognitive processes, such as planning, can be synthesized as an aggregate of the perception-action cycles. Motivational aspects in the LIDA system concern feelings, which have their own valence (positive or negative), associated with satisfaction, or pain (which evidently attributes LIDA also to the emotional developments and solutions).

Stimuli recorded by sensors and pre-processed, are next analyzed in a working/operational memory referring to various types of long-term memory (perceptual, episodic, declarative and procedural). Memory is instrumental in creating a current model of actual circumstances, which constitute an executive groundwork for the process of selecting the desired reaction (using the procedural memory). *Conscious contents* are intended to add an external context to this model, and to enable learning processes. Once selected, the reaction is directly implemented by the actuators.

## 2.2 CLARION

Connectionist Learning with Adaptive Rule Induction On-line, CLARION, represents a cognitive architecture based on theories from cognitive and social psychology [11, 62–64]. CLARION implements several AI results to ensure the effect of creating an *intelligent* system. CLARION's architecture, developed and implemented by Ron Sun, is composed of four units shown in Fig. 2:

- ACS – (procedural) Action Centered Sub-system,
- NACS – Non-Action Centered Sub-system,
- MS – Motivational Sub-system,
- MCS – Meta-Cognitive Sub-system.

In each of the above sub-systems the data and structures are represented dually: at a higher level (overt/explicit) and at a lower level (covert/implicit). This dual representation in CLARION, connected with (different) philosophic theories and with the issue of memory representation [35, 51], enables autonomous learning in two ways: bottom-up (induction) and top-down (deduction). The assumptions applied are fully compliant with the requirements of the embodied intelligence design discussed earlier.

The action oriented sub-system (ACS) is responsible for all kinds of the agent's reactions, both internal and external (concerning the environment). The covert (implicit) part is implemented as a neural network, while the overt (explicit) layer represents a rule base. The non-action centered sub-system (NACS), which mimics the role of the semantic and episodic memories, is responsible for the storage and delivery of knowledge. It is also divided into two parts. Its hidden part takes the form of an associative neural network, while its explicit layer can be described with the use of symbolic notations and rules. The inference performed in this module is founded on similarities.

Motivation means are also important to the design of the cognitive structure of CLARION. Corresponding motivational elements of the MS sub-system are of both the explicit and implicit type. Explicit (higher) elements include targets (explicit goals), such as: belonging, recognition, power, autonomy, respect, and honesty. On the other hand, the lower motivational factors (prime movers) of the CLARION system, realize the idea similar to the concept of *needs* (discussed later), which are of a physiological nature (consider eating, drinking, sleep, security, and reproduction). In addition, CLARION's MS sub-system allows you to program your own *secondary needs* to define a more subtle motivation (in order to achieve a certain goal).

The MCS sub-system is responsible for a meta-cognitive function resembling attention or awareness. It monitors and regulates all other cognitive processes of the agent and fulfills the idea of *consciousness*. More specifically, MCS chooses which goals are most important, with autonomous inferencing and learning, and how to adjust the gain of the learning process. It is also responsible for information filtering and for selecting the method of data interpretation.

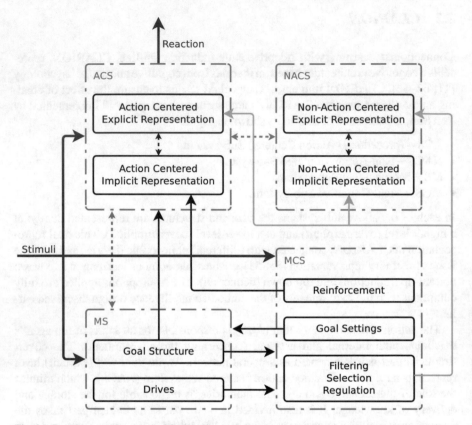

**Fig. 2** Cognitive architecture in Clarion: the orange lines present attention (in general), green lines indicate data exchange, and red lines show interaction with the system's environment

## 2.3 SOAR

State, Operator And Result, SOAR, is a cognitive architecture invented by Laird, Newell, and Rosenbloom [10, 24, 36, 38, 39, 48]. It is one of the earliest systems of this type (its first version is dated back to 1983), whose main purpose is behavior resembling an intelligent agent. Its architecture is suitable for working in varying conditions, from routine tasks to creative, open problems. It requires appropriate forms of knowledge representation, and suitable types of memories (procedural, semantic, episodic and iconic). To be consistent with the assumptions of embodiment, the agent needs to interact with the ambient world, and to learn constantly about its features. The decision making in SOAR is based on the current situation perceived from the environment, whereas the necessary information and knowledge is acquired by suitable dynamic processing of the data gained through the sensors. An internal expert system plays the role of fundamental processing unit.

SOAR's cognitive architecture has several components concerning [37]:

- memory functioning, for the task of knowledge storage,
- processing module of attention, used for extraction, mixing and remembering knowledge,
- semantics and syntax of the language used for storage and processing of knowledge.

Similarly to LIDA, SOAR is based on a certain decision cycle. A perception sub-system manipulates the data stored in a symbolic short-term memory. Deductive rules are used to test the agent's capabilities in the context of possible actions. Another layer of rules is applied to suggest optimal reactions (operations) adequate to the current situation evaluated by perception and motivational sub-systems, and next the agent's preferences are calculated. Finally, according to the perceived state (situation), and given a pre-processed set of possible reactions and preferences, SOAR is ready to select one of the estimated reactions, and then to apply it using the system actuators.

The cognitive structure of SOAR is shown in Fig. 3, where decision cycle is implemented by the block of decision procedure. In the SOAR system, emotions are generated in the appraisal detection block, and next they serve as reinforcement applied in learning processes (indirectly through *mood* and *feelings*). Semantic memory is an essential element in the treatment of procedural and episodic knowledge (using long-term memory). It allows the agent to store information about the environment. On the other hand, the episodic memory contains the knowledge related to the execution and effects of various types of actions, including the degree of fulfillment of the rules and operations performed by the agent (and others). Long-term visual memory as well as imagination assist in the agent's mind operations concerning spatial processing.

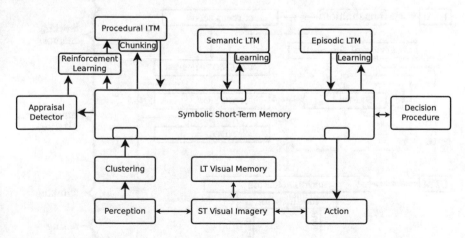

**Fig. 3** Cognitive system of SOAR (ver. 9)

## 2.4 Intelligent System of Decision-Making

Intelligent System of Decision-making, ISD, as presented in the recent papers [12, 26–31, 35], is a control system of an agent that intends to covert and implement the contemporary theory of embodied intelligence and decision theory, and in particular, the models of cognitive psychology and motivation theory. It mimics roughly the way people make decisions, from the arrival of the stimuli to the generation of a reaction. As a consequence, the ultimate design of the ISD unit is the result of a thorough modeling of human psychology embedded in elementary findings of an extensive literature study. In practice, ISD is a universal system which can control robots and unmanned ground vehicles, including cars, as is presented in [12]. A view on ISD is presented in Fig. 4.

ISD is a cognitive decision-making system, which implements all of the stages of decision-making, presented earlier. The main mechanism of decision-making in ISD is based on the concept of needs, which are principal drives for acting. Needs are variables programmable by the user. They can also be possibly created autonomously

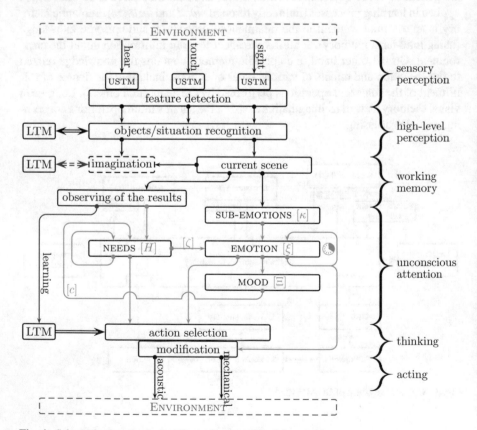

**Fig. 4** Schematic view on the Intelligent System of Decision-making

by the agent and adjusted for certain situations. Thus, different sets of needs may be used to shape the characteristics (personality) of the agent, according to its environmental conditioning. Observed objects and events, and actions performed by the agent (namely their inner and outer results) have impact on the state of the agent's needs.

ISD presents also cognitive abilities with respect to the *understanding* of the environment (in practice, without them the system would not be consistent). It means that from the robotic point of view, the agent is 'conscious' of its environment, it knows its position, and the position of surrounding objects and their definition. Stimuli perceived by the agent's senses (sensors) are stored in an ultra-short-term memory (USTM). Simple features of perceived objects (impressions), such as colors, shapes, textures, etc. (like red flat rectangle), are extracted from USTM, and stored in a short-term memory (STM). To recognize a simple impression, the agent can apply various mechanisms, developed as filters, masks, neural networks, fuzzy systems, decision rules and others. For example, a Haar cascade can be used for recognizing head shapes (impressions). During extraction, certain stimuli may cause an immediate unconscious action of the agent (like: 'step back' in response to pain). On the basis of the observed features (impressions), complete discoveries/objects are 'mentally' created, taking into account the relative location of the features in space. In a simple translation, the discovery consists of impressions in a specific location. Next, they are compared to known objects stored in a long-term memory (LTM). If the 'mind' detects a certain level of similarity between the perceived discovery and a know object from LTM, the discovery is recognized/identified with the object from LTM. A suitable recognition procedure is described in [12]. Some of the discoveries may result in half-conscious activities, previously learned through multiple repetitions.

Recognized objects are transferred to the agent's operational memory that represents the current scene, where they are analyzed from different angles, taking into account:

- the impact of external (environmental) facts/objects, as they may affect the needs or cause sub-emotions, which can, in turn, change the agent's proper emotion; Remember also that both the needs and the sub-emotions must be previously stored as connected to certain discoveries (e.g. a pink blanket from childhood can connect with the need for security), and thus affect the agent's current system of needs;
- the effect of the internal (*body*) facts/states, as they can also modify the agent's system of needs (e.g. an energy sensor connected to the need of energy, can directly change the need of the agent, according to its value).

According to the above, the states of needs are constantly updated, creating, and pointing to, new goals. The agent tries to find (or formulate) a conscious action to be implemented by the system in order to fulfill its most important or painful needs [27–29]. The action undertaken by the ISD unit is then tracked by the part of the thinking process which is referred to as the observer of results. This process always seeks to see a desired effect of the previous action (for instance, in the change of the degree of fulfilment of the agent's needs) by penetrating the contents of the operational memory. It is also related to the learning process in ISD. The achieved results of the previous activities are memorized (for future searches of optimal actions).

In line with the human motivation theory, emotions are one of the most important factors of human behavior. Systems, based on human psychology (both cognitive and motivative), but deprived of emotions would be ineffective. Emotions in ISD perform their function at a higher level of control than the basic ISD control ruled by the system of needs. In our robotics applications, emotions allow us to narrow down the set of possible reactions to those that are most adequate (in the view of the system designer) for the current time moment and the state of the system [30, 34].

Pre-defined sub-emotions (emotions associated with identified objects) do influence the current state of the proper emotion of the agent, which strikes (assumes) one of 24 possibilities, according to the theory of Plutchik. The degree of satisfaction of all the agent's needs, the former emotional state, and the effect of calming down (emotion simply decays with time), all influence the state of the emotion of the agent. Changes in emotion affect the mood, which, in turn, tune the fuzzy parameters of the needs models. As mentioned earlier, emotion effectively preselects (narrows) the set of possible reactions. In addition, it can modify some reactions (for instance, by using additional forms of expression, like wording, gestures, or facial expressions).

There are different types of long-term memory in the ISD system [31]:

- semantic (abstract and realistic),
- episodic,
- procedural.

Knowledge in ISD is stored in the form of (abstract or instance) discoveries, consisting of many different features/impressions (including those related to needs and emotions), labels, and relations to other discoveries [35]. Episodic memory is used to describe events on the time axis, and with reference to respective discoveries stored in the semantic memory. A forgetting phenomenon decays the activity level of remembrances (the events remembered in the episodic memory). Depending on this level, the more frequent the remembrances (memories) are, the faster they can be recalled. Procedural memory contains specifications (declarations) of the agent's actions.

## 3 Comparison

The above-presented systems represent a cognitive approach to the problem of decision-making. All of them are trying to combine the bottom-up and top-down approaches and methods. In practice, however, they are very different in the aspects of implementation and concept. There is no great sense to compare them in terms of parameters such as computational complexity, speed of response, accuracy and performance of individual activities, because of the large variety of implementation and use of these systems. Certainly, there are several useful tests for autonomous

**Table 1** Comparison of cognitive architectures

|  | LIDA | CLARION | SOAR | ISD |
|---|---|---|---|---|
| Structure | Perception-action cycles | Explicit and implicit sub-systems (parallel) | Cycles | Cycles with interruptions |
| Stimuli | Internal and external | External | Dependent on designer | Internal and external |
| Perception memory | Slip-Net (associative) | Connected to working memory | *Not known* | Impressions |
| Basic memory unit | Codlet | Chunk | Rule | Discovery |
| Short-time or working memory | Global workspace theory | Limited (visuospatial, auditory, other) | Symbolic short-term memory | Current scene and imagination with activation levels (limited) |
| Long-time memory structure | Perceptual, episodic, declarative, procedural | Non-action centered subsystem (semantic, associative knowledge) | Procedural, semantic, episodic | Semantic (abstract and instance), episodic, procedural |
| Drivers | *Not known* | Similar to human needs, goals | Emotions | Needs and emotions |
| Emotions | Feelings (positive or negative) | *Not known* | Appraisal (mood and feelings) | Based on plutchik |
| Decision-making | Based on current environmental situation | Rules and neural networks | Rules and reasoning | Motivation driven |
| Programming language | Java | C# | Java and C++ | Python |
| Usage | Medical diagnostic | Simulations concerning wide spectrum of cognition | Simulations from towers of Hanoi to quakebot | Partial simulations |

cognitive systems like user-end tests for *coffee-making* or *student behavior*, but they have a limited use, due to the lack of the necessary actuators. The utility of such one-sided (one goal) tests is also controversial due to their selectivity, at which some cognitive systems appear to be better than the other ones, depending on the particular test task. However, one may always compile a multi-purpose comparison of the cognitive architectures in terms of structure models, driving systems, and implementing concepts, as has been shown in Table 1.

# 4 Synchronization of Cognitive Systems

Each of the presented systems approaches the issue of modeling the human cognitive processes in its own way. They appear to be more or less explanatory, and usually to some extend (partially) support the psychological theories on these processes. This knowledge allows us both to evaluate the psychological theories and generalize or adapt the cognitive processes for autonomous agents. For example, each of presented systems has some basic memory entity, which let the agent to comprehend particular real objects, and an overall semantic memory, necessary for grasping the actual situation by an autonomous robot.

Note that cognitive architectures are primarily designed to make decisions under the circumstances of autonomous work. Nevertheless modeling the environment of the agent appears to be even more difficult than the inferencing itself. Therefore, it is important that the developed systems also indicate how to describe the environment for the purpose of autonomous agents (letting the necessary and inevitable interaction).

For comparative purposes and definite concluding results, each of the presented systems should be implemented on a platform of an autonomous (mobile) robot, and then tested under identical conditions (this would be more effective than partial simulation, certainly). In particular, the cognitive architectures should be tested at different angles, highlighted below:

- perception – estimated in terms of speed and accuracy of environmental recognition,
- attention – to determine the importance of objects due to agent's security and decisions,
- decision-making – adequate for practical uncertainty,
- learning and reasoning – enabling the agent to correct its mistakes and to expand its *knowledge* about the surrounding environment,
- computing power – necessary for proper functioning of the system.

# 5 Summary

The paper discusses the idea of embodied intelligence as an approach that combines both the cognitive modeling of complex systems (top-down approach), as well as the (bottom-up) implementation of systems designed to detect and comprehend the basic characteristics of the environment. Needing a variety of tools, the creation of such architecture principally relies on established theories, and thus results in workable reformulations of several essential definitions concerning intelligence.

The agent that has the ability to actively process cognitive information using its sensors and mechanisms to adapt itself to the changing environment and to achieve its objectives (at least to strive for them), possesses embodied intelligence. In our pursuit of the goal of embodied intelligence, we used a systematic approach to the cognitive decision-making process through the implementation of several major ideas of cognitive psychology and motivation theory, which led us to design of the Intelligent System of Decision-making (ISD).

Though the presented cognitive systems have been developed for different purposes, all of them model the decision-making process in a very interesting, instructive and practicable way, using differently defined motivational aspects. In the near future, such systems will have the opportunity to achieve a high level of sophistication in terms of both the design conception and technical implementation—with great hope to achieve at least some level of intelligence of simple living creatures (like lizards, for instance).

# References

1. Agarwal, M., Goel, S.: Expert system and its requirement engineering process. In: International Conference on Recent Advances and Innovations in Engineering, pp. 1–4. IEEE (2014)
2. Alsop, S.: Beyond Cartesian Dualism: Encountering Affect in the Teaching and Learning of Science, vol. 26, Springer Science Business Media (2005)
3. Anderson, M.L.: Embodied cognition: a field guide. Artif. Intell. **149**(1), 91–130 (2003)
4. Arkin, R.C.: Behavior-Based Robotics. MIT Press, Cambridge, MA (1998)
5. Bennett, C.C., Doub, T.W.: Artificial Intelligence in Behavioral and Mental Health Care. In: Luxton, D.D. (ed.) Artificial Intelligence in Behavioral and Mental Health Care, 2, pp. 27–51. Elsevier (2016)
6. Brim, N., Orville, G., Glass, D.C.: Personality and Decision Processes: Studies in the Social Psychology of Thinking. Stanford University Press (1962)
7. Brooks, R.A.: Intelligence without reason. In: International Joint Conference on Artificial Intelligence, pp. 569–595. Sydney (1991)
8. Brooks, R.A.: Intelligence without representation. Artif. Intell. **47**(1–3), 139–159 (1991)
9. Chen, W., Qu, T., Zhou, Y., Weng, K., Wang, G., Fu, G.: Door recognition and deep learning algorithm for visual based robot navigation. In: IEEE International Conference on Robotics and Biomimetics IEEE, pp. 1793–1798 (2014)
10. Chown, E., Jones, R., Henninger, A.: An architecture for emotional decision-making agents. In: Proceedings of the First International Joint Conference on Autonomous Agents and Multiagent Systems: part 1, pp. 352–353. ACM, Bologna (2002)
11. Coward, L., Sun, R.: Criteria for an effective theory of consciousness and some preliminary attempts. Conscious. Cogn. **13**(2), 268–301 (2004)
12. Czubenko, M., Ordys, A., Kowalczuk, Z.: Autonomous driver based on intelligent system of decision-making. Cogn. Comput. **7**(5), 569–581 (2015)
13. Damjanovic, V., Kravcik, M., Devedzic, V.: eQ: an adaptive educational hypermedia-based BDI agent system for the semantic Web. In: Fifth IEEE International Conference on Advanced Learning Technologies, pp. 421–423. IEEE (2005)
14. De Silva, L., Ekanayake, H.: Behavior-based robotics and the reactive paradigm a survey. In: International Conference on Computer and Information Technology, pp. 36–43. Khulna (2008)
15. Dewey, J.: How We Think. D.C. Heath Company, Mineola, N.Y. (1910)
16. Du, P., Liu, H.y.: Study on air combat tactics decision-making based on Bayesian networks. In: 2nd IEEE International Conference on Information Management and Engineering, pp. 252–256. IEEE, Chengdu (2010)
17. Flemmer, R.C.: A scheme for an embodied artificial intelligence. In: 2009 4th International Conference on Autonomous Robots and Agents, pp. 1–9. IEEE (2010)

18. Franklin, S., Madl, T., D'Mello, S., Snaider, J.: LIDA: a systems-level architecture for cognition, emotion, and learning. IEEE Trans. Auton. Ment. Dev. **6**(1), 19–41 (2014)
19. Goodwin, P., Wright, G.: Decision Analysis for Management Judgment. Wiley (2009)
20. Gottfredson, L.: The general intelligence factor. Sci. Am. Presents **9**(4), 24–29 (1998)
21. Hernandez, A., El Fallah-Seghrouchni, A., Soldano, H.: Distributed learning in intentional BDI multi-agent systems. In: Proceedings of the Fifth Mexican International Conference in Computer Science, pp. 225–232. IEEE (2004)
22. Herve, L.G., Sorin, M.: A model of cooperative agent based on imitation and Maslow's Pyramid of needs. In: International Joint Conference on Neural Networks, pp. 1229–1236. IEEE (2009)
23. Ji, S., Yang, M., Yu, K.: 3D convolutional neural networks for human action recognition. IEEE Trans. Pattern Anal. Mach. Intell. **35**(1), 221–31 (2013)
24. Jones, R., Laird, J.: Constraints on the design of a high-level model of cognition. In: Proceedings of the Nineteenth Annual Conference of the Cognitive Science Society (1997)
25. Korecko, S., Herich, T., Sobota, B.: JBdiEmo OCC model based emotional engine for Jadex BDI agent system. In: 12th International Symposium on Applied Machine Intelligence and Informatics (SAMI), pp. 299–304. IEEE, Herl'any (2014)
26. Kowalczuk, Z., Czubenko, M.: DICTOBOT an autonomous agent with the ability to communicate. In: Zeszyty Naukowe Wydziału ETI Politechniki Gdaskiej. Technologie Informacyjne, pp. 87–92 (2010)
27. Kowalczuk, Z., Czubenko, M.: Interactive cognitive-behavioural decision making system. In: Rutkowski, L. (ed.) Artifical Intelligence and Soft Computing Lecture Notes in Computer Science, Lecture Notes in Artificial Intelligence, vol. 6114 (II), pp. 516–523. Springer-Verlag, Berlin, New York (2010)
28. Kowalczuk, Z., Czubenko, M.: Model of human psychology for controlling autonomous robots. In: 15th International Conference on Methods and Models in Automation and Robotics, pp. 31–36 (2010)
29. Kowalczuk, Z., Czubenko, M.: Intelligent decision-making system for autonomous robots. Int. J. Appl. Math. Comput. Sci. **21**(4), 621–635 (2011)
30. Kowalczuk, Z., Czubenko, M.: xEmotion—a computational model of emotions dedicated for intelligent decision-making systems, in polish (xEmotion obliczeniowy model emocji dedykowany dla inteligentnych systemów decyzyjnych). Pomiary, Automatyka, Robotyka **2**(17), 60–65 (2013)
31. Kowalczuk, Z., Czubenko, M.: Cognitive memory for intelligent systems of decision-making, based on human psychology. In: Korbicz, J., Kowal, M. (eds.) Intelligent Systems in Technical and Medical Diagnostics, Advances in Intelligent Systems and Computing, vol. 230, chap. Cognitive, pp. 379–389. Springer, Berlin, Heidelberg (2014)
32. Kowalczuk, Z., Czubenko, M.: Overview of humanoid robots, in polish (Przegld robotów humanoidalnych). Pomiary, Automatyka, Robotyka **19**(4), 67–75 (2015)
33. Kowalczuk, Z., Czubenko, M.: Computational approaches to modeling artificial emotion an overview of the proposed solutions. Front. Robot. AI **3**(21), 1–20 (2016)
34. Kowalczuk, Z., Czubenko, M.: Interpretation and Modeling of Emotions for the Governance of Autonomous Agent-Robots with the Use of the Paradigm of Scheduling Variable Control in preparation (2016)
35. Kowalczuk, Z., Czubenko, M., Jędruch, W.: Learning Processes in Autonomous Agents using an Intelligent System of Decision-making. In: Kowalczuk, Z. (ed.) Advances in Intelligent Systems and Computing, pp. 301–315. Springer, Berlin, Heidelberg New York (2016)
36. Laird, J.: The Soar Cognitive Architecture. MIT Press (2012)
37. Laird, J.: Extending the Soar cognitive architecture. In: Wang, P., Goertzel, B., Franklin, S. (eds.) Proceedings of the Artificial General Intelligence, vol. 171, pp. 224–235. IOS Press (2008)
38. Laird, J., Mohan, S.: A case study of knowledge integration across multiple memories in Soar. Biologically Inspired Cognitive Archit **8**, 93–99 (2014)

39. Laird, J.E., Newell, A., Rosenbloom, P.S.: SOAR: an architecture for general intelligence. Artif. Intell. **33**(1), 1–64 (1987)
40. Madl, T., Franklin, S.: Constrained incrementalist moral decision making for a biologically inspired cognitive architecture. In: Trappl, R. (ed.) A Construction Manual for Robots' Ethical Systems, pp. 137–153. Springer International Publishing, Cognitive Technologies (2015)
41. Mann, L., Harmoni, R., Power, C.: The GOFER course in decision making. In: Brown, J., Brown, R. (eds.) Teaching Decision Making to Adolescents. Routledge Taylor and Francis Group, New Jersey, London (1991)
42. Marsella, S., Gratch, J., Petta, P.: Computational models of emotion. In: Scherer, K.R., Bänziger, T., Roesch, E.B. (eds.) A Blueprint for Affective Computing: A Sourcebook and Manual, pp. 21–41. Oxford University Press, Oxford, UK (2010)
43. Matsumoto, Y., Nishida, Y., Motomura, Y., Okawa, Y.: A concept of needs-oriented design and evaluation of assistive robots based on ICF. In: International Conference on Rehabilitation Robotics, Zurich (2011)
44. Mintzberg, H., Raisinghani, D., Théorêt, A.: The structure of 'unstructured' decision processes. Adm. Sci. Q. **21**(2), 246–275 (1976)
45. Miwa, H., Itoh, K., Ito, D., Takanobu, H., Takanishi, A.: Introduction of the need model for humanoid robots to generate active behavior. IEEE/RSJ Int Con Intell Robots Syst **2**, 1400–1406 (2003)
46. Moravec, H.: Mind Children. Harvard University Press, The Future of Robot and Human Intelligence (1988)
47. Newell, A., Simon, H.A.: Human Problem Solving. Prentice-Hall, Englewood Cliffs (1972)
48. Nielsen, P., Koss, F., Taylor, G., Jones, R.: Communication with intelligent agents. In: Proceedings of IITSEC, pp. 824–834. Orlando, FL (2000)
49. Norvig, P.: On Chomsky and the two cultures of statistical learning. On-line essay in response to Chomsky's remarks ... (2011)
50. Novak, E.: Toward a mathematical model of motivation, volition, and performance. Comput. Edu. **74**, 73–80 (2014)
51. Paivio, A., Csapo, K.: Picture superiority in free recall: imagery or dual coding? Cogn. Psychol. **5**(2), 176–206 (1973)
52. Pan, Y.T., Tsai, M.S.: Development a BDI-based intelligent agent architecture for distribution systems restoration planning. In: 15th International Conference on Intelligent System Applications to Power Systems, pp. 1–6. IEEE, Curitiba (2009)
53. Pickering, A.: The Cybernetic Brain. The University of Chicago Press (2011)
54. Pijanowski, J.: The role of learning theory in building effective college ethics curricula. J. Coll. Charact. **10**(3), 1–14 (2009)
55. Rasheed, N., Amin, S.H., Sultana, U., Shakoor, R., Zareen, N., Bhatti, A.R.: Theoretical accounts to practical models: Grounding phenomenon for abstract words in cognitive robots. Cogn. Syst. Res. **40**, 86–98 (dec 2016)
56. Ren, L., Liu, W., Liang, X.: The research on the needs model of the China network game. In: IEEE International Conference on Communications Technology and Applications, pp. 255–258. IEEE (2009)
57. Seepanomwan, K., Caligiore, D., Cangelosi, A., Baldassarre, G.: Generalisation, decision making, and embodiment effects in mental rotation: a neurorobotic architecture tested with a humanoid robot. Neural Netw. **72**, 31–47 (2015)
58. Simon, H.A.: The New Science of Managment Decision. Prentice Hall PTR (1960)
59. Spearman, C.: General intelligence objectively determined and measured. Am. J. Psychol. **15**(2), 201–292 (1904)
60. Starzyk, J.: Motivation in Embodied Intelligence (2008)
61. Sternberg, R.J., Salter, W.: Handbook of Human Intelligence. Cambridge University Press, UK, Cambridge (1982)
62. Sun, R.: Moral judgment, human motivation, and neural networks. Cogn. Comput. **5**(4), 566–579 (2013)

63. Sun, R., Helie, S.: Psychologically realistic cognitive agents: taking human cognition seriously. J. Exp. Theor. Artif. Intell. **25**(1), 65–92 (2013)
64. Sun, R., Merrill, E., Peterson, T.: From implicit skills to explicit knowledge: a bottom-up model of skill learning. Cogn. sci. **25**(2), 203–244 (2001)
65. Wang, L., Wang, M.: Modeling of combined Bayesian networks and cognitive framework for decision-making in C2. J. Syst. Eng. Electron. **21**(5), 812–820 (2010)
66. Żurada, J., Barski, M., Jędruch, W.: Artificial Neural Networks, in Polish (Sztuczne sieci neuronowe). Wydawnictwo naukowe PWN, Warszawa (1996)

# Evolutionary Approach for Automatic Design of PID Controllers

Krystian Łapa and Krzysztof Cpałka

**Abstract** In this paper a new approach for automatic design of PID controllers is presented. It is based on meta-heuristic hybrid algorithm which is a combination of the genetic algorithm and the imperialist one. Main characteristic of the proposed approach is capability to design the structure and the structure parameters of a controller. It is a big advantage because it eliminates trial and error process of design the controller structure. Moreover, the proposed approach has been developed in a way that allows to obtain controllers taking different control criteria and a different control object into consideration.

## 1 Introduction

The controller is a main component of the control system. Its purpose is to control specified object in a way to obtain expected (or close to expected) behavior of the object. On the other hand, the automatic control of the object lies on making controller dependent on changes of measurable physical values (feedback signals) from the object (for example: current, voltage, temperature, pressure, velocity, etc.). Furthermore, control process should take different control criteria, which depend usually on control object and control goal.

Many types of controllers can be found in the literature, such as: proportional-integral-differential (PID) controllers, controllers based on computational intelligence (e.g. neural networks [4, 42, 49, 50], fuzzy systems [7–10, 13, 14, 31–37, 45–48], clustering algorithms [11, 16, 26] and hybrid controllers (based on both PID controller and computational intelligence methods). However, the PID controllers correspond to the needs of most automation systems [22] and they are mostly used

K. Łapa · K. Cpałka (✉)
Institute of Computational Intelligence, Czestochowa University of Technology,
Al. Armii Krajowej 36, 42-200 Czestochowa, Poland
e-mail: krzysztof.cpalka@iisi.pcz.pl

K. Łapa
e-mail: krystian.lapa@iisi.pcz.pl

© Springer International Publishing AG 2018
A.E. Gawęda et al. (eds.), *Advances in Data Analysis with Computational Intelligence Methods*, Studies in Computational Intelligence 738,
https://doi.org/10.1007/978-3-319-67946-4_16

in practice [25]. Typical PID controller consists of the following elements: proportional P with reinforcement parameter $K_p$, integral I with time constant $T_i$ (denoted also as $K_i$) and differential D with time constant $T_d$ (denoted also as $K_d$). The purpose of P element is to compensate offset between expected value and real value from the object, the purpose of I element is to compensate offsets from previous time steps and the purpose of D element is to compensate future offsets. To achieve appropriate quality of the controller the elements P, I and D should be properly connected and their parameters should be properly selected.

In the literature many methods for parameters selection (tuning) can be found, such as: Ziegler-Nichols method, methods working in a field of frequency (giving information about the supply gain and supply phase), methods using relay tuning, methods based on optimization of control criteria, methods with inner model, methods seeking roots of closed loop, methods using optimal module critter, methods based on image recognition etc. Additional methods that can be used for parameters tuning are computational intelligence methods, in particular population-based methods [2, 15, 19, 20, 24, 25, 38, 41, 44]. These methods are efficient procedures of searching space and they are based on processing of group (population) of possible solutions (individuals) for the problem under consideration. In these algorithms each individual contain a set of parameters representing single controller.

Among the typical PID controllers similar controllers with modified (or reduced) structures can be found. These include, for example, the following controllers: PI, PI in cascade, PI with feed-forward [6, 17, 21, 28], PI or PID with additional lowpass [21], PID with anti-windup and compensation mechanism [29], pseudo-derivative feedback (PDF), PDF with feed-forward gain (PDFF) [40] etc.

Selection of a proper controller structure is a complex problem that usually needs a priori expert knowledge. In practice a need for methods for automatic selection controller parameters and structure arises. The aim of these methods is to test nontypical combinations of feedback signals from the object, taking into account the actual working conditions of the control system (including non-linearity and distortion) and taking into account deviations from the principles of analytical design of the regulator in order to simplify and improve the quality of control.

In this paper a method for selection the controller structure and for simultaneously tuning of this structure parameters is proposed. We used our experience from the field of hybrid population-based algorithms [27] to create this method. In order to develop this method the following problems had to be solved: (a) problem of generalization of the controller structure (which allows us to obtain all mentioned structures and process any number of feedback signals), (b) problem of proper encoding of the controller (not only its parameters), (c) problem of taking into consideration multiple control criteria, (d) problem of proper processing of the population of encoded controllers. The latter problem arises from the fact that existing population-based algorithms are not suitable for use in the considered problem of the controller selection. For example, the genetic algorithm efficiently tunes encoded binary values, thus it is well suited to the structure selection. In turn, the imperialist algorithm efficiently tunes real values but it is not equipped in mechanism to tune encoded binary values, thus it is well suited to parameters selection. In this paper an algorithm which is a hybridization of both mentioned algorithms is proposed.

In Sect. 2 the proposed approach for designing control systems with possibilities of a new hybrid genetic-imperialist algorithm is described. The proposed hybrid algorithm is described in Sect. 3. Simulation results are presented in Sect. 4 and conclusions are placed in Sect. 5.

## 2 Proposed Generalized Controller Structure

In this paper an attempt for generalization of MISO (multiple input, single output) controller structure is made. To achieve that a new structure of the controller is proposed (see Fig. 1). In case of MIMO system, proposed in Fig. 1 structure should be paralleled. Proposed structure contains CB and NB blocks (see Fig. 2) and each CB block is adequate to the single PID structure. The main advantage of using proposed controller structure is the possibility of selecting the most unusual/non-typical structures of the control system (for example with non-typical feedback). The goal of the structure modification is to obtain simple structure which in the best possible way meets the required control criteria.

The number of CB and NB blocks in a proposed controller is a parameter of algorithm that arises from complexity of considered problem. In practice, the number of controller input signals $e_i$, $i = 1, ..., I$ (including feedback signals from the controlled object) is usually small, so the initial number of CB and NB blocks is also small. The inputs of CB blocks might include: (a) offsets of feedback signals and expected values of signals from the object, (b) feed-forward signals directly from

**Fig. 1** Proposed generalized controller structure based on CB and NB blocks, designed to automatic selection by evolutionary algorithm

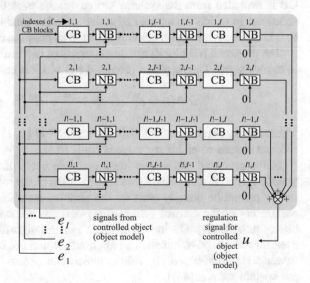

**Fig. 2** Proposed structure of: **a** control block (CB), **b** node block (NB)

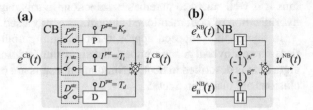

the object (solutions like that are often used in the literature to improve the quality of control), (c) cascading signals from other CB blocks connected with NB blocks.

The proposed CB block structure consists of P, I and D elements, which can be additionally turned off (by elements acting the same way as the switch in an electrical circuit) (see Fig. 2). Thus, the goal of tuning algorithm is to select proper (simple and efficient) structure of controller (by modification of electrical circuit-like switches) and simultaneously to select this structure parameters (by modification real value parameters of P, I and D elements). The output of proposed CB block is calculated as:

$$u^{\text{CB}}(t) = \left( P^{\text{str}} \cdot K_p \cdot e^{\text{CB}}(t) + I^{\text{str}} \cdot \frac{1}{T_i} \cdot \int_0^t e^{\text{CB}}(t)\,dt + D^{\text{str}} \cdot T_d \cdot \frac{de^{\text{CB}}(t)}{dt} \right), \quad (1)$$

where binary parameters (referred as switches) $P^{\text{str}}$, $I^{\text{str}}$, $D^{\text{str}}$ stand for activation (when binary value is equal to 1) of corresponding P, I and D elements. It is worth noting that if necessary, a different initial structure of the controller can be used (for example with filter elements) which, however, does not affect the concept of the proposed approach. If all switches $P^{\text{str}}$, $I^{\text{str}}$, $D^{\text{str}}$ are set to 0, then the whole CB is excluded from the system. Moreover, the evolution process promotes these solutions, in which the number of active switches (and thus reduced the number of P, I, D elements and CB blocks) is as small as possible (see Sect. 3.3).

The proposed NB block structure consists of two multipliers which allow signals to change signs. The output of proposed NB block is calculated as:

$$u^{\text{NB}}(t) = e_{\text{A}}^{\text{NB}}(t) \cdot (-1)^{A^{\text{str}}} + e_{\text{B}}^{\text{NB}}(t) \cdot (-1)^{B^{\text{str}}}, \quad (2)$$

where binary parameters $A^{\text{str}} \in \{0, 1\}$ and $B^{\text{str}} \in \{0, 1\}$ stand for changes of signs of input signals when binary value is equal to 1.

In the proposed method an important role plays proposed hybrid evolutionary genetic-imperialist algorithm (see Fig. 3). Most of the recent population-based algorithms cannot be directly used for the simultaneous selection of the structure and parameters of the control system (they can process only real parameters or only binary parameters). On the other hand, genetic algorithms could theoretically be used for this purpose, but such an approach would not be effective (there are many newer population-based algorithms which can obtain better results on real value parameters-see e.g. [44]).

**Fig. 3** The idea of the proposed method of automatic design of PID controllers

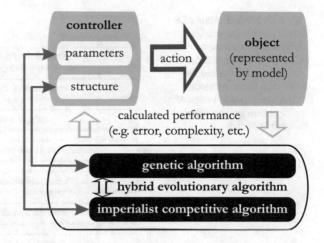

The proposed algorithm is an ensemble of genetic algorithm to process binary parameters and imperialist algorithm to process real parameters. The idea of genetic algorithms is based on biological evolution of species (see e.g. [5]) and the idea of imperialist algorithm (Imperialist Competitive Algorithm, ICA) is based on social evolution (see e.g. [3]). New elements of the algorithm proposed in this paper includes, among the others, adapting it to the processing both the binary and real parameters and the introduction of the modified mutation operator. A detailed description of the proposed algorithm was presented in Sect. 3.

## 3 Proposed Hybrid Genetic-Imperialist Algorithm Description

Each solution (individual) in the imperialist algorithm terminology (see e.g. [3]) is called a colony. On the basis of the best colonies from initial population (colonies characterized by the best values of fitness function) the empires are created (each colony from best colonies creates an empire and became imperialist of this empire). The rest of the colonies are spread among all empires. The colonies are a subject to evolutionary operators which are referring to the human social evolution: the assimilation operator and the revolution operator. Additionally, a binary mutation of colonies takes place (mutation of structure of the controller), derived from the genetic algorithm. After this process all empires are re-evaluated by fitness function and empires compete with each other which results in transfer of the colonies between strongest and weakest empire from the competition. If, due to colonies transfer, empire lost all colonies, it is eliminated (removed) from the whole process. Next, a stopping criterion is checked. This criterion can be based on quality of the best solution in the population or on the number of total iterations defined in the algo-

**Fig. 4** A block diagram of a hybrid genetic-imperialist algorithm for automatic selection of the structure and parameters of the controller based on a linear correction terms. The presented steps of the algorithm are described in detail in Sect. 3

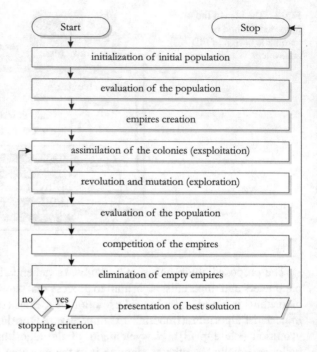

rithm. Therefore, the purpose of the algorithm is to systematize improvement of the solutions in terms of the value of evaluation function. The steps of the algorithm are shown in Fig. 4 and are described in detail in further part of this section.

## 3.1 Encoding of the Structure and Parameters

Solutions encoded in a population are identified as $\mathbf{X}_j, j = 1, \ldots, N$ ($N$ stands for the number of solutions in population). Each solution contains two parts: $\mathbf{X}_j^{\text{str}}$ and $\mathbf{X}_j^{\text{par}}$ ($\mathbf{X}_j = \left\{ \mathbf{X}_j^{\text{str}}, \mathbf{X}_j^{\text{par}} \right\}$). The first part $\mathbf{X}_j^{\text{str}}$ encodes the structure of the controller and it is expressed as follows:

$$
\mathbf{X}_j^{\text{str}} = \begin{bmatrix} P_{j,1,1}^{\text{str}}, I_{j,1,1}^{\text{str}}, D_{j,1,1}^{\text{str}}, A_{j,1,1}^{\text{str}}, B_{j,1,1}^{\text{str}}, \ldots, \\ P_{j,1,I}^{\text{str}}, I_{j,1,I}^{\text{str}}, D_{j,1,I}^{\text{str}}, A_{j,1,I}^{\text{str}}, B_{j,1,I}^{\text{str}}, \ldots, \\ P_{j,I!,1}^{\text{str}}, I_{j,I!,1}^{\text{str}}, D_{j,I!,1}^{\text{str}}, A_{j,I!,1}^{\text{str}}, B_{j,I!,1}^{\text{str}}, \ldots, \\ P_{j,I!,I}^{\text{str}}, I_{j,I!,I}^{\text{str}}, D_{j,I!,I}^{\text{str}}, A_{j,I!,I}^{\text{str}}, B_{j,I!,I}^{\text{str}} \end{bmatrix} = \left[ X_{j,1}^{\text{str}}, \ldots, X_{j,L^{\text{str}}}^{\text{str}} \right], \quad (3)
$$

where each parameter $X_{j,g}^{\text{str}}$, $g = 1, \ldots, L^{\text{str}}$ encodes information about state of the switch ($P^{\text{str}}$, $I^{\text{str}}$ or $D^{\text{str}}$) in the controller structure and about nodes NB parameters

$(A^{\text{str}}, B^{\text{str}})$, $L^{\text{str}} = 5 \cdot I! \cdot I$ stands for amount of parameters of the solution $\mathbf{X}_j^{\text{str}}$. The second part $\mathbf{X}_j^{\text{par}}$ encodes parameters of the controller and it is expressed as follows:

$$\mathbf{X}_j^{\text{par}} = \begin{bmatrix} P_{j,1,1}^{\text{par}}, I_{j,1,1}^{\text{par}}, D_{j,1,1}^{\text{par}}, \dots, \\ P_{j,1,I}^{\text{par}}, I_{j,1,I}^{\text{par}}, D_{j,1,I}^{\text{par}}, \dots \\ P_{j,I!,1}^{\text{par}}, I_{j,I!,1}^{\text{par}}, D_{j,I!,1}^{\text{par}}, \dots, \\ P_{j,I!,I}^{\text{par}}, I_{j,I!,I}^{\text{par}}, D_{j,I!,I}^{\text{par}} \end{bmatrix} = \left[ X_{j,1}^{\text{par}}, \dots, X_{j,L^{\text{par}}}^{\text{par}} \right], \tag{4}$$

where each parameter $X_{j,g}^{\text{par}}$, $g = 1, \dots, L^{\text{par}}$ encodes information about real parameter $K_p$, $T_i$ or $T_d$ of the CB block structure of the controller, $L^{\text{par}} = 3 \cdot I! \cdot I$ stands for amount of parameters of the solution $\mathbf{X}_j^{\text{par}}$.

## 3.2 Initialization of Initial Population

All parameters of the initial population of solutions $\mathbf{X}_j$ are generated randomly. The parameters of the first part of $\mathbf{X}_j^{\text{str}}$ encoding the structure of the controller take the binary values drawn from the set $X_{j,g}^{\text{str}} \in \{0, 1\}$, where index of the parameter $g = 1, \dots, L^{\text{str}}$. The parameters of the second part of $\mathbf{X}_j^{\text{par}}$ encoding parameters of the controller can take real number values and they are randomly generated from ranges selected individually for the problem under consideration (separately for each group of parameters: $P, I, D$).

The example of encoding of controller structure with two feedback signals is shown in Fig. 5a and example with simplified presentation and encoding of controller structure parameters is shown in Fig. 5b.

## 3.3 Evaluation of the Population

In the proposed algorithm the evaluation of all solutions $\mathbf{X}_j$ from the population of algorithm takes an important part. This evaluation is based on properly defined fitness function. Fitness function allows us to evaluate the controller encoded by single solution $\mathbf{X}_j$. This evaluation can not only take into account many criteria but also each criterion can be weighted. Consideration of many criteria requires an appropriate aggregation of them or use of a different approach in the field of multi-criteria optimization (see e.g. [23]). In this paper a modified weighted sum method (WSM) (see e.g. [12]) was used. In this method the fitness function for the solutions $\text{FF}(\mathbf{X}_j)$ is defined as:

$$\text{FF}(\mathbf{X}_j) = \sum_{m=1}^{M} w_m \cdot \left( a_m \cdot \text{ff}_m(\mathbf{X}_j) \right)^2, \tag{5}$$

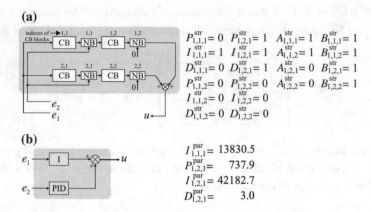

**Fig. 5** Example of controller with two input signals ($I = 2$): **a** controller structure and example parameters encoding this structure, **b** simplified controller structure (according to structure parameters) with only real value parameters presented (this presentation is used in the further part of this paper)

where $ff_m\left(\mathbf{X}_j\right)$ stands for component of the fitness function connected to criterion $m$ ($m = 1, \ldots, M$), $M$ stands for amount of considered criteria, $w_m$ stands for weight correlated to the $m$-th criterion, $a_m$ stands for normalization parameter of the $m$-th criterion. Using normalization parameter (which is not a standard element of WSM) eliminates situation where one of the components of the fitness function $ff_m\left(\mathbf{X}_j\right)$ determines value of the function $FF\left(\mathbf{X}_j\right)$.

## 3.4 Empires Creation

From the solutions $\mathbf{X}_j$ (called also individuals or colonies) created in the initialization process, $Ni$ the best solutions (based on the fitness function value) are used to create empires. The amount of these solutions can be set freely, however in the literature can be found a suggestions to set this number to $Ni = \text{int}\left(\frac{N}{10}\right)$ (int ($\cdot$) is a function which approximates the number to its nearest integer) (see e.g. [3]). Each of the best solutions creates an own empire and becomes an imperialist of it.

Next, the rest of the colonies (in amount of $N - Ni$) are added to the empires. Each empire gets specified amount of colonies chosen randomly from remaining colonies. This amount is based on the power of the empire $P_k$ ($k = 1, \ldots, Ni$) calculated in the following way:

$$Pi_k = \left| \frac{FF\left(\mathbf{Xi}_k\right) - \max\limits_{s=1,\ldots,Ni}\left\{FF\left(\mathbf{Xi}_s\right)\right\}}{\sum\limits_{q=1}^{Ni}\left(FF\left(\mathbf{Xi}_q\right) - \max\limits_{s=1,\ldots,Ni}\left\{FF\left(\mathbf{Xi}_s\right)\right\}\right)} \right|, \tag{6}$$

where numerator and denominator contain normalized value of the fitness function. The power of empire allows us to obtain amount of colonies $Nic_k = \text{int} \left( Ni \cdot Pi_k \right)$. The colonies added into empire will be denoted as $\mathbf{Xic}_{k,r} = \left\{ \mathbf{Xic}_{k,r}^{str}, \mathbf{Xic}_{k,r}^{par} \right\} = \left\{ Xic_{k,r,1}, \ldots, Xic_{k,r,L^{str}+L^{par}} \right\}$ ($k = 1, \ldots, Ni$, $r = 1, \ldots, Nic_k$). The system of empires and their colonies formed in this step will be subject to change as described in Sects. 3.5 and 3.8.

## 3.5 Assimilation of the Colonies

The purpose of making changes in the colonies is to explore the search space of parameters and structure for the considered problem. The purpose of exploitation is to shift colonies closer to the imperialist of their empire (as models with the best value of fitness function). It is made on the basis of the assimilation operator (typical for the imperialist algorithm, see e.g. [3]). The purpose of exploration (global and local exploration) is to make random changes in colonies, which allows to find new, not known solutions. It is made on the basis of the revolution operator (typical for the imperialist algorithm, see e.g. [3]) and mutation operator (typical for the genetic algorithm, see e.g. [30]).

The assimilation operator works on part $\mathbf{Xc}_r^{par}$ of the solutions $\mathbf{Xc}_r$, which encodes real parameters of the controller. It allows colonies to move towards imperialist with using additional small random direction angle. It can be written as follows:

$$Xic_{k,r,g}^{par} := \left( Xi_{k,g}^{par} - Xic_{k,r,g}^{par} \right) \cdot U_r (0,2) \cdot U_g (-\gamma, \gamma), \tag{7}$$

where $U_g (0, 2)$ stands for random number from the range $(0, 2)$ generated for assimilation for each colony $r$, $U_g (-\gamma, \gamma)$ stands for random number from the range $(-\gamma, \gamma)$ generated individually for each gene of each $r$ colony, $\gamma$ is a parameter defining random angle. The assimilation operator allows us to: (1) maintain a strong position of empires, (2) not to introduce such changes in the population of individuals that make impossible to find the optimal solution in terms of the adopted criteria.

## 3.6 Revolution and Mutation

After assimilation, a revolution and mutation of the colonies take place. These operators work only on part of the parameters of solutions. For each parameter a two random numbers are generated from the range $[0, 1]$. If the first number is lower than probability of revolution $p_r$, then the parameter is modified by a revolution operator. If the second number is lower than probability of mutation $p_m$, the parameter is modified by mutation operator. Both probabilities parameters are similar to standard mutation probability operator from the genetic algorithm. The revolution concerns parameters of the controller ($\mathbf{Xc}_r^{\mathrm{par}}$) and the mutation concerns parameters of the structure of the controller ($\mathbf{Xc}_r^{\mathrm{str}}$). Revolution operator effect can be written as follows:

$$Xic_{k,r,g}^{\mathrm{par}} := \underline{Xic}_{k,r,g}^{\mathrm{par}} + \left(\overline{Xic}_{k,r,g}^{\mathrm{par}} - \underline{Xic}_{k,r,g}^{\mathrm{par}}\right) \cdot \mathrm{U}_g\,(0,1), \tag{8}$$

where $\underline{Xic}_{k,r,g}^{\mathrm{par}}$ stands for minimum acceptable value of gene, $\overline{Xic}_{k,r,g}^{\mathrm{par}}$ stands for maximum acceptable value of gene. Values $\underline{Xic}_{k,r,g}^{\mathrm{par}}$ and $\overline{Xic}_{k,r,g}^{\mathrm{par}}$ arise from the specificity of the considered problem.

The mutation concerns parameters of the controller structure. The binary parameters (switches) modified by the mutation parameter are inverted (from 1 to 0 and vice-versa). Since the revolution and mutation interact intensively on colonies, the value of $p_{r/m}$ cannot be too large to not cause degeneration of the population.

## 3.7 Evaluation of the Population

After changes described in Sects. 3.5 and 3.6 all individuals are re-evaluated by fitness function (5). The purpose of this step is to update fitness function values of the individuals before empires competition takes place, which relies on fitness function values of the individuals.

## 3.8 Competition of the Empires

The changes in empires are made in three steps. In the first one each colony $\mathbf{Xi}_k$ in the empire is compared with imperialist of this empire. If the fitness function of the colony is better than fitness function of the imperialist ($\mathbf{Xic}_{k,r}$), the colony takes control over empire and replaces existing imperialist. The second step is based on the imperialist (empires) competition. In this step the weakest empire (taking into account the empire power) losing its weakest colony (taking into account the fitness function of the colonies inside empire). This colony is transferred into empire which won the main competition. The competition is based on the empire power and probability. Total empire power is defined as follows:

$$C_k = \mathrm{FF}\left(\mathbf{Xi}_k\right) + \xi \cdot \frac{\sum_{r=1}^{Nci_k} \mathrm{FF}\left(\mathbf{Xic}_{k,r}\right)}{Nci_k}, \tag{9}$$

where $\xi \in [0, 1]$ stands for importance of the colonies in the empire (it is a static algorithm parameter). The probability of acquisition of the weakest colonies can be calculated using total empire power which is defined as (similarly as in formula (6)):

$$Pic_k = \left| \frac{C_k - \max\limits_{s=1,\dots,Ni} \{C_s\}}{\sum\limits_{q=1}^{Ni} \left( C_q - \max\limits_{s=1,\dots,Ni} \{C_s\} \right)} \right|. \tag{10}$$

The sum of probability of wining competition is equal to 1. For the strongest empire this value is the highest, for the weakest empire this value is equal to 0. A virtual roulette wheel is obtained by allocating each empire on a segment of the wheel, which size is proportional to the probability of wining the competition by the considered empire. Next, a single number is drawn from the unit interval. It indicates the empire placed on the roulette wheel which has won the competition. Thus, the process is analogous to the selection by the roulette wheel used in the genetic algorithm (see e.g. [30]).

## 3.9 Elimination of Empty Empires

The rotation of the colonies between empires allows us to eliminate (step three) the weakest empires. It is realized in such a way that the empires which do not have any colonies are removed. Moreover, rotation of the weakest colonies between empires causes that the strongest empires become weaken. Due to this process, the algorithm is less sensitive to local minima (which is a big advantage).

## 3.10 Stopping Criterion

The last step of the algorithm is based on checking the number of the algorithm iterations. If this number reaches specified value, the best solution is presented and algorithm stops, otherwise the algorithm goes back to the step described in Sect. 3.5.

## 4 Simulations

In this section a simulation problem, simulation method and simulation results are presented.

## 4.1  Simulation Problem

In the simulations a problem of automatic selection of the structure and the structure parameters for quarter car active suspension system [1, 18, 39] was considered. The main idea of this system is shown in Fig. 6. In the controlled object a following stands are used: $m_u$ denotes unsprung mass, $m_s$ denotes sprung mass, $k_t$ denotes tire stiffness, $k_s$ denotes sprung stiffness, $d_s$ denotes sprung damping, $z_r$ denotes road profile, $z_t$ denotes tire compression, $z_u$ denotes displacement of unsprung mass, $z$ denotes suspension travel, $z_s$ denotes displacement of sprung mass. Parameters of active suspension controller were set as follows: $m_u = 48.3$ kg, $m_s = 395.3$ kg, $k_s = 30010$ N/m, $k_t = 340000$ N/m, $d_s = 145$ Ns/m. Controlled object is modelled as follows:

$$\dot{\mathbf{x}} = \mathbf{Ax} + \mathbf{Bu} + \mathbf{f}, \tag{11}$$

where $\mathbf{A}$ is a state matrix in the form:

$$\mathbf{A} = \begin{bmatrix} 0 & 1 & 0 & 0 \\ -\frac{k_s}{m_s} & -\frac{d_s}{m_s} & \frac{k_s}{m_s} & \frac{d_s}{m_s} \\ 0 & 0 & 0 & 1 \\ \frac{k_s}{m_u} & \frac{d_s}{m_s} & -\frac{k_s+k_t}{m_u} & -\frac{d_s}{m_s} \end{bmatrix}, \tag{12}$$

$\mathbf{x}$ is a state vector (initial values of the state vector were set to zero) described as follows:

$$\mathbf{x} = \begin{bmatrix} x_1 & x_2 & x_3 & x_4 \end{bmatrix}^T = \begin{bmatrix} z_s & \dot{z}_s & z_u & \dot{z}_u \end{bmatrix}^T, \tag{13}$$

$\mathbf{B}$ is an input matrix represented by the formula:

$$\mathbf{B} = \begin{bmatrix} 0 & \frac{1}{m_s} & 0 & -\frac{1}{m_u} \end{bmatrix}^T, \tag{14}$$

**Fig. 6**  Active suspension controller ($e_1 = fb_1$, $e_2 = fb_2$, $e_3 = fb_3$)

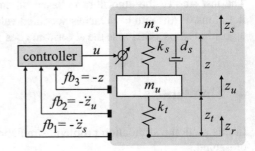

**u** is a vector of output signals obtained from the controller in a size equal to one (**u** = [$u$], see Fig. 6), **f** is an input vector from kinematic extortion described by the following equation:

$$\mathbf{f} = \left[0\ 0\ 0\ -\frac{k_t}{m_u}\right]^T \cdot z_r. \tag{15}$$

For the purposes of simulation, we used the discrete form of Eqs. (11)–(15), which were discretized with time step $T_s$. It is worth to mention that in the practical part of implementation, in the microprocessor system the Eq. (1) is also subject to discretization of step $T_r$ (Table 1).

## 4.2 Simulation Method

In our simulations two cases were considered. In the first case, learning phase of the system (in evolution process) without inclusion of signals drift and in the second case inclusion of signals drift was used. Drift of the signals is a time constant value which should be added to the signals' values. Simulation for both cases was made in a few configurations using additionally included noise of the signals and different road profile to test the system (see Fig. 7 and Table 2). Both the drift and the noise

**Table 1** Parameters of the simulations

| Description | Value |
|---|---|
| Range $\left[\underline{Xic}_{k,r,g}^{par}, \overline{Xic}_{k,r,g}^{par}\right]$ for $P$ ($K_p$) parameters of the CB | [0, 2000] |
| Range $\left[\underline{Xic}_{k,r,g}^{par}, \overline{Xic}_{k,r,g}^{par}\right]$ for $I$ ($1/T_i$) parameters of the CB | [0, 50000] |
| Range $\left[\underline{Xic}_{k,r,g}^{par}, \overline{Xic}_{k,r,g}^{par}\right]$ for $D$ ($T_d$) parameters of the CB | [0, 20] |
| Range, of the control signal $u$ of the controller (see [39]) | [−1000, 1000] |
| Time step of the discretization of the controller in time domain | $T_s = 0.1$ ms |
| Quantization resolution for signals $u, e_i$, $i = 1, \ldots, n$ | 0.0001 |
| Value of optional drift of the input signals $e_1$ and $e_2$ | 0.01 |
| Optional noise range of input signals $e_1$ and $e_2$ | [−0.004, 0.004] |
| Simulation length | $T = 8s$ |
| The number of samples of a single simulation | $Z = \frac{T}{T_s} = 80000$ |
| Interval between subsequent controller activations | $T_r = 5 \cdot T_s = 0.5$ ms |

**Fig. 7** Considered road profiles: **a** trapezoidal, **b** sinusoidal

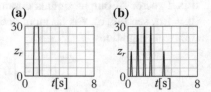

**Table 2** Simulation cases with used road profile (t-stands for trapezoidal shape of road profile, s-stands for sinusoidal shape of road profile-see Fig. 7), signals noise and signals drift

| # | Active system | Learning phase | | Testing phase | | |
|---|---------------|----------------|-------|---------------|-------|-------|
|   |               | Road profile   | Drift | Road profile  | Drift | Noise |
| (a) | No  | t | –   | t | No  | No  |
| (b) | Yes | t | No  | t | No  | No  |
| (c) | Yes | t | No  | t | Yes | Yes |
| (d) | Yes | t | Yes | t | No  | No  |
| (e) | Yes | t | Yes | t | Yes | Yes |
| (f) | No  | t | –   | s | No  | No  |
| (g) | Yes | t | Yes | s | No  | No  |
| (h) | Yes | t | Yes | s | Yes | Yes |

of the signals results from tolerance of the used sensors and should be provided by specifications of the hardware manufacturer. In our simulations a drift signal with value 0.01 and random noise with amplitude $[-0.004, 0.004]$ were used (Table 1).

Both simulation cases start from general structure of the controller shown in Fig. 1. In this structure $I = 3$, therefore, the initial number of CB blocks (Fig. 2) was 18. The variants (simulation configurations) for both simulation cases are shown in Table 2. The variants (b) and (c) relate to tests of structures obtained without taking into account the signals drift in the evolution. The variants (d), (e), (g) and (h) relate to tests of structures obtained taking into account the signals drift in the evolution. The variants (a) and (f) relate to test of the system with open regulation loop (without using a controller) and were taken into account in order to compare the results.

For both simulation cases the same parameters of the simulations (Table 1) and the same following parameters of the algorithm were used: intensity of shifts of the assimilation operator $\gamma = \langle -0.15, 0.15 \rangle$, revolution probability $p_r = 0.25$, mutation probability $p_m = 0.15$, colonies importance factor $\xi = 0.1$, population size $N = 100$, algorithm iteration number was set to 1000.

In the process of selection of the structure and the parameters of the controller we used hybrid genetic-imperialist algorithm, whose detailed description is presented in Sect. 3. The aim of the considered problem was to select structure and parameters of the controller taking into account the following criteria: passenger comfort, car handling, etc. It was realized by properly defined fitness function (5). The components of the fitness function $\mathrm{ff}_m\left(\mathbf{X}_j\right)$, $m = 1, \ldots, 7$ $(M = 7)$ are presented in the Table 3. It

**Table 3** The obtained values of the fitness function and its components (5)

| $m$ | Name | $w_m$ | $a_m$ | $ff_m(\mathbf{X})$ definition | $ff_m(\mathbf{X})$ values for considered simulation variants | | | | | | | |
|---|---|---|---|---|---|---|---|---|---|---|---|---|
| | | | | | (a) | (b) | (c) | (d) | (e) | (f) | (g) | (h) |
| 1 | Passenger comfort | 1.00 | 5 | $\sqrt{\dfrac{1}{Z}\cdot\sum_{z=1}^{Z} z_{s_z}^2}$ | $4.2\times10^{-1}$ | $1.3\times10^{-1}$ | $1.3\times10^{-1}$ | $1.5\times10^{-1}$ | $1.6\times10^{-1}$ | $28.8\times10^{-1}$ | $15.4\times10^{-1}$ | $15.8\times10^{-1}$ |
| 2 | Car handling | 0.25 | 2000 | $\sqrt{\dfrac{1}{Z}\times\sum_{z=1}^{Z} z_{t_z}^2}$ | $6.5\times10^{-4}$ | $4.4\times10^{-4}$ | $4.5\times10^{-4}$ | $4.4\times10^{-4}$ | $4.7\times10^{-4}$ | $33.3\times10^{-4}$ | $21.5\times10^{-4}$ | $21.4\times10^{-4}$ |
| 3 | Suspension max. travel | 0.10 | 20 | $\max_{z=1,\dots Z}\{|z_z|\}$ | $21.9\times10^{-3}$ | $29.1\times10^{-3}$ | $33.8\times10^{-3}$ | $28.1\times10^{-3}$ | $30.5\times10^{-3}$ | $111.4\times10^{-3}$ | $79.9\times10^{-3}$ | $83.5\times10^{-3}$ |
| 4 | Suspension travel | 0.10 | 20 | $\sqrt{\dfrac{1}{Z}\cdot\sum_{z=1}^{Z} z_z^2}$ | $4.8\times10^{-3}$ | $6.2\times10^{-3}$ | $23.3\times10^{-3}$ | $6.7\times10^{-3}$ | $7.2\times10^{-3}$ | $32.4\times10^{-3}$ | $20.6\times10^{-3}$ | $21.5\times10^{-3}$ |
| 5 | Complexity | 0.50 | 2 | $\sum_{g=1}^{L^{str}} X_{j,g}^{str}$ | – | $31.2\times10^{-3}$ | $31.2\times10^{-3}$ | $62.5\times10^{-3}$ | $62.5\times10^{-3}$ | – | $62.5\times10^{-3}$ | $62.5\times10^{-3}$ |
| 6 | Control force | 0.10 | $2\times10^{-3}$ | $\sqrt{\dfrac{1}{Z}\cdot\sum_{z=1}^{Z} u_z^2}$ | – | $19.3\times10^{2}$ | $70.6\times10^{2}$ | $19.9\times10^{2}$ | $21.7\times10^{2}$ | – | $47.3\times10^{2}$ | $49.0\times10^{2}$ |
| 7 | Oscillations of controller | 0.25 | $2\times10^{-4}$ | $\sum_{o=1}^{O-1}\left|u(t_o)-u(t_{o+1})\right|$ | – | $4.0\times10^{3}$ | $3.7\times10^{3}$ | $4.3\times10^{3}$ | $165.1\times10^{3}$ | – | $19.3\times10^{3}$ | $150.0\times10^{3}$ |
| | | | | $FF(\mathbf{X}_j)=$ | $4.77$ | $1.04$ | $1.47$ | $1.21$ | $5.44$ | $2.19\times10^{2}$ | $0.71\times10^{2}$ | $5.16\times10^{2}$ |

**Fig. 8** Minimums and maximums of the output signal $u$ for fitness function component $ff_7(\mathbf{X})$-see Table 3

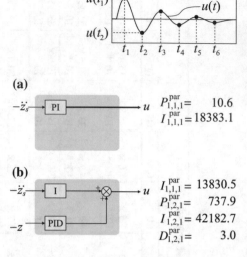

**Fig. 9** Obtained controllers (with simplified presentation of their structures and their parameters) for: **a** case without using drift and noise in the learning phase (variants **b**) and **c** in Table 2), **b** case with using drift and noise of the signals in the learning phase (variants **d**), **e**, **g** i **h** in Table 2)

is worth to mention that the definition of criteria can be very elastic. For example, a criterion applying to oscillations of the control signal points the absolute difference between the values in successive amplitude oscillations until their disappearance (see Fig. 8).

## 4.3 Simulation Results

The conclusions of the simulations can be summarized as follows:

- Controllers obtained from the evolution process are shown in Fig. 9. As might be expected, the structure obtained without taking into account drift of the signals (see Fig. 9a) is less complex than the structure obtained with drift of the signals (see Fig. 9b). In such structure, signals from accelerator sensors were needed to obtain satisfactory quality of the controller. Direct measure of the suspension travel was also not necessary (a similar proposal was formulated in [43]).
- The structure obtained without taking into account drift of the signals works well only in conditions similar to the ones from the learning phase (see Fig. 10b), which are ideal. At the same time, this structure cannot cope with the tests, which take into account drift of the signals (in real conditions) (see Fig. 10c). In particular, the control signal $u$ and signal $z$ (suspension travel) reached the limit value other than 0. The reason for this behavior is the reaction of the controller on the integral component to the presence of a signal drift. The obtained results disqualified the first controller for its practical use. The results obtained with the use of open loop (without using a controller) are shown in Fig. 10a for comparison.

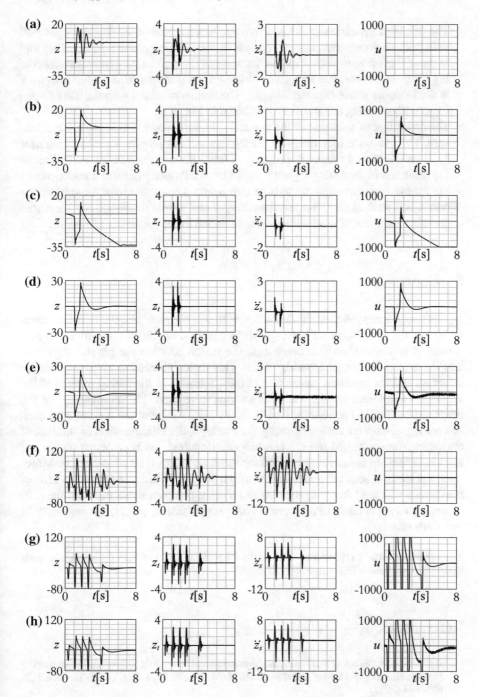

**Fig. 10** Signals obtained in the simulation for cases **a–h** considered in Table 2

- The obtained structure which takes into account the drift of the signals copes well with both environments: similar to those in the learning phase (see Fig. 10d) and in the real conditions with drift and the noise of the signals taken into account (see Fig. 10e). Moreover, this structure works well on different road profile than used in the learning phase (see Fig. 10g, h). The results obtained with the use of open loop (without using a controller) are shown in Fig. 10f.
- The procedure of selection of the controller initially takes into account three sensors. The evolutionarily obtained controllers did not require signal from sensor $\ddot{z}_u$. It was achieved thanks to properly defined fitness function for evaluation of the solutions in the tuning algorithm population. Properly defined components of the fitness function allow us to take into account relevant characteristics of the controller, such as, e.g. the reaction time of the sensor to change the measured quantity, the price of used components included in the controller and so on.

## 5 Conclusions

In this paper a new method for designing both the structure and the structure parameters of the controllers with using hybrid genetic-imperialist algorithm was presented. It is a new algorithm created on the fusion between the genetic algorithm and the imperialist competitive algorithm. This fusion allowed us to obtain both the structure and the structure parameters of the controller. In the tuning process of the algorithm a different criteria and their weights were used. The disadvantage of the proposed method is the need for a suitably accurate model of the controlled object. A very significant advantage of using the model is the minimization of the risk of damaging the controlled object. The proposed method has been tested on the car active suspension system problem and the results confirmed its effectiveness. Moreover, the method selected a simpler controller for the problem in which the drift of the signals was not taken into account and a more complex structure when the drift of the signals was taken into account. This proves, among others, the flexibility of our method.

**Acknowledgements** The project was financed by the National Science Centre (Poland) on the basis of the decision number DEC-2012/05/B/ST7/02138.

## References

1. Agharkakli, A., Sabet, G.S., Barouz, A.: Simulation and analysis of passive and active suspension system using quarter car model for different road profile. Int. J. Eng. Trends Technol. **3**(5), 636–644 (2012)
2. Ali, S.R., Aldair, A.A., Almousawi, A.K.: Design an optimal PID controller using artificial bee colony and genetic algorithm for autonomous mobile robot. Int. J. Comput. Appl. **100**(1), 6 (2014)

3. Atashpaz-Gargari, E., Lucas, C.: Imperialist competitive algorithm: an algorithm for optimization inspired by imperialistic competition. IEEE Congr. Evolutionary Comput. **7**(4661), 4666 (2007)
4. Bas, E.: The training of multiplicative neuron model based artificial neural networks with differential evolution algorithm for forecasting. J. Artif. Intell. Soft Comput. Res. **6**, 5–12 (2016)
5. Binitha, S., Sathya, S.S.: A survey of bio-inspired optimization algorithms. Int. J. Soft Comput. Eng. (IJSCE) **2**(2), 137–151 (2012)
6. Boiko, I.: Variable-structure PID controller for level process. Control Eng. Pract. **21**(5), 700–707 (2013)
7. Cpałka, K.: A Method for Designing Flexible Neuro-fuzzy systems. Lecture Notes in Artificial Intelligence, Springer **4029**, 212–219 (2006)
8. Cpałka, K.: Design of Interpretable Fuzzy Systems. Springer (2017)
9. Cpałka, K., Rebrova, O., Nowicki, R., Rutkowski, L.: On design of flexible neuro-fuzzy systems for nonlinear modelling. Int. J. Gen. Syst. **42**(6), 706–720 (2013)
10. Cpałka, K., Rutkowski, L.: Flexible Takagi-Sugeno. Fuzzy systems. In: Neural Networks, Proceedings of the 2005 IEEE International Joint Conference on IJCNN '05, vol. 3, pp. 1764–1769 (2005)
11. Duda, P., Jaworski, M., Pietruczuk, L.: On pre-processing algorithms for data stream. International Conference on Artificial Intelligence and Soft Computing. Lecture Notes in Artificial Intelligence, vol. 7268, pp. 56–63. Springer (2012)
12. Eckenrode, R.T.: Weighting multiple criteria. Manag. Sci. **12**, 19–180 (1965)
13. Gabryel, M., Cpałka, K., Rutkowski, L.: Evolutionary strategies for learning of neuro-fuzzy systems. In: Proceedings of the I Workshop on Genetic Fuzzy Systems, Granada, pp. 119–123 (2005)
14. Gaweda, A.E., Scherer, R.: Fuzzy number-based hierarchical fuzzy system. ICAIS, pp. 302–307 (2004)
15. Ghorbani, R., Wu, Q., Wang, G.G.: Nearly optimal neural network stabilization of bipedal standing using genetic algorithm. Eng. Appl. Artif. Intell. **20**, 473–480 (2007)
16. Jaworski, M., Pietruczuk, L., Duda, P.: On resources optimization in fuzzy clustering of data streams. In: International Conference on Artificial Intelligence and Soft Computing. Lecture Notes in Artificial Intelligence, vol. 7268, pp. 92–99. Springer (2012)
17. Leva, A., Papadopoulos, A.V.: Tuning of event-based industrial controllers with simple stability guarantees. J. Process Control **23**, 1251–1260 (2013)
18. Lin, J., Lian, R.: Intelligent control of active suspension systems. IEEE Trans. Ind. Electron. **58**(2), 618–628 (2010)
19. Łapa, K., Cpałka, K., Wang, L.: New method for design of fuzzy systems for nonlinear modelling using different criteria of interpretability. Lect. Notes Comput. Sci. **8467**, 217–232 (2014)
20. Łapa, K., Szczypta, J., Venkatesan, R.: Aspects of structure and parameters selection of control systems using selected multi-population algorithms. Lect. Notes Comput. Sci. **9120**, 247–260 (2015)
21. Maggio, M., Bonvini, M., Leva, A.: The PID+p controller structure and its contextual autotuning. J. Process Control **22**, 1237–1245 (2012)
22. Malhotra, R., Sodh, R.: Boiler flow control using PID and fuzzy logic controller. IJCSET **1**(6), 315–31 (2011)
23. Marler, R.T., Arora, J.S.: Survey of multi-objective optimization methods for engineering. Struct. Multidiscip. Optim. **26**, 369–395 (2004)
24. Marwala, T.: Control of complex systems using Bayesian networks and genetic algorithm. IJES **5**, 28–37 (2004)
25. Perng, J.-W., Chen, G.-Y., Hsieh, S.-C.: Optimal PID controller design based on PSO-RBFNN for wind turbine systems. Energies **7**, 191–209 (2014)
26. Pietruczuk, L., Rutkowski, L., Jaworski, M., Duda, P.: How to adjust an ensemble size in stream data mining. Inf. Sci. **381**, 46–54 (2017)

27. Przybył, A., Łapa, K., Szczypta, J., Wang, L.: The method of evolutionary designing the elastic controller structure. Lect. Notes Comput. Sci. **9692**, 476–492 (2016)
28. Rasoanarivo, I., Brechet, S., Battiston, A., Nahid-Mobarakeh, B.: Behavioral analysis of a boost converter with high performance source filter and a fractional-order PID controller. In: IEEE Industry Applications Society Annual Meeting (IAS), pp. 1–6 (2012)
29. Ribića, A.I., Mataušek, M.R.: A dead-time compensating PID controller structure and robust tuning. J. Process Control **22**, 1340–1349 (2012)
30. Rutkowski, L.: Computational Intelligence. Springer (2007)
31. Rutkowski, L., Cpałka, K.: A general approach to neuro-fuzzy systems. In: The 10th IEEE International Conference on Fuzzy Systems, 2001, Melbourne, pp. 1428–1431 (2001)
32. Rutkowski, L., Cpałka, K.: A neuro-fuzzy controller with a compromise fuzzy reasoning. Control Cybern. **31**(2), 297–308 (2002)
33. Rutkowski, L., Cpałka, K.: Compromise approach to neuro-fuzzy systems. In: 2nd Euro-International Symposium on Computation Intelligence Location: KOSICE, SLOVAKIA Date: 16–19 Jume 2002, vol. 76, pp. 85–90 (2002)
34. Rutkowski, L., Cpałka, K.: Flexible weighted neuro-fuzzy systems. In: Proceedings of the 9th International Conference on Neural Information Processing (ICONIP'02), Orchid Country Club, Singapore, vol. 4, pp. 1857–1861 (2002)
35. Rutkowski, L., Cpałka, K.: Neuro-fuzzy systems derived from quasi-triangular norms. In: Proceedigns of the IEEE International Conference on Fuzzy Systems, Budapest, July 26–29, vol. 2, pp. 1031–1036 (2004)
36. Rutkowski, L., Przybył, A., Cpałka, K.: Novel online speed profile generation for industrial machine tool based on flexible neuro-fuzzy approximation. IEEE Trans. Ind. Electron. **59**(2), 1238–1247 (2012)
37. Rutkowski, L., Przybył, A., Cpałka, K., Er, M.J.: Online speed profile generation for industrial machine tool based on neuro-fuzzy approach. Lect. Notes Artif. Intell. **114**, 645–650 (2010)
38. Saad, M.S., Jamaluddin, H., Sodh, I.Z.M.: Implementation of PID controller tuning using differential evolution and genetic algorithms. Int. J. Innov. Comput. Inf. Control **8**(11), 7761–7779 (2012)
39. Sande, T.P.J., Gysen, B.L.J., Besselink, I.J.M., Paulides, J.J.H., Lomonova, E.A., Nijmeijer, H.: Robust control of an electromagnetic active suspension system: simulations and measurements. Mechatronics **23**, 2 (2013)
40. Stone, C., Chi-Wei, L.: Fuzzy PDFF-IIR controller for PMSM drive systems. Control Eng. Pract. **19**, 828–835 (2011)
41. Szczypta, J., Łapa, K., Shao, Z.: Aspects of the selection of the structure and parameters of controllers using selected population based algorithms. Lect. Notes Comput. Sci. **8467**, 440–454 (2014)
42. Teng, T.H., Tan, A.H., Żurada, J.M.: Self-Organizing neural networks integrating domain knowledge and reinforcement learning. IEEE Trans. Neural Netw. Learn. Syst. **26**(5), 889–902 (2015)
43. Van de Wal, M., Philips, P., De Jager, B.: Actuator and sensor selection for an active vehicle suspension aimed at robust performance. Int. J. Control **70**(5), 703–720 (1998)
44. Yazdani, A.M., Ahmadi, A., Buyamin, S., Rahmat, M.F., Davoudifar, F., Rahim, H.A.: Imperialist competitive algorithm-based fuzzy PID control methodology for speed tracking enhancement of stepper motor. Int. J. Smart Sens. Intell. Syst. **5**, 3 (2012)
45. Zalasiński, M.: New algorithm for on-line signature verification using characteristic global features. Adv. Intell. Syst. Comput. **432**, 137–146 (2016)
46. Zalasiński, M., Cpałka, K.: New algorithm for on-line signature verification using characteristic hybrid partitions. Adv. Intell. Syst. Comput. **432**, 147–157 (2016)
47. Zalasiński, M., Cpałka, K., Hayashi, Y.: A new approach to the dynamic signature verification aimed at minimizing the number of global features. Lect. Notes Comput. Sci. **9693**, 218–231 (2016)
48. Zalasiński, M., Cpałka, K., Rakus-Andersson, E.: An Idea of the dynamic signature verification based on a hybrid approach. Lect. Notes Comput. Sci. **9693**, 232–246 (2016)

49. Żurada, J.M., Jedruch, W., Barski, M.: Neural Networks. Polish Scientific Publishers, Warsaw, Poland (1996)
50. Żurada, J.M.: Introduction to Artificial Neural Systems. Jaico Publishing House (2005)

. . . . . . . . . . . . . . . . . . . . . . . . . . . . . . . . . . . . . . . . . . . . . . . . . . . . . . . . . . . . . . . . . . . .

. . . . . . . . . . . . . . . . . . . . . . . . . . . . . . . . . . . . . . . . . . . . . . . . . . . . . . . . . . . . . . . . . . . .

# Fuzzy-Genetic Approach to Identity Verification Using a Handwritten Signature

**Marcin Zalasiński, Krzysztof Cpałka and Leszek Rutkowski**

**Abstract** Verification of the dynamic signature is an important issue of biometrics. There are many methods for the signature verification using dynamics of the signing process. Many of these methods are based on the so-called global features. In this paper we propose a new approach to the signature verification using global features. The proposed approach can be characterized as follows: (a) Classification of the signature is performed using a fuzzy-genetic system. (b) We select an individual set of features for each signer. (c) In the procedure of features selection we use a genetic algorithm with appropriately designed evaluation function. It works without access to the signatures called skilled forgeries (this is a major advantage of the proposed approach). (d) We determine weights of importance for evolutionarily selected features. (e) The weights are taken into account in the classification process. (f) An additional advantage of the proposed classifier is the possibility of its work interpretation and possibility of an analytical determination of its parameters without machine learning. In this paper we present the simulation results for the BioSecure signature database, distributed by the BioSecure Association.

## 1 Introduction

Signature is a biometric characteristic (see e.g. [13, 17, 83–87]) which is easy to acquire and socially acceptable, so it is often used to develop effective systems for identity verification. In the literature there are two main types of the signatures. The

M. Zalasiński · K. Cpałka (✉) · L. Rutkowski
Institute of Computational Intelligence, Czestochowa University of Technology,
Al. Armii Krajowej 36, 42-200 Czestochowa, Poland
e-mail: krzysztof.cpalka@iisi.pcz.pl

M. Zalasiński
e-mail: marcin.zalasinski@iisi.pcz.pl

L. Rutkowski
Information Technology Institute, Academy of Social Sciences,
Ul. Sienkiewicza 9, 90-113 łódź, Poland
e-mail: leszek.rutkowski@iisi.pcz.pl

© Springer International Publishing AG 2018
A.E. Gawęda et al. (eds.), *Advances in Data Analysis with Computational Intelligence Methods*, Studies in Computational Intelligence 738,
https://doi.org/10.1007/978-3-319-67946-4_17

first is called static signature (off-line). Analysis of this type of signature is based on its geometric features, such as shape, size ratios, etc. (see e.g. [3, 4, 43]). The second is called dynamic signature (on-line) and it contains information about dynamics of the signing process. The most commonly used signals, which are the basis of the dynamic signature analysis, include a signal of pen pressure on the tablet surface and a signal of pen velocity. The second one is determined indirectly on the basis of the signals describing a position of the pen on the tablet surface. There are also other types of available signals, but the method of their processing is analogous. Dynamic signature verification is much more effective than a static signature verification because: (a) dynamics of the signature is very individual characteristic of the signer, (b) it is difficult to forge, (c) waveforms describing the dynamics of the signature are difficult to translate into the process of signing, (d) waveforms describing the dynamics of the signature can be easily analyzed.

## 1.1 Approaches to the Dynamic Signature Analysis Proposed in the Literature

In the literature four main approaches to the analysis of the dynamic signature have been presented: **(a) global feature based approach** (see e.g. [28, 46, 53–55, 82, 88]), **(b) function based approach** (see e.g. [24, 36, 40, 42, 49, 56]), **(c) regional based approach** (see e.g. [11, 12, 25, 27, 34, 41, 61, 65]), **(d) hybrid approach** (see e.g. [16, 52, 57]). It should also be emphasized that the algorithms for analysis of the dynamic signature can be relatively easily used in other areas of biometric applications, which are based on the analysis of dynamic behavior (see e.g. [15, 21]). Among the four mentioned approaches to analyze the dynamic signature, the methods using global features deserve special attention (see e.g. [28, 44, 58, 59]). The literature in this field contains, among others, definitions of the global features, description of the features selection and classification algorithms based on the features. We encourage you to read the more detailed review of the literature on the dynamic signature verification, which has been presented in our previous papers (see e.g. [11, 12]).

## 1.2 Our Approach to the Dynamic Signature Analysis

In this paper we propose a new method for the dynamic signature verification based on global features, which stands out from the methods of other authors by the following characteristics:

- It uses a genetic algorithm (see e.g. [1, 22, 50, 62, 67, 75, 79, 81]) for the individual selection of the features (for each signer), which among others eliminates the features decreasing the accuracy of the verification (we use our previous experiences on evolutionary algorithms, see e.g. [6]). Genetic algorithm belongs to the computational intelligence methods (see e.g. [19, 20, 23, 38, 39, 63, 64]) In the papers of other authors different methods for the global features selection have been described, but the selection has not been realized individually for each user (see e.g. [28, 53, 54]). The method proposed in this paper realizes this type of selection.

- It determines individually for each signer weights of importance of the features and takes them into account in the process of the signature verification (we use the triangular norms with the weights of arguments, proposed by us earlier, see e.g. [71]). In the papers of other authors different methods for the determination of weights have been described, but it has not been realized individually for each user (see e.g. [28, 46, 55]).
- It takes advantage of the fuzzy set theory and fuzzy systems in the process of the signature verification (we use our previous experiences in the field of the flexible fuzzy systems, see e.g. [9, 10, 48, 68–70, 72–74]). In this paper we propose a new way to use that system to the dynamic signature verification and a new method of its parameters selection. This method allows to avoid the so-called iterative machine learning (see e.g. [90]), which we used in our previous papers (these papers are not related to the dynamic signature verification, but they concerned different structures of the system, applications and methods of automatic selection of the structure and parameters). In the papers of other authors in the field of the dynamic signature verification we have not found this solution.
- It allows to interpret the knowledge accumulated in the system used to the signature verification (we use our previous experiences in the field of interpretability of knowledge of fuzzy systems, see e.g. [5, 7, 8, 47, 48, 66, 76, 78, 89]). In the papers of other authors different methods for the dynamic signature verification have been described, but they were mainly focused on speed and accuracy. The algorithm proposed in this paper works in such a way that the processing method of the signatures and determination of the signatures descriptors (based on the values of global features) could be easily interpreted. This is an important advantage of the algorithm.
- It does not require so-called skilled forgeries and reference signatures of other signers in the training phase (this is a big advantage in the considered group of methods). This is a consequence of properly designed evaluation function in used genetic algorithm. Some methods proposed by other authors requires reference signatures of other users or false signatures (so-called skilled forgeries) in the learning phase. This causes that the accuracy of the algorithm depends on the number of users stored in the database and the effectiveness of the so-called skilled forgers (false signatures created by them are available in popular databases of the signatures, which are used to compare efficiency of the verification methods). Moreover, it causes problems during practical implementation. The proposed method does not depend on the number of users in the database. It uses false signatures only in the testing phase. This is achieved through appropriately structured flexible fuzzy system, which is the one-class classifier.

- It is distinguished by the independence of the used set of features which can be arbitrarily reduced or expanded. In other words, the proposed algorithm is flexible because it is not sensitive to the selection of the initial set of features. Methods of other authors are often highly dependent on the used set of features.

In the simulations we have used paid signature database BioSecure, distributed by the BioSecure Association (see [32]).

This paper is organized into four sections. In Sect. 2 we present description of the proposed algorithm for the signature verification based on global features. In Sect. 3 simulation results are presented. Conclusions are drawn in Sect. 4.

## 2  Description of the Fuzzy-Genetic Approach for Signature Verification

The proposed method consists of two phases: learning (training on the basis of the reference signatures) and testing (verification of the test signature). In the first phase the selection of features is performed individually for each signer, descriptors of features and weights of importance of features are determined. They are needed for a proper work of the classifier in the test phase. These parameters are stored in a database. In the second phase parameters stored for each signer in the learning phase are downloaded from the database. Next, verification of signatures is realized on the basis of these parameters. In the remainder of this section, learning procedure (Sect. 2.1) and signature verification procedure (Sect. 2.2) have been described.

### 2.1  Description of the Learning Phase

This section describes steps of the algorithm executed in the learning phase.

**Step 1** The learning phase starts by acquiring $J$ reference signatures of the signer $i$. Different types of tablets may have a different sampling frequency thus acquired signatures should be normalized. In the normalization procedure for each user the most typical reference signature, called base signature, is selected. It is one of the reference signatures collected in the acquisition phase, for which a distance to the other reference signatures is the smallest. The distance is calculated according to the adopted distance measure (e.g. Euclidean). Training or testing signatures are matched to the base signature using the Dynamic Time Warping algorithm (see e.g. [2, 26, 77]), which operates on the basis of matching velocity and pressure signals. The result of matching of two signatures is a map of their corresponding points. On the basis of the map, trajectories of the signatures are matched. Matching using DTW could not be done directly with the use of trajectories, because this would remove the differences between the shapes of the signatures. It would have a very negative impact on training. Elimination of differences in rotation of signatures is performed by the PCA algorithm which in the literature is commonly used to make the images rotation invariant (see e.g. [31]). A more detailed description of the normalization techniques can be found in the literature (see e.g. [35, 45, 60]).

**Step 2** In this step of the algorithm, the matrix $\mathbf{G}_i$ is determined. The matrix contains values of all global features which describe the dynamics of the reference signatures of the signer $i$. It has the following structure:

$$\mathbf{G}_i = \begin{bmatrix} g_{i,1,1} & \cdots & g_{i,N,1} \\ \vdots & & \vdots \\ g_{i,1,J} & \cdots & g_{i,N,J} \end{bmatrix}, \tag{1}$$

where $I$ is a number of the signers, $J$ is a number of the signatures created by the signer in the acquisition phase, $N$ is a number of used global features, and $g_{i,n,j}$ is a value of the global feature $n, n = 1, \ldots, N$, determined for the signature $j, j = 1, \ldots, J$, created by the signer $i, i = 1, \ldots, I$. Method of determining values of 85 global features used by us in the simulations has been described in detail in [28] and it will not be considered in this paper.

**Step 3** In this step of the algorithm, the vector $\bar{\mathbf{g}}_i = \left[ \bar{g}_{i,1}, \ldots, \bar{g}_{i,N} \right]$ is determined, where $\bar{g}_{i,n}$ is an average value of $n$-th global feature of all $J$ reference signatures of the signer $i$:

$$\bar{g}_{i,n} = \frac{1}{J} \cdot \sum_{j=1}^{J} g_{i,n,j}. \tag{2}$$

**Step 4** In this step of the algorithm, evolutionary selection of subset of global features takes place. The subset contains features which are the most characteristic for the signer $i$ (procedure `Evolutionary Features Selection`$(\mathbf{G}_i, \bar{\mathbf{g}}_i)$). Evolutionary algorithm is a method modelled on natural evolution for solving problems, mainly optimization ones. It is the search procedure based on the mechanisms of natural selection and inheritance. It uses the evolutionary principle of survival of the fittest individuals. Evolutionary algorithms differ from traditional optimization methods, among others, in that: (a) they do not process the task parameters directly, but their encoded form, (b) they start a search not from a single point, but from the population of points, (c) they use only the objective function, not its derivatives, (d) they use probabilistic rather than deterministic selection rules. As a result, they have the advantage over other optimization techniques, for example analytical methods, random methods, etc. (see e.g. [67]). Procedure `Evolutionary Features Selection` $(\mathbf{G}_i, \bar{\mathbf{g}}_i)$ randomly generates an initial set of so-called chromosomes, which form a population of abundance $Ch$. Each of them specifies other subset of features. The chromosome is denoted as the vector $\mathbf{x}_{i,ch} = \left[ x_{i,ch,1}, \ldots, x_{i,ch,N} \right]$, where $x_{i,ch,g} \in \{0, 1\}$ indicates whether feature $g$ ($g = 1, \ldots, N$) encoded in the chromosome $ch$ ($ch = 1, \ldots, Ch$) will be used to verify the signature of the signer $i$ (1-it will be used, 0-it will not be used). Next, the evaluation of the chromosomes adaptation is performed and operators of crossing and mutation are applied to the chromosomes.

These genetic operators provide exploitation and exploration of the searching space of the features. This action is repeated within the next steps, so-called generations (number of generations is a parameter of the algorithm). Thanks to the use of genetic operators, chromosomes in each subsequent generation have got a better value of the evaluation function (a way of its determination is given in the Sect. 2.1.2). This means that encoded subset of features is becoming more characteristic for the considered signer $i$. From the population of chromosomes, in the latest generation chromosome with the smallest value of the evaluation function is selected (the best for minimization function). The selected chromosome encodes an evolutionarily selected subset of features. It is rewritten to the vector $\mathbf{x}'_i$.

**Step 5** In this step of the algorithm, determination of the reduced matrix of global features $\mathbf{G}'_i$ and reduced vector $\bar{\mathbf{g}}'_i$ of average values of global features is performed. They are created taking into account the vector $\mathbf{x}'_i$, therefore they contain only information about those features which have been evolutionarily selected for the signer $i$. A number of columns of the vector $\bar{\mathbf{g}}'_i$ and the matrix $\mathbf{G}'_i$ is $N'_i$, where $N'_i \leq N$ is a number of features selected for the signer $i$.

**Step 6** In this step of the algorithm, calculation of the classifier parameters used in the test phase is performed. This procedure is called `Classifier Determination` $(i, \mathbf{x}'_i, \mathbf{G}'_i, \bar{\mathbf{g}}'_i)$ and it has been described in the Sect. 2.1.3. In particular, distances $maxd_{i,n}$ and weights $w_{i,n}$ $(i = 1, \ldots, I, n = 1, \ldots, N'_i)$ are determined individually for the signer $i$. Each parameter $maxd_{i,n}$ determines instability of signing of the signer $i$ in the context of the feature $n$. Its value is dependent on the variability of the feature. Each weight $w_{i,n}$ describes importance of the global feature $n$.

**Step 7** In the last step of the algorithm, the following information about the signer $i$ are stored into a database: the vector $\mathbf{x}'_i$, the vector $\bar{\mathbf{g}}'_i$, and parameters of the classifier $maxd_{i,n}$ and $w_{i,n}$. Training phase for the signer $i$: (a) proceeds similarly to all signers, but for each signer regardless, (b) in practice is performed once for each signer.

### 2.1.1 Evolutionary Features Selection

A purpose of the procedure `Evolutionary Features Selection` $(\mathbf{G}_i, \bar{\mathbf{g}}_i)$ is the choice of such a subset of features whose values determined for the reference signatures of the signer $i$ are similar to each other. This is not an easy task, because e.g. for 85 features (the number of features which we used in the simulations) the number of combinations is over $38 \times 10^{24}$ (exactly it is $\sum_{n=1}^{N} N! / (n! \cdot (N - n)!)$). It is expected that the evolutionary algorithm finds a subset of the features close to the optimum in acceptable time. Considered procedure works according to the algorithm shown in Fig. 1. At the beginning, random initialization of the vectors $\mathbf{x}_{i,ch}$ takes place. The vectors are interpreted as chromosomes in the population encoding subsets of the features. Next, evaluation of chromosomes by determining the values of their adaptation function is performed (see Sect. 2.1.2). Having the values of

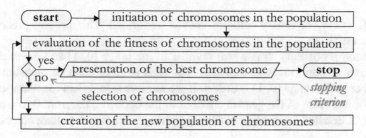

**Fig. 1** Scheme of the procedure Evolutionary Features Selection ($\mathbf{G}_i$, $\bar{\mathbf{g}}_i$)), consistent with the scheme of the genetic algorithm

the adaptation function the stop condition of the algorithm is checked. It takes into account the achievement of the threshold value by the function or execution by the algorithm a certain number of generations. If the stop condition is satisfied, then the evolutionary feature selection procedure terminates and returns the best chromosome from the population. It rarely takes place immediately after the initialization of the population, so the population must be processed in a process of evolution. Its first step is a draw of the individuals in order to apply genetic operators to them. A typical method of individuals selection is e.g. the tournament selection (see e.g. [51, 67]). In this method a few chromes are drawn from the entire population. These chromosomes create so-called tournament group and the chromosome having the best fitness function value is selected from them. Then, another tournament group is created and one chromosome from it is selected. This process is repeated until a new population is created. Next, pairs of chromosomes exchange genes (crossing is applied) at random points and finally some randomly selected genes of the chromosomes mutate (their value changes from 0 to 1 or vice versa). The algorithm takes into account a probability of crossover and mutation, which are its parameters. In this way, the parent population form descendant population, which again is evaluated and the process is repeated.

Operation of the procedure Evolutionary Features Selection ($\mathbf{G}_i$, $\bar{\mathbf{g}}_i$) is dependent on the following parameters:

- Size of the population (number of chromosomes). It specifies the number of features subsets processed in a single step of the algorithm (so-called single generation).
- Number of generations. It specifies the maximum number of steps $S$ in the evolutionary feature selection algorithm for a single user.
- Crossover probability. It is a real number in the range [0, 1] and determines the intensity of the crossing (gene exchange) between chromosomes. For each randomly selected pair of chromosomes selected in the tournament method, a real number in the range [0, 1] is drawn. If the number is less than the crossover probability, an exchange of genes between the chromosomes is performed. Moreover, the number of the crossing points is also associated with this operation. At these

points a "cut" of binary chromosomes is performed. This process precedes the genes exchange.
- Mutation probability. It is a real number in the range [0, 1] and determines the intensity of chromosomes mutation. For each gene of each chromosome a real number in the range [0, 1] is drawn. If the number is less than the mutation probability, the value of the gene is changed to the opposite, i.e. from 0 to 1 and vice versa. A detailed description of the algorithm can be found, among others, in [51, 67].

We would like also to emphasize that the originality of the proposed approach results from a specific way of determining the evaluation function of chromosomes from the population (Calculate Ff $(\mathbf{G}_i, \bar{\mathbf{g}}_i, \mathbf{x}_{i,ch})$). Evaluation of the chromosomes is based on the similarity of features for the reference signatures created in the training phase (described in Sect. 2.2).

### 2.1.2 Determination of Fitness Function

In the determination of the fitness function of the chromosome, the following parameters are taken into account:

- $\mathbf{G}_i$—a matrix of all global features values, determined for all reference signatures of the signer $i$,
- $\bar{\mathbf{g}}_i$—a vector of average values of global features, averaged in the context of all reference signatures of the signer $i$,
- $\mathbf{x}_{i,ch}$—a chromosome with index $ch$ in the population associated with the signer $i$, for which the value of the evaluation function is calculated. In the considered procedure (and only in this procedure) will be used reduced versions of the mentioned parameters: $N^*$, $\mathbf{G}^* = \left[ \mathbf{g}^*_{j=1}, \ldots, \mathbf{g}^*_{j=J} \right]$, and $\bar{\mathbf{g}}^*$. They were created on the basis of the values of the vector $\mathbf{x}_{i,ch}$ in the same way as previously described parameters: $N'_i$, $\mathbf{G}'_i$, and $\bar{\mathbf{g}}'_i$ (on the basis of the vector $\mathbf{x}'_i$).

Considered method Calculate Ff $(\mathbf{G}_i, \bar{\mathbf{g}}_i, \mathbf{x}_{i,ch})$ starts by determination of the covariance matrix for the matrix of all global features (**Step 1**). Covariance cov $(\mathbf{G}^*)$ is a measure of the linear correlation between global features values of the reference signatures $\mathbf{G}^*$ of the signer $i$ (created in the acquisition phase). In the **Step 2** of the algorithm, determination of the vector of Mahalanobis distances (see e.g. [14]) $\mathbf{m}$ is performed. It contains distances between the vector of average values of the global features $\bar{\mathbf{g}}^*$ and the matrix of the global features values $\mathbf{G}^*$ represented by the vectors $\mathbf{g}^*_j, j = 1, \ldots, J$:

$$m_j = \sqrt{ \left( \mathbf{g}^*_j - \bar{\mathbf{g}}^* \right) \left( \text{cov}(\mathbf{G}^*) \right)^{-1} \left( \mathbf{g}^*_j - \bar{\mathbf{g}}^* \right)^T }. \tag{3}$$

Mahalanobis distance well defines the similarity of the selected features vector of the reference signature $j$ (features indicated by the tested chromosome) $\mathbf{g}^*_j$ to the vector of average values of these features $\bar{\mathbf{g}}^*$. It takes into account their mutual correlation and individual variance (expressed by the arithmetic mean of the squared deviations from the arithmetic mean). It should be noted that for each subset of features $J$ distances are determined. The subset of features associated with the lowest distance is the most valuable for the signer $i$ in the training phase. In the last step of the algorithm (**Step 3**), determination of the evaluation function of the chromosome $\mathbf{x}_{i,ch}$ is performed:

$$\mathrm{ff}\left(\mathbf{x}_{i,ch}\right) = \frac{1}{J} \cdot \sum_{j=1}^{J} m_j. \tag{4}$$

Lower value of the fitness function $\mathrm{ff}\left(\mathbf{x}_{i,ch}\right)$ means that the chromosome $\mathbf{x}_{i,ch}$ is "better" (subset of global features encoded in the chromosome $\mathbf{x}_{i,ch}$ is the most characteristic for the signer $i$).

### 2.1.3 Determination of the Classifier Parameters

In the procedure described in this section only individually selected (for the signer $i$) dynamic signature features are considered (there are $N'_i$ features). It means that in determination of the classifier parameters only the matrix $\mathbf{G}'_i$ and the vector $\bar{\mathbf{g}}'_i$ are taken into account.

Procedure `Classifier Determination` $(i, \mathbf{x}'_i, \mathbf{G}'_i, \bar{\mathbf{g}}'_i)$ starts by determination of Euclidean distances $d_{i,n,j}$ between each global feature $n$ encoded in the chromosome $\mathbf{x}'_i$ and average value of the global feature for all $J$ signatures of the signer $i$ (**Step 1**):

$$d_{i,n,j} = \left| \bar{g}_{i,n} - g_{i,n,j} \right|. \tag{5}$$

In the **Step 2** of the considered procedure, selection of maximum distance for each global feature $n$ is performed (from distances determined in the Step 1):

$$maxd_{i,n} = \max_{j=1,\dots,J} \left\{ d_{i,n,j} \right\}. \tag{6}$$

If reference signatures are more similar to each other, the tolerance of our classifier is lower, because $maxd_{i,n}$ takes smaller values. In the **Step 3** of the considered procedure, computation of weights $w_{i,n}$ is performed. Each weight is calculated on the basis of standard deviation of $n$-th global feature of the signer $i$ and average value of distances for $n$-th feature of the signer $i$:

$$w_{i,n} = 1 - \frac{\sqrt{\frac{1}{J} \cdot \sum_{j=1}^{J} d_{i,n,j}^2}}{\frac{1}{J} \cdot \sum_{j=1}^{J} d_{i,n,j}}. \tag{7}$$

It should be emphasized that the distances and the weights are used in the classification process of the signature.

## 2.2 Description of the Signatures Verification Phase

The purpose of the signatures verification phase is to determine whether the tested signature, which belongs to a signer claiming to be the signer $i$, in fact belongs to the signer $i$. In the **Step 1** of the procedure a signer, whose identity should be verified, creates one test signature. In this step he also claims his identity as $i$. As in the case of the learning phase, the signature has to be geometrically pre-processed. In the **Step 2** of the procedure, the following information are downloaded from the database: information about selected features of the signer $i$ ($\mathbf{x}'_i$), average values of this features calculated during training phase ($\bar{\mathbf{g}}_i$) and classifier parameters of the signer $i$ ($maxd_{i,n}$, $w_{i,n}$). In the **Step 3** of the procedure, determination of the values of the global features $gtst_{i,n}$, $n = 1, \ldots, N'_i$, for the test signature is performed. The values refer to the features which have been selected as the most characteristic for the signer $i$ in the training phase. In the **Step 4** of the procedure, similarities of global features values of the test signature to the average values of the global features for the reference signatures are determined:

$$dtst_{i,n} = \left| \bar{g}_{i,n} - gtst_{i,n} \right|. \tag{8}$$

In the last step (**Step 5**) of the procedure, the verification of the test signature using one-class flexible fuzzy classifier of the Mamdani type (Sect. 2.2.2) is performed. Its structure is described in the next section. Values of the signals $dtst_{i,n}$ determined in the Step 4 are given at the input of the system.

### 2.2.1 A New One-Class Flexible Fuzzy Classifier

In the signature verification value of the variable $dtst_{i,n}$ is considered. It refers to the similarity between values of the test signature global features and average values of these features determined for the reference signatures. It has an imprecise nature and it is difficult to describe with classical theory of sets and two-valued logic. Therefore, we have used the theory of fuzzy sets and we described values the "high similarity" and "low similarity" using fuzzy sets. Then we have formulated clear fuzzy rules and used approximate inference. As a result, we have obtained a complete fuzzy system which for values of similarities $dtst_{i,n}$ ($n = 1, \ldots, N'_i$) given on inputs determines the similarity of the values of evolutionary selected features of the test signature to the values of the reference signatures global features. In the proposed method it is the basis for evaluation of the reliability of the signature in Sect. 2.2.2. Our system for signature verification works on the basis of two fuzzy rules in the form:

$$
\left\{
\begin{bmatrix}
\text{IF } \left( dtst_{i,1} \text{is} A_{i,1}^1 \right) \Big| w_{i,1} \text{AND} \dots \\
\dots \text{AND } \left( dtst_{i,N'_i} \text{is} A_{i,N'_i}^1 \right) \Big| w_{i,N'_i} \text{THEN} y_i \text{is} B^1 \\
\text{IF } \left( dtst_{i,1} \text{is} A_{i,1}^2 \right) \Big| w_{i,1} \text{AND} \dots \\
\dots \text{AND } \left( dtst_{i,N'_i} \text{is} A_{i,N'_i}^2 \right) \Big| w_{i,N'_i} \text{THEN} y_i \text{is} B^2
\end{bmatrix}
\right\},
\tag{9}
$$

where:

- $dtst_{i,n}$, $i = 1, \dots, I$, $n = 1, \dots, N'_i$, are input linguistic variables (see e.g. [18, 30]) indicating the "similarity between the values of the global feature $n$ of the test signature and the average values of the global feature defined for the reference signatures of the signer $i$". Values "high" and "low" assumed by these variables are Gaussian fuzzy sets $A_{i,1}^1, \dots, A_{i,N'_i}^1$ and $A_{i,1}^2, \dots, A_{i,N'_i}^2$ (see Fig. 2), described by the membership functions $\mu_{A_{i,n}^1}$ and $\mu_{A_{i,n}^2}$. In the case when a fuzzification of the singleton type is used, input linguistic variables can be considered as input signals of the system, which are determined using the formula (8).
- $y_i$, $i = 1, \dots, I$, is output linguistic variable "similarity between the values of the selected evolutionary global features of the test signature and the features of the reference signatures of the signer $i$". Value "high" assumed by this variable is the fuzzy set $B^1$ of the $\gamma$ type, value "low" is the fuzzy set $B^2$ of the $L$ type (see Fig. 2). Sets $B^1$ and $B^2$ are described by the membership functions $\mu_{B^1}$ and $\mu_{B^2}$ (see e.g. [67]).
- $maxd_{i,n}$, $i = 1, \dots, I$, $n = 1, \dots, N'_i$, can be equated with the border values of features of individual signers (calculated by the formula (6)) and $w_{i,n}$ are weights of importance related to the global feature number $n$ of the signer $i$ (calculated by the formula (7)).

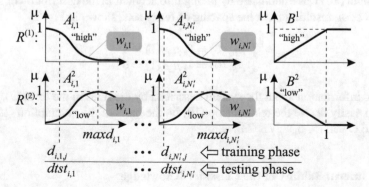

**Fig. 2** Input and output fuzzy sets of the one-class flexible fuzzy classifier of the Mamdani type for signature verification of the signer $i$

### 2.2.2 Signature Verification

In the proposed method, the test signature is recognized as belonging to the signer $i$ (genuine) if the assumption $\bar{y}_i > cth_i$ is satisfied, where $\bar{y}_i$ is the value of the output signal of fuzzy system described by the rules (9):

$$\bar{y}_i = \frac{T^* \left\{ \begin{array}{c} \mu_{A^1_{i,1}}\left(dtst_{i,1}\right), \ldots, \mu_{A^1_{i,N'_i}}\left(dtst_{i,N'_i}\right); \\ w_{i,1}, \ldots, w_{i,N'_i} \end{array} \right\}}{\left[ \begin{array}{c} T^* \left\{ \begin{array}{c} \mu_{A^1_{i,1}}\left(dtst_{i,1}\right), \ldots, \mu_{A^1_{i,N'_i}}\left(dtst_{i,N'_i}\right); \\ w_{i,1}, \ldots, w_{i,N'_i} \end{array} \right\} + \\ T^* \left\{ \begin{array}{c} \mu_{A^2_{i,1}}\left(dtst_{i,1}\right), \ldots, \mu_{A^2_{i,N'_i}}\left(dtst_{i,N'_i}\right); \\ w_{i,1}, \ldots, w_{i,N'_i} \end{array} \right\} \end{array} \right]}, \tag{10}$$

where:

- $T^* \{\cdot\}$ is the algebraic weighted t-norm (see [6, 71]) in the form:

$$T^* \left\{ \begin{array}{c} a_1, a_2; \\ w_1, w_2 \end{array} \right\} = T \left\{ \begin{array}{c} 1 - w_1 \cdot (1 - a_1), \\ 1 - w_2 \cdot (1 - a_2) \end{array} \right\}$$

$$\overset{e.g.}{=} . (1 - w_1 \cdot (1 - a_1)) \cdot (1 - w_2 \cdot (1 - a_2)), \tag{11}$$

where t-norm $T \{\cdot\}$ is a generalization of the usual two-valued logical conjunction (studied in classical logic), $w_1$ and $w_2 \in [0, 1]$ mean weights of importance of the arguments $a_1, a_2 \in [0, 1]$. Please note that $T^* \{a_1, a_2; 1, 1\} = T \{a_1, a_2\}$ and $T^* \{a_1, a_2; 1, 0\} = a_1$.

- $cth_i \in [0, 1]$—coefficient determined experimentally for each signer to eliminate disproportion between FAR (False Acceptance Rate) and FRR (False Rejection Rate) error (see e.g. [80]).

Formula (10) was established by taking into account in the description of system simplification resulting from the spacing of fuzzy sets, shown in Fig. 2:

$$\left\{ \begin{array}{l} \mu_{B^1}(0) = 0, \mu_{B^1}(1) = 1 \\ \mu_{B^2}(0) = 1, \mu_{B^2}(1) = 0 \end{array} \right. . \tag{12}$$

Detailed information about the system described by the rules in the form (9), which allow to easily derive the relationship (10) on the basis of the assumption (12), can be found e.g. in [5, 6, 8, 71, 73].

### 2.2.3 Interpretability of the Classifier Knowledge

In the literature one can find the conditions that must be met by the rules of the fuzzy systems, which cause that the rules are clear. For example, in the paper [29] 4 interpretability levels have been presented (complexity at the rule base level, complexity at the level of fuzzy partitions, semantics at the rule base level, semantics at the fuzzy partition level). The rules in the form (9) meet defined levels. Moreover, it is worth to note that in the proposed method: (a) all parameters of the rules are determined analytically and they have their own interpretation, (b) the rules have the same form for all signers but different values of the parameters.

## 2.3 Description of the Computational Complexity

In practice, the learning phase of the algorithm is performed once for each user and the testing phase (signature verification) can be performed multiple times. A decisive influence on the computational complexity of the learning phase has a complexity of used genetic algorithm (see Table 2). In turn, a way of determining the global features has a decisive influence on the computational complexity of the testing phase (minimal in practice) (see Table 2). Implementation details of the proposed algorithm have not been considered in the paper, but a need to start the process of evolution once for each user in the learning phase should not be a problem in the practical implementation of the algorithm. However, if there is a need of processing a very large number of users registering to the system at the same time, the algorithm could be run in a parallel server environment. Another solution could be queuing of tasks associated with an automatic evolutionary selection of features.

**Table 1** Performance comparison of our method with other methods using BioSecure database

| Method | Average FAR (%) | Average FRR (%) | Average error (%) |
|---|---|---|---|
| Methods used in signature evaluation campaign 2009 [33] | – | – | 1.71–27.76 |
| Horizontal partitioning [12] | 2.94 | 4.45 | 3.70 |
| Vertical partitioning [11] | 3.13 | 4.15 | 3.64 |
| Evolutionary selection with PCA [88] | 5.29 | 6.01 | 5.65 |
| Our method without evolutionary selection | 3.29 | 3.82 | 3.56 |
| Our method with evolutionary selection | 2.32 | 2.48 | 2.40 |

**Table 2**  Computational complexity of the proposed algorithm

| Step | Learning phase 1 | Testing phase |
|------|------------------|---------------|
| 1 | $J$ | 1 |
| 2 | $J \cdot \sum_{n=1}^{N} c_n$ | $4 \cdot N'$ |
| 3 | $J \cdot N$ | $\sum_{n=1}^{N'} c_n$ |
| 4 | $S \cdot$ $\left( N + 9 + N \cdot (2 + J) + \frac{2 \cdot N^{*3} + 9 \cdot N^{*2} + 13 \cdot N^*}{6} \right)$ | $N'$ |
| 5 | $2 \cdot N$ | $1 + 2 \cdot N'$ |
| 6 | $4 \cdot J \cdot N$ | – |
| 7 | $4 \cdot N$ | – |

## 3  Simulation Results

Simulations were performed using commercial BioSecure database which contains signatures of 210 signers. The signatures were acquired in two sessions using the digital graphic tablet. Each session contains 15 genuine signatures and 10 skilled forgeries per person. During training phase we used 5 randomly selected genuine signatures of each signer. During test phase we used 10 remaining genuine signatures and all 10 skilled forgeries of each signer. The process was performed five times, and the results were averaged. The described method is commonly used in evaluating the effectiveness of methods for the dynamic signature verification and it corresponds to the standard cross validation procedure. The test was performed using the authorial testing environment implemented in C# language. During the simulations the following assumptions have been adopted: (a) population contains 100 chromosomes, (b) algorithm stops after the lapse of a determined number of 1000 generations, (c) during selection of chromosomes tournament selection method is used, (d) crossover is performed with probability equal to 0.8 at three points, (e) mutation is performed for each gene with probability equal to 0.02. Details concerning the interpretation of these parameters can be found, among others, in [51, 67].

Conclusions of the simulations can be summarized as follows:

- The proposed method for the considered BioSecure database works with high accuracy in comparison with the methods presented in the Table 1 and in the paper [33]. The comparison criterion was the value of the error EER (Equal Error Rate), which is commonly used to evaluate the accuracy of biometric methods (see e.g. [24, 42]). In practice, also other measures, such as e.g. $d'$, can be used in assessing the effectiveness of the biometric systems (see e.g. [37]). The $d'$ measures the separation between the means of the genuine and impostor probability distributions in standard deviation units. Its mean value, averaged for five test sessions and all signers, is equal to 7.58 for the BioSecure database.

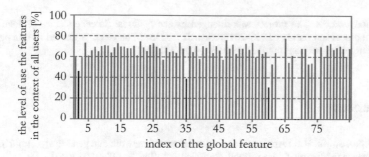

**Fig. 3** Percentage frequency of selection of the global features of the signature in the context of all signers for BioSecure database

- In simulations a common value of $cth_i = 0.45$ was used for all signers. We adopted the assumption that the number of false acceptance should be close to the number of false rejection. If the algorithm working in practice has to be e.g. more sensitive to false acceptance (e.g. in high security systems), value of $cth_i$ should be higher than 0.45.
- The considered set of features does not contain features selected to verify signature of all signers (see Fig. 3). However, there are those which were not selected at all. Their names are not given, because the verification of a usefulness of the features in the context of the database BioSecure was not our goal. It should be noted that use of all available features causes increasing of ERR value to 3.56%.

## 4 Conclusions

In this paper we have proposed a new fuzzy-genetic biometric method for the dynamic signature verification using global features. It is based on the appropriately designed evaluation function of the genetic algorithm. It is used for individual choice of a subset of the global features which are the most characteristic for the reference signatures of the considered signer. Moreover, the proposed method determines the weights of importance of the evolutionarily selected global features and uses them in the classification process. It is also worth noting that the proposed algorithm works independently of the initial set of features, works without access to the so-called skilled forgeries and uses the capabilities of the fuzzy one-class classifier, whose knowledge can be interpreted. We would also like to emphasize that the proposed method worked with very high accuracy for the BioSecure signature database in comparison to the methods of other authors (described in the available positions of the literature).

In our further research in the field of the dynamic signature verification we are planning to take care of, among others, research about the relationship between the dynamic signature verification accuracy and the number of the global features used in the verification.

**Acknowledgements** The project was financed by the National Science Centre (Poland) on the basis of the decision number DEC-2012/05/B/ST7/02138. The work presented in this paper was also supported by the grant number BS/MN 1-109-301/16/P.

# References

1. Arabgol, S., Ko, H.S.: Application of artificial neural network and genetic algorithm to health-carewaste prediction. J. Artif. Intell. Soft Comput. Res. **3**, 243–250 (2013)
2. Banko, Z., Janos, A.: Correlation based dynamic time warping of multivariate time series. Expert Syst. Appl. **39**, 12814–12823 (2012)
3. Batista, L., Granger, E., Sabourin, R.: Dynamic selection of generative discriminative ensembles for off-line signature verification. Pattern Recogn. **45**, 1326–1340 (2012)
4. Bhattacharya, I., Ghosh, P., Biswas, S.: Offline signature verification using pixel matching technique. Proc. Technol. **10**, 970–977 (2013)
5. Cpałka, K.: A new method for design and reduction of neuro-fuzzy classification systems. IEEE Trans. Neural Netw. **20**, 701–714 (2009)
6. Cpałka, K.: On evolutionary designing and learning of flexible neuro-fuzzy structures for nonlinear classification. Nonlinear Anal. Ser. A Theory Methods Appl. **71**, 1659–1672 (2009)
7. Cpałka, K.: Design of Interpretable Fuzzy Systems. Springer (2017)
8. Cpałka, K., Łapa, K., Przybył, A., Zalasiński, M.: A new method for designing neuro-fuzzy systems for nonlinear modelling with interpretability aspects. Neurocomputing **135**, 203–217 (2014)
9. Cpałka, K., Rebrova, O., Nowicki, R., Rutkowski, L.: On design of flexible neuro-fuzzy systems for nonlinear modelling. Int. J. General Syst. **42**(6), 706–720 (2013)
10. Cpałka, K., Rutkowski, L.: Flexible Takagi-Sugeno fuzzy systems. In: Neural Networks, Proceedings of the 2005 IEEE International Joint Conference on IJCNN '05 vol. 3, pp. 1764–1769. (2005)
11. Cpałka, K., Zalasiński, M.: On-line signature verification using vertical signature partitioning. Expert Syst. Appl. **41**, 4170–4180 (2014)
12. Cpałka, K., Zalasiński, M., Rutkowski, L.: New method for the on-line signature verification based on horizontal partitioning. Pattern Recogn. **47**, 2652–2661 (2014)
13. Cpałka, K., Zalasiński, M., Rutkowski, L.: A new algorithm for identity verification based on the analysis of a handwritten dynamic signature. Appl. Soft Comput. **43**, 47–56 (2016)
14. De Maesschalck, R., Jouan-Rimbaud, D., Massart, D.L.: The Mahalanobis distance. Chemom. Intell. Lab. Syst. **50**, 1–18 (2000)
15. Dean, D., Sridharan, S.: Dynamic visual features for audio-visual speaker verification. Comput. Speech Lang. **24**, 136–149 (2010)
16. Doroz, R., Porwik, P., Orczyk, T.: Dynamic signature verification method based on association of features with similarity measures. Neurocomputing **171**, 921–931 (2016)
17. Duch, W., Korbicz, J., Rutkowski, L., Tadeusiewicz, R.: Biocybernetics and biomedical engineering. EXIT, Warszawa (2013)
18. Duch, W., Setiono, R., Zurada, J.M.: Computational intelligence methods for rule-based data understanding. Proc. IEEE **92**, 771–805 (2004)
19. Duda, P., Hayashi, Y., Jaworski, M.: On the strong convergence of the orthogonal series-type kernel regression neural networks in a non-stationary environment. In: Artificial Intelligence and Soft Computing, vol. 7267, pp. 47–54. Springer (2012)
20. Duda, P., Jaworski, M., Pietruczuk, L.: On pre-processing algorithms for data stream. In: International Conference on Artificial Intelligence and Soft Computing. Lecture Notes in Artificial Intelligence, vol. 7268, pp. 56–63. Springer (2012)

21. Ekinci, M., Aykut, M.: Human gait recognition based on kernel PCA using projections. J. Comput. Sci. Technol. **22**, 867–876 (2007)

22. El-Samak, A.F., Ashour, W.: Optimization of traveling salesman problem using affinity propagation clustering and genetic algorithm. J. Artif. Intell. Soft Comput. Res. **5**, 239–246 (2015)

23. Er, M.J., Duda, P.: On the weak convergence of the orthogonal series-type kernel regresion neural networks in a non-stationary environment. In: International Conference on Parallel Processing and Applied Mathematics. Lecture Notes in Computer Science, vol. 7203, pp. 90–98. Springer (2012)

24. Faundez-Zanuy, M.: On-line signature recognition based on VQ-DTW. Pattern Recogn. **40**, 981–992 (2007)

25. Faundez-Zanuy, M., Pascual-Gaspar, J.M.: Efficient on-line signature recognition based on multi-section vector quantization. Form. Pattern Anal. Appl. **14**, 37–45 (2011)

26. de Canetea, Fernandez, J., Garcia-Cerezoa, A., Garcia-Morala, I., Del Saza, P., Ochoa, E.: Correlation based dynamic time warping of multivariate time series. Expert Syst. Appl. **40**, 5648–5660 (2013)

27. Fierrez, J., Ortega-Garcia, J., Ramos, D., Gonzalez-Rodriguez, J.: HMM-based on-line signature verification: feature extraction and signature modeling. Pattern Recogn. Lett. **28**, 2325–2334 (2007)

28. Fierrez-Aguilar, J., Nanni, L., Lopez-Penalba, J., Ortega-Garcia, J., Maltoni, D.: An on-line signature verification system based on fusion of local and global information. In: Audio-and Video-based Biometric Person Authentication. Lecture Notes in Computer Science, vol. 3546, pp. 523–532 (2005)

29. Gacto, M.J., Alcala, R., Herrera, F.: Interpretability of linguistic fuzzy rule-based systems: an overview of interpretability measures. Inf. Sci. **181**, 4340–4360 (2011)

30. Gaweda, A.E., Zurada, J.M.: Data-driven linguistic modeling using relational fuzzy rules. IEEE Trans. Fuzzy Syst. **11**, 121–134 (2003)

31. Gonzalez, R.C., Woods, R.E.: Digital Image Processing. Pearson Education Inc., London (2008)

32. Homepage of Association BioSecure. http://biosecure.it-sudparis.eu. Accessed 22 July 2016

33. Houmani, N., Garcia-Salicetti, S., Mayoue, A., Dorizzi, B.: BioSecure signature evaluation campaign 2009 (BSEC'2009): Results. http://biometrics.it-sudparis.eu/BSEC2009/downloads/BSEC2009_results.pdf. Accessed 22 July 2016 (2009)

34. Huang, K., Hong, Y.: Stability and style-variation modeling for on-line signature verification. Pattern Recogn. **36**, 2253–2270 (2003)

35. Ibrahim, M.T., Khan, M.A., Alimgeer, K.S., Khan, M.K., Taj, I.A., Guan, L.: Velocity and pressure-based partitions of horizontal and vertical trajectories for on-line signature verification. Pattern Recogn. **43**, 2817–2832 (2010)

36. Jain, A.K., Griess, F.D., Connell, S.D.: On-line signature verification. Pattern Recogn. **35**, 2963–2972 (2002)

37. Jain, A.K., Ross, A.: Introduction to Biometrics. In: Flynn, P., Ross, A.A., Jain, A.K. (eds.) Handbook of Biometrics, pp. 1–22. Springer, US (2008)

38. Jaworski, M., Er, M.J., Pietruczuk, L.: On the application of the parzen-type kernel regression neural network and order statistics for learning in a non-stationary environment. In: International Conference on Artificial Intelligence and Soft Computing. Lecture Notes in Artificial Intelligence, vol. 7267, pp. 90–98. Springer (2012)

39. Jaworski, M., Pietruczuk, L., Duda, P.: On resources optimization in fuzzy clustering of data streams. In: International Conference on Artificial Intelligence and Soft Computing. Lecture Notes in Artificial Intelligence, vol. 7268, pp. 92–99. Springer (2012)

40. Jeong, Y.S., Jeong, M.K., Omitaomu, O.A.: Weighted dynamic time warping for time series classification. Pattern Recogn. **44**, 2231–2240 (2011)

41. Khan, M.A.U., Khan, M.K., Khan, M.A.: Velocity-image model for online signature verification. IEEE Trans. Image Process. **15**, 3540–3549 (2006)

42. Kholmatov, A., Yanikoglu, B.: Identity authentication using improved online signature verification method. Pattern Recogn. Lett. **26**, 2400–2408 (2005)

43. Kumar, R., Sharma, J.D., Chanda, B.: Writer-independent off-line signature verification using surroundedness feature. Pattern Recogn. Lett. **33**, 301–308 (2012)
44. Lee, L.L., Berger, T., Aviczer, E.: Reliable on-line human signature verification systems. IEEE Trans. Pattern Anal. Machine Intell. **18**:643–647 (1996)
45. Lei, H., Govindaraju, V.: A comparative study on the consistency of features in on-line signature verification. Pattern Recogn. Lett. **26**, 2483–2489 (2005)
46. Lumini, A., Nanni, L.: Ensemble of on-line signature matchers based on overcomplete feature generation. Expert Syst. Appl. **36**, 5291–5296 (2009)
47. Łapa, K., Cpałka, K., Wang, L.: New method for design of fuzzy systems for nonlinear modelling using different criteria of interpretability. Lect. Notes Comput. Sci. **8467**, 217–232 (2014)
48. Łapa, K., Szczypta, J., Venkatesan, R.: Aspects of structure and parameters selection of control systems using selected multi-population algorithms. Lect. Notes Comput. Sci. **9120**, 247–260 (2015)
49. Maiorana, E.: Biometric cryptosystem using function based on-line signature recognition. Expert Syst. Appl. **37**, 3454–3461 (2010)
50. Mazurowski, M.A., Habas, P.A., Zurada, J.M., Tourassi, G.D.: Decision optimization of case-based computer aided decision systems using genetic algorithms with application to mammography. Phys. Med. Biol. **53**, 895–908 (2008)
51. Michalewicz, Z.: Genetic Algorithms+Data Structures=Evolution Programs. Springer, Berlin, Heidelberg (1998)
52. Moon, J.H., Lee, S.G., Cho, S.Y., Kim, Y.S.: A hybrid online signature verification system supporting multi-confidential levels defined by data mining techniques. Int. J. Intell. Syst. Technol. Appl. **9**, 262–273 (2010)
53. Nanni, L.: An advanced multi-matcher method for on-line signature verification featuring global features and tokenised random numbers. Neurocomputing **69**, 2402–2406 (2006)
54. Nanni, L., Lumini, A.: Ensemble of Parzen window classifiers for on-line signature verification. Neurocomputing **68**, 217–224 (2005)
55. Nanni, L., Lumini, A.: Advanced methods for two-class problem formulation for on-line signature verification. Neurocomputing **69**, 854–857 (2006)
56. Nanni, L., Lumini, A.: A novel local on-line signature verification system. Pattern Recogn. Lett. **29**, 559–568 (2008)
57. Nanni, L., Maiorana, E., Lumini, A., Campisi, P.: Combining local, regional and global matchers for a template protected on-line signature verification system. Expert Syst. Appl. **37**, 3676–3684 (2010)
58. Nelson, W., Kishon, E.: Use of dynamic features for signature verification. In: Proceedings of the IEEE International Conference on Systems, Man, and Cyber, vol. 1, pp. 201–205 (1991)
59. Nelson, W., Turin, W., Hastie, T.: Statistical methods for on-line signature verification. Int. J. Pattern Recogn. Artif. Intell. **8**, 749–770 (1994)
60. O'Reilly, Ch., Plamondon, R.: Development of a Sigma-Lognormal representation for on-line signatures. Pattern Recogn. **42**, 3324–3337 (2009)
61. Pascual-Gaspar, J.M., Faúndez-Zanuy, M., Vivaracho, C.: Fast on-line signature recognition based on VQ with time modelling. Eng. Appl. Artif. Intell. **24**, 368–377 (2011)
62. Peteiro-Barral, D., Guijarro-Berdias, B., Pérez-Sánchez, B.: Learning from heterogeneously distributed data sets using artificial neural networks and genetic algorithms. J. Artif. Intell. Soft Comput. Res. **2**, 5–20 (2012)
63. Pietruczuk, L., Duda, P., Jaworski, M.: Adaptation of decision trees for handling concept drift. In: International Conference on Artificial Intelligence and Soft Computing. Lecture Notes in Artificial Intelligence, vol. 7894, pp. 459–473. Springer (2013)
64. Pietruczuk, L., Rutkowski, L., Jaworski, M., Duda, P.: How to adjust an ensemble size in stream data mining. Inf. Sci. **381**, 46–54 (2017)
65. Razzak, M.I., Alhaqbani, B.: Multilevel fusion for fast online signature recognition using multi-section VQ and time modelling. Neural Comput. Appl. **26**, 1117–1127 (2015)

66. Rigatos, G.G., Siano, P.: Flatness-based adaptive fuzzy control of spark-ignited engines. J. Artif. Intell. Soft Comput. Res. **4**, 231–242 (2014)
67. Rutkowski, L.: Computational Intelligence. Springer, Berlin, Heidelberg (2008)
68. Rutkowski, L., Cpałka, K.: A neuro-fuzzy controller with a compromise fuzzy reasoning. Control Cybern. **31**(2), 297–308 (2002)
69. Rutkowski, L., Cpałka, K.: Compromise approach to neuro-fuzzy systems. In: 2nd Euro-International Symposium on Computation Intelligence, Kosice, Slovakia, June 16–19, vol. 76, pp. 85–90 (2002)
70. Rutkowski, L., Cpałka, K.: Flexible weighted neuro-fuzzy systems. In: Proceedings of the 9th International Conference on Neural Information Processing (ICONIP'02), Orchid Country Club, Singapore, vol. 4, pp. 1857–1861 (2002)
71. Rutkowski, L., Cpałka, K.: Flexible neuro-fuzzy systems. IEEE Trans. Neural Netw. **14**, 554–574 (2003)
72. Rutkowski, L., Cpałka, K.: Neuro-fuzzy systems derived from quasi-triangular norms. In: Proceedings of the IEEE International Conference on Fuzzy Systems, Budapest, July 26–29, vol. 2, pp. 1031–1036 (2004)
73. Rutkowski, L., Cpałka, K.: Designing and learning of adjustable quasi triangular norms with applications to neuro-fuzzy systems. IEEE Trans. Fuzzy Syst. **13**, 140–151 (2005)
74. Rutkowski, L., Przybył, A., Cpałka, K.: Novel online speed profile generation for industrial machine tool based on flexible neuro-fuzzy approximation. IEEE Trans. Ind. Electron. **59**(2), 1238–1247 (2012)
75. Sivanandam, S.N., Deepa, S.N.: Introduc. Genet. Algorithms. Springer, Berlin, Heidelberg (2008)
76. Stanovov, V., Semenkin, E., Semenkina, O.: Self-configuring hybrid evolutionary algorithm for fuzzy imbalanced classification with adaptive instance selection. J. Artif. Intell. Soft Comput. Res. **6**, 173–188 (2016)
77. Svalina, I., Galzina, V., Lujić, R., Šimunović, G.: Correlation based dynamic time warping of multivariate time series. Expert Syst. Appl. **40**, 6055–6063 (2013)
78. Theodoridis, D.C., Boutalis, Y.S., Christodoulou, M.A.: Robustifying analysis of the direct adaptive control of unknown multivariable nonlinear systems based on a new neuro-fuzzy method. J. Artif. Intell. Soft Comput. Res. **1**, 59–80 (2011)
79. Yang, C.H., Moi, S.H., Lin, Y.D., Chuang, L.Y.: Genetic algorithm combined with a local search method for identifying susceptibility genes. J. Artif. Intell. Soft Comput. Res. **6**, 203–212 (2016)
80. Yeung, D.Y., Chang, H., Xiong, Y., George, S., Kashi, R., Matsumoto, T., Rigoll, G.: SVC2004: first international signature verification competition. Lect. Notes Comput. Sci. **3072**, 16–22 (2004)
81. Yin, Z., O'Sullivan, C., Brabazon, A.: An analysis of the performance of genetic programming for realised volatility forecas. J. Artif. Intell. Soft Comput. Res. **6**, 155–172 (2016)
82. Zalasiński, M.: New algorithm for on-line signature verification using characteristic global features. Adv. Intell. Syst. Comput. **432**, 137–146 (2016)
83. Zalasiński M, Cpałka, K.: A New Method Of On-line Signature Verification Using A Flexible Fuzzy One-class Classifier, pp. 38–53. Academic Publishing House EXIT (2011)
84. Zalasiński, M., Cpałka, K.: New algorithm for on-line signature verification using characteristic hybrid partitions. Adv. Intell. Syst. Comput. **432**, 147–157 (2016)
85. Zalasiński, M., Cpałka, K., Hayashi, Y.: New method for dynamic signature verification based on global features. In: Artificial Intelligence and Soft Computing. Lecture Notes in Computer Science, vol. 8467, pp. 251–265. Springer (2014)
86. Zalasiński, M., Cpałka, K., Hayashi, Y.: A new approach to the dynamic signature verification aimed at minimizing the number of global features. Lect. Notes Comput. Sci. **9693**, 218–231 (2016)
87. Zalasiński, M., Cpałka, K., Er, M.J.: New method for dynamic signature verification using hybrid partitioning. in: Artificial Intelligence and Soft Computing. Lecture Notes in Computer Science, vol. 8467, pp. 236–250. Springer (2014)

88. Zalasiński, M., Łapa, K., Cpałka, K.: New algorithm for evolutionary selection of the dynamic signature global features. Lect. Notes Artif. Intell. **7895**, 113–121 (2013)
89. Zhao, W., Lun, R., Espy, D.D., Reinthal, M.A.: Realtime motion assessment for rehabilitation exercises: integration of kinematic modeling with fuzzy inference. J. Artif. Intell. Soft Comput. Res. **4**, 267–286 (2014)
90. Żurada, J.M.: Introduction to Artificial Neural Systems. Jaico Publishing House (2005)

# A Method of Design and Optimization for SiC-Based Grid-Connected AC-DC Converters

S. Piasecki, R. Szmurlo, J. Rabkowski and M.P. Kazmierkowski

**Abstract** This chapter presents a method of design and optimization dedicated for three-phase AC-DC converters. The main idea of presented work is to provide a tool which supports design process and helps to achieve desired properties: efficiency, volume, weight and system cost. The proposed design method is described in the chapter with special attention to calculations regarding power section of the converter. Newly introduced technology of SiC power devices is in scope of author's analysis. Features of proposed method are illustrated by three SiC-based laboratory models rated at 10 an 20 kVA respectively. Each model is a result of the optimization process performed at different input requirements related to volume and efficiency. Finally, performance all models is verified during operation with $3 \times 400$ V AC grid.

## 1 Introduction

There is no doubt that a technology of Silicon Carbide power devices has become a permanent part of the power electronics picture. Today, statements about lower on-state resistances and higher switching speeds in comparison to Silicon devices sound quite obvious. New possibilities in a number of power electronics applications allow to achieve higher efficiency and power density of the applied converters [1, 2]. However, a cost of SiC transistors and diodes is still higher than Si

S. Piasecki · R. Szmurlo · J. Rabkowski · M.P. Kazmierkowski (✉)
Warsaw University of Technology, Institute of Control and Ind. Electronics,
Warsaw, Poland
e-mail: mpk@isep.pw.edu.pl

S. Piasecki
e-mail: szymon.piasecki@ee.pw.edu.pl

R. Szmurlo
e-mail: robert.szmurlo@ee.pw.edu.pl

J. Rabkowski
e-mail: jacek.rabkowski@ee.pw.edu.pl

© Springer International Publishing AG 2018
A.E. Gawęda et al. (eds.), *Advances in Data Analysis with Computational Intelligence Methods*, Studies in Computational Intelligence 738,
https://doi.org/10.1007/978-3-319-67946-4_18

counterparts, but may be compensated by strongly improved parameters of power converters. A common knowledge is also a fact that a simple replacement of the Si devices by SiC counterparts is not the best move. In most cases power electronic converters should be completely redesigned when a new technology is being introduced. Especially, when passive components—parts of the filters—contribute in the volume and, especially, power losses. This is exactly the case of the grid-connected AC-DC converter (see Fig. 1), which contains power section (three-phase, fully controlled bridge) and three-phase filter. Relations between all components of the converter are complex and a design process contains number of different variables. Therefore, additional support to the designer seems to be interesting option and multi-objective analysis might be considered in this case. The design process shows a decisive impact on the expected properties of the converter and obtained functionalities. Typically, high power quality as well as high efficiency are required, moreover, low price and volume of the converter should be also maintained. This means that conflicting objectives need to be combined during a design and production process [3, 4].

Solving of the multi-dimensional design problems with conflicting objectives can be supported by Multi-Objective Optimization (MOO) methods, successfully implemented in power electronics [5, 6]. In the design process of power electronic system, as the result of performed optimization, the sets of "the best" design parameters are expected. Usually, there is no one, optimal solution, for analysed and optimized problem but a set of different solutions, different trade-offs.

One of the key components of the AC-DC converter, specifically power section, are power devices. Operation conditions related to performance of the switching devices (nominal current, switching speed, on-state resistance, surface for heat dissipation) determine the number of system parameters, such as cooling section, grid filter, DC-link voltage level and others, as is presented in Fig. 1.

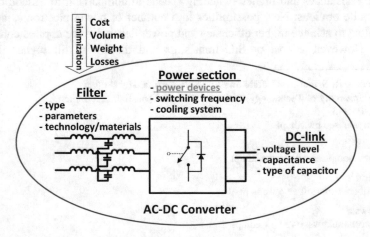

**Fig. 1** Main parameters of the AC-DC converter and expected design objectives of the system

In the presented paper a MOO is applied to assist process of the AC-DC converter design with special attention paid to analysis of SiC based power transistors. The developed optimization tool, allows to analyze how changes of one or more design variables will affect system parameters and desired properties of the converter and finally achieve the design parameters which any change would bring no benefit according to assumed criteria—so called Pareto optimum [7]. The implementation of proposed methodology enables this analysis to be performed in early stage of the design process, giving the engineer a general overview of the available possibilities and choices.

Selected optimization criteria (design objectives) for the AC-DC converter are general properties of this system: volume, efficiency, weight, power quality and price. The optimization parameters are design variables (see Fig. 1): grid filter (type of the filter, values of elements, type of used material and element), type of power switches, cooling system, switching frequency, DC-link capacitance, type of DC-link capacitor and DC-link voltage level (see Fig. 1).

## 2 Design and Optimization Methodology

Several methodologies have been proposed to design and optimize power electronic circuits [8–13]. Complexity level of applied models and mathematical equations are different, usually they rise with precision of the obtained results. Actually, whole optimization process is more complicated with advanced models complicate, time required for calculations is also increased. Moreover, some approaches require implementation of various simulation environments for analysis of different physical phenomena (thermal, electrical, etc). Therefore, an objective of the presented methodology is to support process of the AC-DC converter design utilizing parameters of available on the market components in order to achieve fast tool, suitable for industrial applications. The methodology is composed by three main parts: the design procedure, the database with parameters of available on the market components and the multi-objective evolutionary optimization block, as is presented in Fig. 2.

The first part of the process is the design procedure. Based on experts knowledge in electrical engineering, especially in power electronics, this part allows to analyse surface of available solutions for various operation conditions of the converter. Several mathematical scripts are employed to obtain general system parameters (as currents, voltages, grid filter parameters, DC capacitance, etc.) for different switching frequencies, DC voltages, thermal resistances of the heatsink and performance of the cooling system. All obtained parameters are collected as matrix of general system parameters—available designs. The described methodology is dedicated for 2-level converter, but by modification of applied scripts may be extended to multilevel topologies.

Detailed parameters of the system components: inductors, semiconductors and capacitors are used in optimization calculations. Here only existing on the market,

**Fig. 2** Simplified block
diagram of proposed design
and optimization
methodology of AC-DC
converter

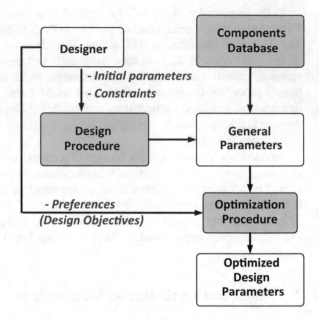

commercially available component are considered. Selected parameters of components are implemented in proposed Components Database (see Fig. 3) on the base of datasheets provided by the producers. Thus, the designer selects elements which are considered in optimization process by their implementation in the Database. At current stage of procedure's development capacitors, inductors and semiconductors are analysed, however, proposed methodology can be applied for other system components.

The next step of analysis is the optimization process. The discussed methodology treats design and optimization of the AC-DC converter as a problem with finite number of alternatives, therefore, the discrete optimization methods, in particular Genetic Algorithms (GAs) are employed. GAs use mechanisms inspired by biological evolution, such as reproduction, mutation, recombination, and selection, and for the all GA techniques the main idea is to select from given population the fittest individuals as in case of natural selection (survival of the fittest). The selection is carried out based on given criteria (cost function), while the fittest measure can be expressed by performance indices. The whole optimization process is based on populations which evolve during generations. In each generation, the individuals from the population are evaluated according to established criteria. The fittest are selected to the next generation and create new population. The new population is subjected to evolution (evolutionary operations, e.g. mutation, crossover) and whole process is repeated till termination conditions are fulfilled.

Based on several scripts components from database are matched with particular designs from matrix of general parameters according to current/voltage limitations. Employing genetic algorithm the known operation conditions (general system

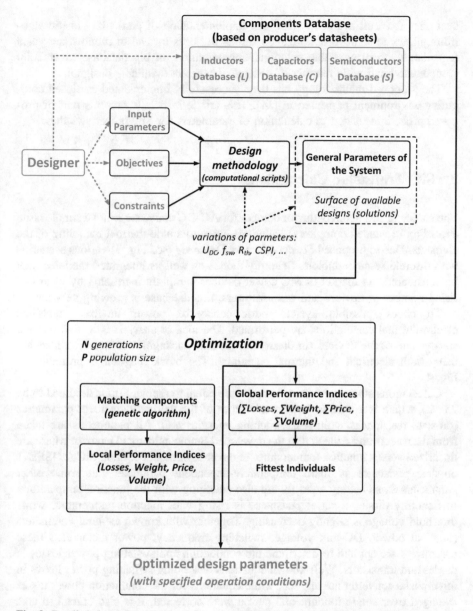

**Fig. 3** Detailed signal flow diagram of proposed design and optimization methodology

parameters) are combined with particular component's parameters. Using dedicated mathematical models and formulas Local Performance Indices related to volume, weight, losses and price are obtained (see Fig. 3). Finally, after n-generations fittest individuals (according to establish by the designer objectives) are selected. Due to relatively high number of generations (>100 000), the calculated parameters are the best ones from all available solutions. Implementation of particular genetic algorithm allows to achieve optimization results in hours instead of months and years (calculation of all available solutions) for operation on big database (over 1000 components) and wide number of general parameters (available designs).

The proposed methodology has been successfully implemented as closed computing environment presented in [14]. This article in details presents part of proposed procedure related to calculation of parameters for SiC power switches.

## 3 SiC Transistor Calculations

The choice of SiC semiconductor devices for AC-DC converter is not a simple issue according to needed complex calculations, which include thermal modeling of the elements. Due to proposed *Semiconductors Database* (see Fig. 3) various scenarios with discrete semiconductors, Schottky diodes as well as integrated modules may be considered. As market of SiC power devices is rapidly increasing by means of rising number of devices and manufacturers, this database is growing very fast.

To check possible system performance of power devices combined electro-thermal calculations are performed. The idea of analysis is to find thermal steady-state of the devices (in discrete or module package) mounted on a heatsink using both electrical and thermal equations. The block diagram is presented in Fig. 4.

Calculations start for junction temperature equal to ambient (here assumed to be 25 °C), which is a base to find actual values of the thermal-dependent parameters (on-state resistances, voltages, switching energies etc). All parameters are taken from datasheets and collected in the developed Semiconductors Database where are linearized versus junction temperature. In the case of most common SiC MOSFETs on-state resistance is main temperature-dependent parameter but some other parameters as switching energies are also changing with the junction temperature. In Schottky diodes a series resistance is rising with junction temperature while threshold voltage is slightly decreasing. Together with known General Parameters (nominal power, DC-link voltage, switching frequency, power factor etc.) these parameters are applied to determine the conduction and switching power losses in diodes and transistors. Well-known equations describing switching power losses in three-phase converter are applied, while equations for the conduction power losses averaged over single fundamental period were corrected. It is also crucial to take into account a reverse conduction of SiC MOSFETs, which is better explained in [15]. In the next step a calculation of the case and junction temperatures takes place using parameters such as thermal resistances of the power modules and the heatsink. This procedure is repeated until the analyzed system reaches a thermal steady-state, which is expected to reflect a situation in the real circuit for given circuit parameters. Calculated power losses for a 10 kVA AC-DC converter for 25 mΩ and 80 mΩ SiC MOSFETs have been presented in Fig. 5.

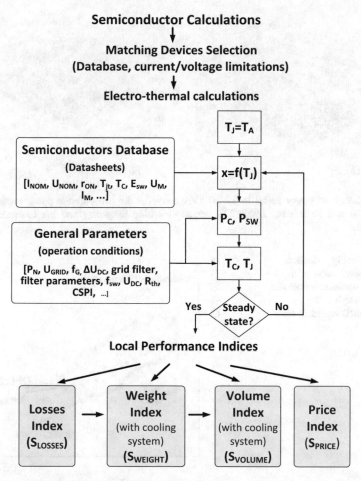

**Fig. 4** Block diagram of the procedure for power transistors selection, electro-thermal modelling and local performance indices estimation

## 4 Power Semiconductors Optimization

The proposed optimization methodology has been implemented as standalone system to optimize the grid-connected AC-DC converter. The application was implemented in Java using Grails framework and calculation scripts are executed by GNU Octave. The system runs on a virtual machine based on Linux Ubuntu and has allocated 8 virtual processors (Intel Xeon X5460) clocked at 3.16 GHz and 4 GB of RAM. The database of components uses the MySQL database server on the same machine. Applied evolutionary algorithms are implemented from the MOEA Framework. The selected optimization algorithm is executed with specified by the designer size of the population (P) and a maximum number of evaluations of the objective function (N) performed during calculation of EA.

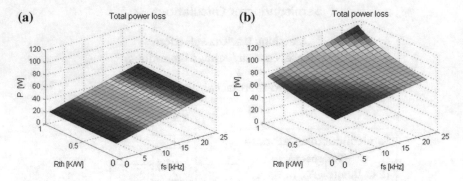

**Fig. 5** Calculated power losses in the 10 kVA converter for two different power modules with 80 mΩ (**a**) and 25 mΩ (**b**) MOSFETs versus switching frequency and the heatsink thermal resistance

**Fig. 6** Quality indicators used for evaluation of the obtained optimization results —two objective (2 dimensional) representation

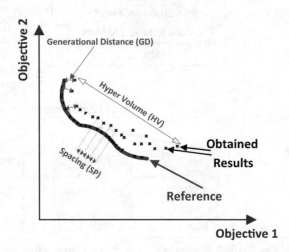

Operation and performance of the proposed methodology with 5 selected genetic algorithms (OMOPSO, NSGAIII, SPEA 2, SMPSO, eMOEA) has been verified through series of numerical analysis. For investigation of the system performance a four objectives are included: volume, weight, price and losses with the same weight coefficients. Achieved results are compared with reference according to 4 quality indicators: Spacing, Generational Distance, Hyper Volume and Elapsed Time to evaluate the obtained 4-dimensional space. Indicators, illustrated in Fig. 6, are defined as follows:

(1) *Spacing (SP)*—indicator gives information how evenly are distributed the results along the known Pareto front;
(2) *Hyper Volume (HV)*—gives information about the volume (in the objective space) covered by non-dominated set of solutions for problem where all objectives need to be minimized. Larger values of the HV indicator are required;

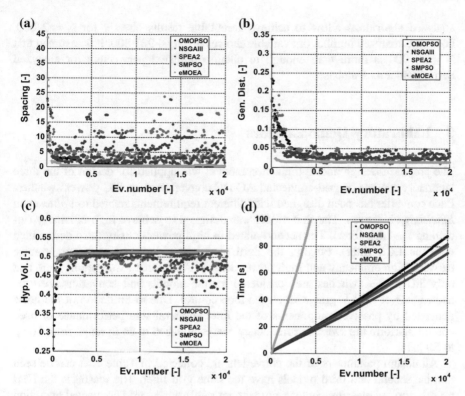

**Fig. 7** Performance of the optimization algorithms. The quality indicators obtained for N = 20 000 evaluations for 5 analyzed evolutionary algorithms and constant population (P = 50); **a** Spacing; **b** Generational distance; **c** Hyper volume **d** Elapsed time, all versus number of evaluations (Ev. Number)

(3) *Generational Distance (GD)*—this indicator gives information how far (on average) are obtained results from true Pareto front. A value of GD equal zero indicates that all calculated elements are on true Pareto front.

(4) *Elapsed Time*—time calculated from the beginning of the optimization process.

For performance analysis a Semiconductors Database with 300 records combined with 16426 General Parameters Vectors was used. In this case the number of all possible combinations is equal 300 * 16426 = 4 927 800. As a reference, the result obtained with 200 000 evaluations (N) of the NSGAIII algorithm with initial population size (P) equal 50 has been selected. Optimization process involves four discussed objectives: volume, weight, losses and price with assumption that all criteria should be minimized with the same weight coefficient. Because the graphical visualization of 4 objectives would be unreadable, the results with quality indicators are presented. In Fig. 7 quality indicators (obtained based on reference) are presented for analysis with N = 20 000 evaluations. It can be observed, that all analyzed algorithms allow to achieve acceptable results already for N = 2 000, however, analyzed number of available choices is only 4 927 800. It is assumed that N = 20 000 is more than enough to obtain optimized parameters for analyzed Semiconductors Database.

## 5  Laboratory Demonstrators

The proposed design and optimization method was applied for design of the three laboratory models of grid-connected AC-DC converters with SiC power switches. Each converter has been designed with different requirements related to *Volume* and *Efficiency* objectives. The first model was designed to achieve high efficiency (including the LCL filter). The second aimed in high power density (also determined with the LCL filter). Finally, the third model was designed as a compromise between high efficiency and high power density. Additional assumptions were that only SiC power switches are considered for the models and converters are connected to 3 × 400 V grid with closed loop control. Due to limitation of the noise generated by passive components of the filter and available performance of used control platform the switching frequency was established to be in range from 16 up to 80 kHz.

All design parameters of the converters are collected in Table 1. It can be seen that the second and third models have the same grid filter. The reason is the EMI noise—applied measurement system was not stable during grid connected operation at 80 kHz switching frequency with converter side inductance ($L_C$) lower than 250 uH, despite the fact that it resulted from the analytical calculations. This issue will be investigated in future works of the authors.

Efficiency of the discussed models has been evaluated by an experimental investigation. System configuration during experiments is presented in Fig. 8a, obtained efficiency versus output power characteristics of the models are presented in Fig. 8b. For efficiency measurements the Yokogawa WT1806 Power Analyzer was used. All converters are connected to the grid through LCL filters, with parameters as presented in Table 1, measured efficiency includes losses in power section of the converter and passive components of the filter (see Fig. 8a).

Analyzed models achieve high efficiency (99.1% with LCL filter) or high power density (5.23 kW/dm3), according to established design objectives. Experimentally obtained performance space related to two main objectives: *Volume* (expressed by *1/Power Density*) and *Efficiency* (expressed by summarized losses) is presented in Fig. 9. Moreover, figure presents main parameters of constructed models. Obtained results are compared to theoretical parameters calculated by the procedure. The main source of obtained, around 20%, error are calculations related to magnetic components. For estimation of losses in the inductors a Modified Steinmetz Equation [16] is used, while parameters for the equation are selected based on material's datasheet provided by producers and laboratory measurements. Volume

**Table 1** Parameters of the analyzed laboratory AC-DC converter models

| Parameter | High efficiency | High efficiency/power density | High power density |
|---|---|---|---|
| Rated power ($P_N$) | 10 [kVA] | 10 [kVA] | 20 [kVA] |
| AC nominal voltage ($U_{AC}$) | 230 [V RMS] | 230 [V RMS] | 230 [V RMS] |
| AC nominal current ($I_{AC}$) | 14.5 [A RMS] | 14.5 [A RMS] | 28.9 [A RMS] |
| DC nominal voltage ($U_{DC}$) | 580–700 [V DC] | 580–700 [V DC] | 580–700 [V DC] |
| DC nominal current ($I_{DC}$) | 14.3–17.3 [A DC] | 14.3–17.3 [A DC] | 28.5–34.4 [A DC] |
| Switching frequency ($f_{sw}$) | 16–24 [kHz] | 40 [kHz] | 80 [kHz] |
| Parameters of LCL filter | | | |
| Converter side inductor ($L_C$) | $L_C$ = 1.5 [mH] | $L_C$ = 250 [µH] | $L_C$ = 250 [µH] |
| $L_C$ core type | 1f, ferrite | 1f, E64/15 3F3 | 1f, ferrite |
| Filter capacitance ($C_{LCL}$) | $C_{LCL}$ = 5 [µF] | $C_{LCL}$ = 5 [µF] | $C_{LCL}$ = 5 [µF] |
| Grid side inductor ($L_G$) | $L_G$ = 100 [µH] | $L_G$ = 100 [µH] | $L_G$ = 100 [µH] |
| $L_G$ core type | 1f, ferrite | 1f, ferrite, E64/10 | 1f, ferrite |
| DC-link capacitance ($C_{DC}$) | 162 [µF] | 100 [µF] | 118 [µF] |
| Power devices | CCS050M12CM2 | 6 × C2M0025120D | 12 × C2M0080120D |
| | | 6 × C4D20120D | 6 × C4D20120A |
| Heatsink | 1 × Fisher SK92 220 mm ($R_{TH}$ = 0.9 K/W) | 2 × Fisher LAM-5-150 ($R_{TH}$ = 0.25 °C/W) | 2 × Fisher LAM-5-150 ($R_{TH}$ = 0.25 °C/W) |

of the inductors is estimated based on peak energy stored in magnetic components, according to equation [17]:

$$L_{VOLUME} = \frac{1}{2} \cdot sf_{L\_Volume} \cdot \sum (L \cdot I_{L\_MAX}^2), \tag{1}$$

where L—inductance, $I_{L\_MAX}$—maximum current of the inductor, $sf_{L\_Volume}$—volume scaling factor (related to the material of the core and applied technology), selected to 0.6 dm$^3$/J, based on [17] and laboratory measurements. This part of the procedure is under development and will be extended by dedicated scripts for magnetic components design.

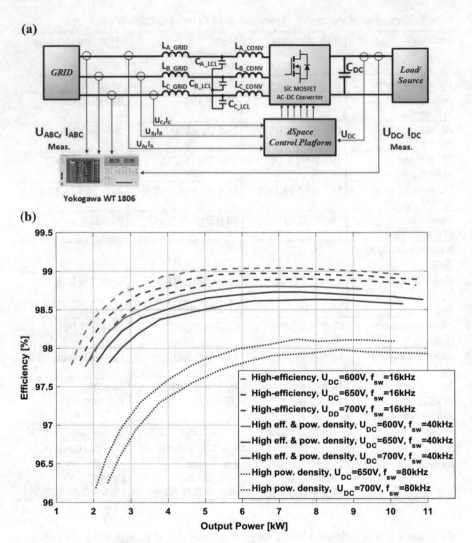

**Fig. 8** Efficiency evaluation of analyzed SiC based AC-DC models for various operation conditions; **a** system configuration during experiment; **b** efficiency versus output power characteristics

# 6  Experimental Investigation

In the further step of experimental investigation the SiC-based demonstrators were verified during grid-connected mode with closed loop control. As a control method the Direct Power Control with Space Vector Modulator (DPC-SVM) is used. The well-known DPC scheme [18] has been extended by additional Phase Locked Loop algorithm, positive and negative Voltage Sequence Extraction module (based on

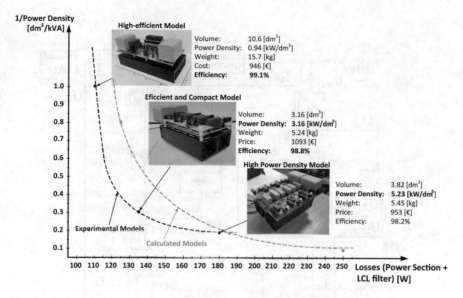

**Fig. 9** The 2 objectives (*1/Power Density vs Losses*) performance space and main parameters of SiC based laboratory models

DSGOGI [19]) and Harmonics Compensation block, as presented in Fig. 10 [19]. Harmonic compensation functionality was realized based on band-pass filters, as described in [19]. This algorithm was implemented on dSpace 1006 platform and used to control each of investigated converters in various operation conditions.

Experimental investigations allow to confirm properties of the SiC based converters in terms of high-efficiency but also improved functionalities of the demonstrators during operation with distorted grid voltage.

In Fig. 11 steady state operation of the high-efficient model as an inverter with nominal power and $U_{DC} = 700$ V is presented. High efficiency and high quality of processed power (Ithd1, Uthd1) are illustrated by screen from Yokogawa Analyzer. Figure 12 illustrates additional functionality of the control method which allows for stable and uninterrupted operation of the converter under grid voltage disturbances. Despite voltage distorted by harmonics and dip grid side currents are controlled, balanced and close to sinusoidal.

Figure 13 presents steady state operation of the high-efficient and compact model as active rectifier under nominal load (10 kW) and $U_{DC} = 700$ V with 40 kHz switching frequency. High efficiency (98.57%), as well as high quality of processed power (Ithd1, Uthd1) are also illustrated by screen from Yokogawa Analyzer. In Fig. 14 harmonics compensation of the grid side current functionality is presented when converter operates with grid voltage distorted by 5% of 5th, 7th and 11th harmonics. Despite distortion grid side current THD is 1% (without compensation 30%).

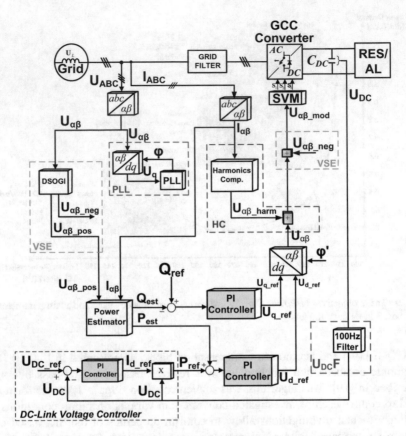

**Fig. 10** Block diagram of the control method implemented in the SiC based laboratory demonstrators

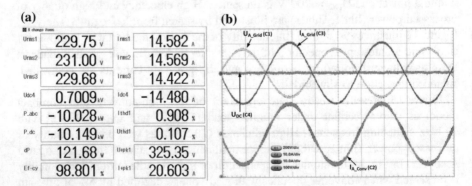

**Fig. 11** Inverting operation of the high-efficient model during steady state, $U_{DC}$ = 700 V, $f_{sw}$ = 16 kHz, $P_{OUT}$ = 10 kW; **a** screen from the Yokogawa power analyzer, **b** current and voltage waveforms, from the top: grid voltage of phase A ($U_{A\_Grid}$), grid side current of the phase A ($I_{A\_Grid}$), DC-link voltage ($U_{DC}$), converter side current of the phase A ($I_{A\_Conv}$)

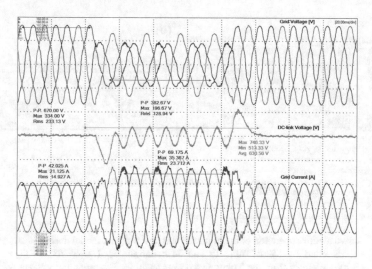

**Fig. 12** Active rectifying operation of the high-efficient model under grid voltage distorted by 40% voltage dip in two phases (voltage unbalance in third phase is caused by separating grid transformer) and 5% of 5th and 7th harmonics. From the top: grid voltage, DC-link voltage and grid current

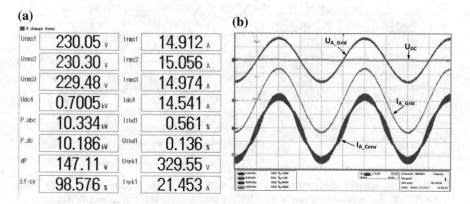

**Fig. 13** Active rectifying operation of the high-efficient and compact during steady state, $U_{DC} = 700$ V, $f_{sw} = 40$ kHz, $P_{OUT} = 10$ kW; **a** screen from the Yokogawa power analyzer, **b** current and voltage waveforms, from the top: grid voltage of phase A ($U_{A\_Grid}$), grid side current of the phase A ($I_{A\_Grid}$), DC-link voltage ($U_{DC}$), converter side current of the phase A ($I_{A\_Conv}$)

Finally, Fig. 15 presents operation of the high power density model as active rectifier under 10 kW load with switching frequency equal 80 kHz during steady state. Obtained efficiency is 97.9% for 700 V in DC-link.

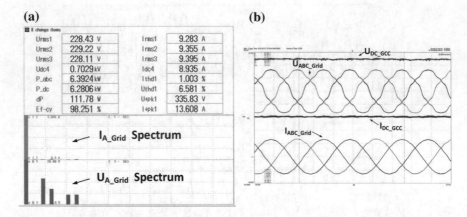

**Fig. 14** Active rectifier operation of the high-efficient and compact model during steady state with grid voltage distorted by 5% of 5th, 7th and 11th harmonics, $U_{DC} = 700$ V, $f_{sw} = 40$ kHz, $P_{OUT} = 6.3$ kW; **a** screen from the Yokogawa power analyzer, **b** current and voltage waveforms, from the top: DC-link voltage ($U_{DC\_GCC}$), grid voltage ($U_{ABC\_Grid}$), DC-link current ($I_{DC\_GCC}$), grid side current ($I_{ABC\_Grid}$)

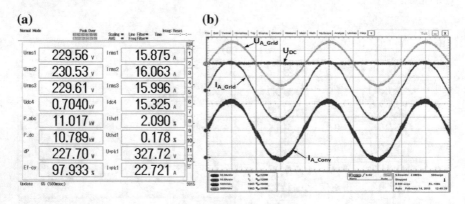

**Fig. 15** Active rectifier operation of the model with high power density during steady state, $U_{DC} = 700$ V, $f_{sw} = 80$ kHz, $P_{OUT} = 10.5$ kW; **a** screen from the Yokogawa power analyzer, **b** current and voltage waveforms, from the top: grid voltage of phase A ($U_{A\_Grid}$), grid side current of the phase A ($I_{A\_Grid}$), DC-link voltage ($U_{DC}$), converter side current of the phase A ($I_{A\_Conv}$)

## 7  Conclusions

The chapter presents the dedicated method for design and optimization of AC-DC converters with special effort on the calculations of SiC power semiconductor devices. For the optimization a dedicated system utilizing genetic algorithms is used. Due to proposed Components Database performed calculations are limited to the existing on the market components, while parameters of the components are implemented in database using producers datasheets. Methodology for the power semiconductor calculations is presented in details and illustrated by three laboratory models which utilize SiC power switches. High efficiency and high power density design scenarios are analysed, moreover, the model which is a compromise between power density and efficiency is presented. Series of experiments confirms properties of the designed models and possibilities offered by dedicated control method. Thus, the presented methodology was verified in practice and provides results, which are very close to initial assumptions. The authors believes that after necessary improvements it may be applied to support designers of the grid-connected AC-DC converters.

**Acknowledgements** This work has been supported by the National Science Center, Poland, based on decision DEC-2012/05/B/ST7/01183.

# References

1. Millan, J., Godignon, P., Perpina, X., Perez-Tomas, A., Rebollo, J.: A survey of wide bandgap power semiconductor devices. IEEE Trans. Power Electron. **29**(5), 2155–2163 (2014)
2. Dimarino, C.M., Burgos, R., Boroyevich, D.: High-temperature silicon carbide: characterization of state-of-the-art silicon carbide power transistors. IEEE Ind. Electron. Mag. **9**(3), 19–30 (2015)
3. Larouci, C., Boukhnifer, M., Chaibet, A.: Design of power converters by optimization under multiphysic constraints: application to a two-time-scale AC/DC–DC converter. IEEE Trans. Ind. Electron. **57**(11), 3746–3753 (2010)
4. Ramachandran, R., Nymand, M.: Design and analysis of an ultra-high efficiency phase shifted full bridge GaN converter. In: IEEE Applied Power Electronics Conference and Exposition (APEC), pp. 2011–2016 (2015)
5. Marler, R.T., Arora, J.S.: Survey of multi-objective optimization methods for engineering. Struct. Multidiscip. Optim. **26**(6), 369–395 (2004)
6. Chan, R.R., Sudhoff, S.D., Lee, Y., Zivi E.L.: Evolutionary optimization of power electronics based power systems. In: 22nd Annual IEEE Applied Power Electronics Conference and Exposition (APEC), pp. 449–456 (2007)
7. Kolar, J.W., Biela, J., Waffler, S., Friedli, T., Badstuebner, U.: Performance trends and limitations of power electronic systems. In: 6th International Conference on Integrated Power Electronics Systems (CIPS), pp. 1–20 (2010)
8. Friedli, T., Round, S.D., Hassler, D., Kolar, J.W.: Design and performance of a 200-kHz all-SiC JFET current DC-link back-to-back converter. IEEE Trans. Ind. Appl. **45**(5), 1868–1878 (2009)
9. Biela, J., Badstuebner, U., Kolar, J.W.: Design of a 5-kW, 1-U, 10-kW/dm3 resonant DC–DC converter for telecom applications. IEEE Trans. Power Electron. **24**(7), 1701–1710 (2009)
10. Boillat, D.O., Krismer, F., Kolar, J.W.: Design space analysis and $\rho$-$\eta$ pareto optimization of LC output filters for switch-mode AC power sources. IEEE Trans. Power Elec. **30**(12), 6906–6923 (2015)
11. Ejjabraoui, K., Larouci, C., Lefranc, P., Marchand, C.: Presizing methodology of DC–DC converters using optimization under multiphysic constraints: application to a buck converter. IEEE Trans. Ind. Electron. **59**(7), 2781–2790 (2012)
12. Busquets-Monge, B.Y.S., et al.: Power converter design optimization. a GA-based design approach to optimization of power electronics circuits. IEEE Ind. Appl. Mag. **10**(1), 32–39 (2004)

13. Muhlethaler, J., Schweizer, M., Blattmann, R., Kolar, J.W., Ecklebe, A.: Optimal design of LCL harmonic filters for three-phase PFC rectifiers. IEEE Trans. Power Electron. **28**(7), 3114–3125 (2013)
14. Piasecki, S.: Research and development of multi-objective optimization procedures for AC-DC grid converters in particular for renewable/distributed energy systems. In: PhD Thesis. Warsaw University of Technology (2016)
15. Piasecki S., Rabkowski J., Experimental Investigations on the Grid-connected AC/DC Converter Based on Three-phase SiC MOSFET Module, proc. of 17th European Conference on Power Electronics and Applications (EPE ECCE Europe), (2015), 1–10
16. Reinert, J., Brockmeyer, A., De Doncker, R.W.: Calculation of losses in ferro- and ferrimagnetic materials based on the modified Steinmetz equation. IEEE Trans. Ind. Appl. **37** (4), 1055–1061 (2001)
17. Bloemink, J.M., Green, T.C.: Reducing passive filter sizes with tuned traps for distribution level power electronics. In: 14th European Conference on Power Electronics and Applications, pp. 1–9 (2011)
18. Kazmierkowski, M.P., Jasinski, M., Wrona, G.: DSP-based control of grid-connected power converters operating under grid distortions. IEEE Trans. Ind. Informatics **7**(2), 204–211 (2011)
19. Jasinski, M., Wrona, G., Piasecki, S.: Control of Grid Connected Converter (GCC) Under Grid Voltage Disturbances. In: Advanced and Intelligent Control in Power Electronics and Drives, Chap. 3, vol. 531. Cham: Springer International Publishing (2014)